Modeling and Practice of Erosion and Sediment Transport under Change

Modeling and Practice of Erosion and Sediment Transport under Change

Special Issue Editors

Hafzullah Aksoy
Gil Mahé
Mohamed Meddi

MDPI • Basel • Beijing • Wuhan • Barcelona • Belgrade

Special Issue Editors

Hafzullah Aksoy
Istanbul Technical University
Turkey

Gil Mahé
IRD, UMR HSM, Univ. Montpellier, CNRS
France

Mohamed Meddi
Ecole Nationale Supérieure d'Hydraulique
Blida
Algeria

Editorial Office
MDPI
St. Alban-Anlage 66
4052 Basel, Switzerland

This is a reprint of articles from the Special Issue published online in the open access journal *Water* (ISSN 2073-4441) from 2018 to 2019 (available at: https://www.mdpi.com/journal/water/special_issues/erosion_sediment).

For citation purposes, cite each article independently as indicated on the article page online and as indicated below:

LastName, A.A.; LastName, B.B.; LastName, C.C. Article Title. *Journal Name* **Year**, *Article Number*, Page Range.

ISBN 978-3-03921-431-0 (Pbk)
ISBN 978-3-03921-432-7 (PDF)

Cover image courtesy of Gil Mahé.

© 2019 by the authors. Articles in this book are Open Access and distributed under the Creative Commons Attribution (CC BY) license, which allows users to download, copy and build upon published articles, as long as the author and publisher are properly credited, which ensures maximum dissemination and a wider impact of our publications.

The book as a whole is distributed by MDPI under the terms and conditions of the Creative Commons license CC BY-NC-ND.

Contents

About the Special Issue Editors vii

Preface to "Modeling and Practice of Erosion and Sediment Transport under Change" ix

Hafzullah Aksoy, Gil Mahe and Mohamed Meddi
Modeling and Practice of Erosion and Sediment Transport under Change
Reprinted from: *Water* **2019**, *11*, 1665, doi:10.3390/w11081665 1

Joanna Szilo and Robert Józef Bialik
Grain Size Distribution of Bedload Transport in a Glaciated Catchment (Baranowski Glacier, King George Island, Western Antarctica)
Reprinted from: *Water* **2018**, *10*, 360, doi:10.3390/w10040360 10

Yang Ho Song, Eui Hoon Lee and Jung Ho Lee
Functional Relationship between Soil Slurry Transfer and Deposition in Urban Sewer Conduits
Reprinted from: *Water* **2018**, *10*, 825, doi:10.3390/w10070825 25

Faiza Hallouz, Mohamed Meddi, Gil Mahé, Samir Toumi and Salah Eddine Ali Rahmani
Erosion, Suspended Sediment Transport and Sedimentation on the Wadi Mina at the Sidi M'Hamed Ben Aouda Dam, Algeria
Reprinted from: *Water* **2018**, *10*, 895, doi:10.3390/w10070895 41

Mahrez Sadaoui, Wolfgang Ludwig, François Bourrin, Yves Le Bissonnais and Estela Romero
Anthropogenic Reservoirs of Various Sizes Trap Most of the Sediment in the Mediterranean Maghreb Basine
Reprinted from: *Water* **2018**, *10*, 927, doi:10.3390/w10070927 73

Yung-Chieh Wang and Chun-Chen Lai
Evaluating the Erosion Process from a Single-Stripe Laser-Scanned Topography: A Laboratory Case Study
Reprinted from: *Water* **2019**, *10*, 956, doi:10.3390/w10070956 92

Shanshan Guo, Zhengru Zhu and Leting Lyu
Effects of Climate Change and Human Activities on Soil Erosion in the Xihe River Basin, China
Reprinted from: *Water* **2018**, *10*, 1085, doi:10.3390/w10081085 108

Russell Adams, Paul Quinn, Nick Barber and Sim Reaney
The Role of Attenuation and Land Management in Small Catchments to Remove Sediment and Phosphorus: A Modelling Study of Mitigation Options and Impacts
Reprinted from: *Water* **2018**, *10*, 1227, doi:10.3390/w10091227 122

Necati Erdem Unal
Shear Stress-Based Analysis of Sediment Incipient Deposition in Rigid Boundary Open Channels
Reprinted from: *Water* **2018**, *10*, 1399, doi:10.3390/w10101399 140

Yi-Chiung Chao, Chi-Wen Chen, Hsin-Chi Li and Yung-Ming Chen
Riverbed Migrations in Western Taiwan under Climate Change
Reprinted from: *Water* **2018**, *10*, 1631, doi:10.3390/w10111631 150

Gergely T. Török, János Józsa and Sándor Baranya
A Shear Reynolds Number-Based Classification Method of the Nonuniform Bed Load Transport
Reprinted from: *Water* **2019**, *11*, 73, doi:10.3390/w11010073 . . . . **163**

Jiří Jakubínský, Vilém Pechanec, Jan Procházka and Pavel Cudlín
Modelling of Soil Erosion and Accumulation in an Agricultural Landscape—A Comparison of Selected Approaches Applied at the Small Stream Basin Level in the Czech Republic
Reprinted from: *Water* **2019**, *11*, 404, doi:10.3390/w11030404 . . . . **179**

About the Special Issue Editors

Hafzullah Aksoy received a BA (1990) in civil engineering and a master's degree (1993) and PhD degree (1998) in hydraulics and water resources from Istanbul Technical University, Turkey. He worked in Germany and the USA during his post-doctoral studies. He was appointed as an assistant professor in 2002 and an associate professor in 2003 at Istanbul Technical University. He spent one year (2009) in Germany as an Alexander von Humboldt guest researcher. Currently, he is a professor of hydrology and water resources in the Hydraulics Unit of the Civil Engineering Department at Istanbul Technical University. His research topics range from statistical methods to process-based modeling techniques for rainfall-runoff-sediment transport in hydrological watersheds and rivers. He was awarded the 2006 Best Paper and 2009 Best Discussion Awards of the ASCE *Journal of Hydrologic Engineering* and the 2011 Engineering Incentive Award of the Scientific and Technological Council of Turkey (TUBITAK). Prof. Aksoy is the universities' representative to the National Hydrology Commission of Turkey and has served as vice president (2007–2011), president elect (2011–2013), president (2013–2017), and past president (2017–2019) of International Commission of Surface Water (ICSW), International Association of Hydrological Sciences (IAHS). He was recently awarded a Fulbright Visiting Researcher Scholarship.

Gil Mahé received a BA (1986) in geology and geophysics from Paris XI University and a master's degree (1987) in hydrology. He then studied hydrology, meteorology, climate, and oceanography with the ORSTOM Institute (now IRD, Research Institute for Developing Countries) in France and Senegal to obtain a PhD from Paris XI University in 1992. In 1993, his PhD received the French national prize of "Georges Hachette" from the Société de Géographie of Paris. He is now a research director at IRD. He spent five years in Mali working on the hydroclimatology and sediment transport of the Niger River. He then spent one year in the Climatic Research Unit of University of East Anglia in Norwich, UK, working on the hydrological modeling and impact of climate change on the Niger River. He went back to West Africa for four more years, working in Burkina Faso at the International School of Engineers on the Sahel paradox: less rainfall and more runoff. He then developed research in Maghreb on the impact of climate and human activities on river regimes and sediment transport, with a more recent focus on the impact of dams on sediment transport to the sea. He is currently the president of the International Commission of Surface Water (ICSW), International Association of Hydrological Sciences (IAHS). He is the chair of the International Committee of the FRIEND program of UNESCO (International Hydrological Program) and is responsible for the Mediterranean FRIEND group. He is the coordinator of the French research program SICMED of the MISTRALS CNRS/INSU program for the study of the Mediterranean environment.

Mohamed Meddi received his PhD from the University of Strasbourg in France. He is a professor of hydrology and a scientific manager of doctoral training at the National High School of Hydraulics of Blida. Prof. Meddi is a member of the editorial board of several journals. His research focuses on hydrology and water science, and he has written more than 50 publications. Prof. Meddi has organized several international conferences in the field of water science. In addition, he is a member of the National Commission for the implementation of the PNR topics. Prof. Meddi is responsible for several national research and international programs. Currently, he is reviewer for several scientific journals. He is a member of the MEDFRIEND (one of the eight FRIEND programs of the International Hydrological Program of UNESCO). Most recently, he was a member of the WASER (World Association for Sedimentation and Erosion Research) Council.

Preface to "Modeling and Practice of Erosion and Sediment Transport under Change"

This book titled "Modeling and Practice of Erosion and Sediment Transport under Change" is composed of five keywords. Two out of the five keywords—"erosion" and "sediment transport"—are related to processes. They are two consecutive processes; one, the former, comes after the other, the latter. Sediment particles are first detached from their original location by erosion under the effect of different agents, such as wind, rainfall, and runoff, and are then transported. Another two keywords are "modeling" and "practice", which are mutually linked—both based on each other. Good practice is not possible without a theoretically well-established model nor is modeling valuable when it is not applicable to practice. Finally, the crucial keyword, "change", is a kind of mirror to the future of the above four keywords. Erosion and sediment processes should be modeled and practiced considering hydrological conditions under continuous change. "Change" is the most challenging and uncertain keyword of the book title, as it is born not only of natural reasons but also due to anthropogenic disturbances. It is a common belief of the editors that all the papers in this book should be considered a contribution to addressing the erosion and sediment transport problem in hydrological watersheds at different scales of space and time and also under any kind of change. This book is expected to attract the interest of hydrologists, water resources researchers, practitioners, and decisionmakers.

Hafzullah Aksoy, Gil Mahé, Mohamed Meddi
Special Issue Editors

Editorial

Modeling and Practice of Erosion and Sediment Transport under Change

Hafzullah Aksoy [1,*], Gil Mahe [2] and Mohamed Meddi [3]

1. Department of Civil Engineering, Istanbul Technical University, 34469 Istanbul, Turkey
2. IRD, UMR HSM IRD/ CNRS/ University Montpellier, Place E. Bataillon, 34095 Montpellier, France
3. Ecole Nationale Supérieure d'Hydraulique, LGEE, Blida 9000, Algeria
* Correspondence: haksoy@itu.edu.tr

Received: 27 May 2019; Accepted: 9 August 2019; Published: 12 August 2019

Abstract: Climate and anthropogenic changes impact on the erosion and sediment transport processes in rivers. Rainfall variability and, in many places, the increase of rainfall intensity have a direct impact on rainfall erosivity. Increasing changes in demography have led to the acceleration of land cover changes from natural areas to cultivated areas, and then from degraded areas to desertification. Such areas, under the effect of anthropogenic activities, are more sensitive to erosion, and are therefore prone to erosion. On the other hand, with an increase in the number of dams in watersheds, a great portion of sediment fluxes is trapped in the reservoirs, which do not reach the sea in the same amount nor at the same quality, and thus have consequences for coastal geomorphodynamics. The Special Issue "Modeling and Practice of Erosion and Sediment Transport under Change" is focused on a number of keywords: *erosion* and *sediment transport*, *model* and *practice*, and *change*. The keywords are briefly discussed with respect to the relevant literature. The papers in this Special Issue address observations and models based on laboratory and field data, allowing researchers to make use of such resources in practice under changing conditions.

Keywords: Anthropocene; climate change; deposition; erosion; modeling; practice; sedimentation; sediment transport; watershed

---

1. Keywords of the Special Issue

This Special Issue entitled *Modeling and Practice of Erosion and Sediment Transport under Change* is focused on five keywords. Two out of the five keywords are related to processes, which are *erosion* and *sediment transport*. Sediment particles are first detached from their original location by erosion under the effect of different agents, such as wind, rainfall, and runoff, followed by transportation. Another two keywords are the *modeling* and *practice*, which are mutually linked and both based on one other. Good practice is possible without a theoretically well-established model nor is modeling valuable when it is not applicable to practice. Finally, the crucial keyword, *change*, comes in the title of the Special Issue and serves as a kind of mirror to the future of the previous four keywords. Erosion and sediment processes should be modeled and practiced considering the hydrological conditions under continuous change. The keyword *change* is the most challenging and unclear of the Special Issue as it is born not only of natural reasons but also as a result of anthropogenic disturbances.

2. Erosion and Sediment Transport

Erosion is the process of detachment and transportation of soil materials by erosive agents [1] such as wind, rainfall, or runoff. In cold regions, snowmelt can cause significant erosion due to freshet and ice jams during the period of spring break-up [2–5]. Earthy or rock materials are loosened or dissolved, and they are removed from any part of the Earth's surface by erosion [6]. Erosion is a very important natural phenomenon that ends in soil loss. It causes also storage volume loss in river reservoirs where eroded sediment deposits. Water erosion is affected by the climate, topography, soil, vegetation, and anthropogenic activities such as tillage systems and soil conservation measures [7]. Land use and land cover characteristics are also important in the erosion process [8].

Four different processes that accomplish sediment removal and transport are the detachment by raindrop impact and runoff, and transport by raindrop splash and runoff. Flow and raindrop detachment rates are not simple and are, therefore, given by empirical formulas [9]. Once eroded, sediment is transported. Sediment particles have become detached are transported within flow as long as sediment load in the flow is smaller than its sediment transport capacity. The shear stress exerted by flow should also be greater than the critical value such that sediment particles are removed from their current locations. Otherwise, deposition takes place, which is the actual rate of mass temporarily reaching the overland surface. Deposition might exceed soil entrainment, which is called the net deposition; the opposite case is called net erosion, where the entrainment is higher than deposition. In the case of an equality of entrainment with sedimentation, equilibrium is reached in terms of erosion [10]. Sediment transport is influenced not only by hydraulic properties of flow but also by the physical properties of soil and surface characteristics.

3. Modeling and Practice

There are numerous methods are available in the literature for quantifying sediment transport over watersheds or in streams. The usual practice is to analyze the streamflow and sediment discharge time series statistically [11,12] and to correlate the available streamflow and sediment discharge data [13,14]. Empirical approaches [15] are also frequently used in practice. Statistical analysis and stochastic modeling techniques have always been attractive alternatives because the monitoring and sampling are not easy tasks and, in most cases, the simultaneous streamflow and sediment discharge records are not available for comparatively long periods to enable conclusive deterministic relationships between sediment concentration and streamflow discharge. Therefore, time series models [16], in particular, were found to be useful in dealing with sediment transport problems. Among the many available studies, a number [17–20] are based on the time series modeling technique. However, traditional equations [21] as well as soft computational techniques [22] have also received great attention. In the meantime, monitoring, sampling, and surveying [23] as well as remote sensing and the use of geographical information systems [24] have continued to be important in sediment transport modeling.

As far as modeling is concerned, the Universal Soil Loss Equation (USLE) was designed as a tool to be used in the management practices of agricultural lands. It is one of the earliest attempts to compute the sediment yield in a catchment. Its development is based on data from the United States though it has subsequently been widely used all over the world. The USLE, along with some modifications and revisions (MUSLE and RUSLE), is still a useful tool in watershed management. A large number of the existing erosion and sediment transport models are based on the USLE. Their applications are, however, limited to the environmental circumstances from which the USLE was generated. The WEPP model in the United States and the SHESED and EUROSEM models in Europe were derived based on physical description of the erosion and sediment transport processes. Although preparation of data for physically based models is a hard task, they have been extensively used. It is obvious that a physically based model has much more detail than either USLE or its derivatives. Therefore, there has been a great effort towards developing physically based erosion and sediment transport models.

Due to their importance, hydrological models accommodate erosion and sediment transport modules in which the movement of sediment—that has been eroded by wind, rainfall, or flow, and transported—through the existing river channel to the reservoir is predicted. The SHESED [25], WEPP [26], SWAT [27], and WEHY [28,29] models are some of the well-known examples.

Erosion and sediment transport models are extensions of hydrological models. Therefore, erosion and sediment transport equations are coupled to existing hydrological algorithms. In such a coupling, the output of the hydrological model becomes the input for the erosion part of the model. In the same sense, an erosion and sediment transport model can be easily extended to a nutrient transport model, as it is known that nutrients are mainly transported by sediment particles. It is much easier to extend a multi-size erosion and sediment transport model to a nutrient transport model since nutrient transport is a size-selective process. For an extensive review on erosion and sediment transport, please refer to [30,31].

4. Climatological and Anthropogenic Changes

The world's climate exhibits natural and unnatural variations and changes, with time scales ranging from millions of years down to one or two years. Over periods of several years, fluctuations of a few tenths of a degree in surface air temperatures over continents are common. These changes are related to global, regional, and even local scale events. Not only the climatological changes but also anthropogenic activities that could be superimposed on this varied hydrometeorological basis affect a wide variety of water use and management approaches. The individual components of the hydrological cycle are affected by the change. They might cause hydrological extremes such as droughts and floods, which are environmental issues turning back as direct consequences of natural or anthropogenic changes. Similarly, erosion and sediment transport over hydrological watersheds are heavily affected by such changes. Any change in either of the components of the hydrological cycle affects the hydrological behavior of the watershed and, thus, the erosion and sediment transport from the watershed surface.

A particularly important issue linked to the change in hydrology could be linked to stochastic modeling, which is based on the assumption that the time series is stationary. Generated sequences of any hydrological process under investigation, sediment discharge for instance, are considered stationary; i.e., they have the same statistical characteristics as the observed sequence. On the one hand, the stationarity is considered dead [32] with the change in hydrological processes [33–36]. On the other hand, however, it is still alive and inevitably useful in modeling hydrological processes [37–40] because it is convenient to use for making reliable predictions in engineering design. Also, the modeling concept based on the observed data is a useful practice due to the fact that the past is representative of the future [40]. Yet, any gradual and sudden changes due either to natural variation of the process under investigation or to any anthropogenic intervention should not be ignored. Such changes that could arise in the form of either a trend or a jump could be considered through the existing trend detection mechanisms and segmentation tools [41–44] to eliminate the nonstationarity of the process. Any jump or trend, when they exist, are determined and added into the modeled hydrological variables such that the nonstationary behavior of the process is taken into account.

5. Summary of the Special Issue

Eleven papers have been published in the Special Issue among a much larger collection of submissions. The published papers deal with different aspects of erosion and sediment transport by using different methodologies, through different practices, implementing different scale datasets, collecting in situ data from the field or gathering experimental data, modeling the process with different approaches, and performing case studies from Europe and the Mediterranean North Africa to typhoon-dominated Asia and even to polar Antarctica. The Special Issue therefore represents highly diverse research activities.

A summary of the papers in the Special Issue follows, given in the order in which they appear.

The grain size distribution of bedload transported within a given water discharge has been investigated for two polar catchments in Antarctica [45], aimed at determining how the grain size distribution is modified during times of peak discharge, determining the relationship between the grain size distribution and its parameters, and the amount of transported material in gravel-bed rivers, and examining whether the modification of grain size distribution during efficient bedload transport events allow for the identification of the development stage of river throughs during changing meteorological conditions. These questions were answered after a measurement campaign in two creeks with proglacial gravel-bed channels. At the end it is confirmed that the variability and modification of grain size distributions are strongly related to the daily variability of bedload transport dynamics. The increased proportion of medium and coarse gravel was strictly proportional to the increase in water discharge. An efficient erosive process, which is confirmed by the general conditions of both streams, is significantly influenced by bedload transport.

Urban sewer conduits prevent flooding in urban areas by discharging runoff generated during rainfall. With an improper design of urban sewer conduits, severe soil sedimentation can occur in the conduits which threatens the capacity and ability of the system to discharge the flood during the rainy season or at localized heavy rainfalls. Soil slurry deposited on the surface of the Earth during rainfall that flows into urban sewer conduits was investigated [46] to propose a functional relationship between critical tractive force in urban sewer conduits and the physical properties of particles in a conduit bed which are taken as the inlet flow velocity of the soil slurry mixture, the volume concentration of the soil, and its particle size. For the two-phase soil slurry flow, a numerical analysis was performed based on various flow conditions to conclude that the findings of this study may be helpful to prevent conduit clogging or conduit damage that may occur during heavy rainfall events. As this study considered only the inflow of a large amount of soil slurry at the beginning of rainfall events, it is open for further investigation to introduce various types of rainfall events by using data from real conduit systems.

Algeria, in northern Africa, has always been confronted with severe periodic droughts as well as catastrophic floods, both being major constraints against the economic and social development of the country [47]. In order to combat both natural hazards, the total number of dams increased quite quickly, from 14 in 1962 to 65 in 2014. The scale of aggradations and the raising of the bottom of these dam impoundments by successive deposits of sediments—brought by the watercourses and the wind—are serious problems whose negative consequences are considerably felt in the agriculture, farming, fishing, electricity, and navigation fields. The sedimentation in North African dams is very high in relation to what is noted in the watershed of Wadi Mina in semiarid northwest Algeria, and it has been the subject of the study of Hallouz et al. [47]. Also, with a trend starting in the late 1980s, the annual production of sediment became seven times larger than the previous period, with a four times greater increase in the rates of contribution in the dry season. By comparing the results, it is observed that the upstream basin is the greatest sediment producer towards the dam.

The Mediterranean Maghreb Basin (MMB), that extends over Morocco, Algeria, and Tunisia, is a region where both mechanical erosion rates and the anthropogenic pressure on surface water resources are high and were subject to analysis [48]. Based on sediment trapping, calculated by the models using information and limited data from 470 out of 670 reservoirs in the area, it is confirmed that natural sediment yields are clearly above the world average, with the largest being in Morocco and the smallest in Tunisia. Trapping rates have an opposite order, being the highest in the Tunisian part of the basin, followed by the Algerian and the Moroccan parts. Trapping of the sediment in their reservoirs greatly reduces the natural sediment flux of the dams in the entire Mediterranean Maghreb Basin to the sea; only slightly more than one-third of the natural river sediment fluxes reaches the coastal Mediterranean waters of the Maghreb [49,50]. The effect of small reservoirs and hillside reservoirs should not be ignored in the interception of sediment compared to large reservoirs, although they have shorter life spans than large reservoirs, and their economic exploitation is limited in time. Understanding the impact of dams and related water infrastructures on riverine sediment dynamics is key in arid zones such as the Mediterranean Maghreb Basin, where global warming is predicted to cause important changes in the climatic conditions and the water availability.

Small-scale laboratory rainfall simulator experimental data based on high-precision DEM are evaluated to provide accurate, but affordable, soil loss estimates [51]. Laser-scanned topography and sediment yields were collected every 5 min in each test. The difference between the DEMs from laser scans of different time steps gives the eroded soil volumes and the corresponding estimates of soil loss in mass. It is seen that sediment yield and eroded soil volume increased with rainfall duration and slope. It is demonstrated that the stripe laser-scanning method is applicable in soil loss prediction and erosion evaluation in laboratory case studies, and could be taken for further case studies of larger scales such that a method that is useful, in practice, can be generated.

Climate change and human activities are two major factors affecting runoff and sediment load in hydrological basins [52]. Gradual or sudden changes in hydrometeorological characteristics, such as annual rainfall, air temperature, runoff, and sediment load, are important in simulating the watershed hydrological cycle. These gradual and sudden changes are mainly linked to the contributions of climate change and human activities to runoff and sediment load under change. Results showed that both rainfall and air temperature increased whilst runoff and sediment load decreased. The air temperature experienced a sudden increase and sediment load decrease in 1988. Soil erosion was found to be worse in the upper part of the basin than other parts, and it is the highest in cultivated land. Climate change exacerbates runoff and sediment load with overall contribution to the total change while, on the other hand, human activities decreased runoff and sediment load with overall contribution to the total change. The conclusion of the study is that the variation of runoff and sediment load in the Xihe River Basin in China is largely caused by human activities.

Soil, hillslopes, and watercourses in small catchments possess a degree of natural attenuation that affects both the shape of the outlet hydrograph and the transport of nutrients and sediments [53]. The headwaters of such catchments are expected to add additional attenuation primarily through increasing the amount of new storage available to accommodate flood flows. The actual types of so-called natural-based solutions include swales, ditches, and small ponds (acting as sediment traps). A modeling study was performed on a small subcatchment of 1.25 km^2 in order to address the impacts of land management by altering hydrological flow paths and the overall catchment attenuation capacity on flow rates and nutrient losses. The model results implied that a small decrease in the order of 5%–10% in the peak concentrations of suspended sediment and nutrients was observed with an increase in the catchment storage.

An indoor laboratory-scale experimental study is quite beneficial to the sediment transport problem as the urban drainage and sewer systems are final reaching points for any type of sediment to be washed into the channels or conduits [54]. Using the self-cleansing concept, different cross-section channels were tested to better clarify the fuzziness between the incipient deposition and incipient motion of sediment particles moving within the flow. With this aim, an experimental study carried out in trapezoidal, rectangular, circular, U-shape, and V-bottom channels for four different sizes of sand as the sediment in the experiments was performed in a tilting flume under nine different longitudinal channel bed slopes. The shear stress approach was considered for the analysis in which the well-known Shields and Yalin methods were used. The circular channel was found to be the second most efficient after the rectangular channel, and the V-bottom channel the least in transporting sediment within the drainage system. The outputs of this study are expected to be useful for practical use in the design of urban drainage and sewer systems that collect the sediment load together with nutrients.

Extreme weather phenomena which have been a part of daily hydrometeorological practice worldwide bring catastrophic disasters under the impact of climate change; and Taiwan has no exemption, due to an increase in the frequency of extreme rainfall events [55]. Extreme rainfall events might cause riverbed migrations, such as degradation and aggradation. The extrapolation of typhoon events—observed in the past by the CCHE1D model—under the climate change scenario of representative concentration pathways 8.5 (RCP8.5) and dynamical downscaling of rainfall data in Taiwan shows that the average peak flow during extreme rainfall events will increase by 20% relative to the base period, but the time required to reach the peak will be 8 h shorter than that in the base period. It is expected that the aggradation of the riverbed will increase by the end of the 21st century. It is also foreseen that anthropogenic activities, such as blocking upstream sediment by structures such as a weir, will clearly increase the severity of scouring downstream. The study finally indicates that not only will larger floods occur within a shorter time duration, but that the catchment will also face more severe degradation and aggradation in the future.

In addition to sand-dominated sediment being transported within flow, gravel may also be transported in rivers with mixed-size bed material when dealing with large rivers with complex hydrodynamics and morphodynamics [56]. This is the collapse of the uniform sediment assumption, which then requires a suitable approach to estimate the nonuniform behavior of the sediment transport. Utilizing the shear Reynolds number, a decision criterion which is based on the combined use of several formulas, each of them having a certain application range, was verified with the field and laboratory data of nonuniform bed material compositions. The proposed approach is able to predict domination of sand or gravel transport with an uncertainty of less than 5%.

In central Czech Republic, a small stream catchment with an agricultural and forest–agricultural landscape and relatively rugged topography and riverbed slope [57] make the terrain very vulnerable to water erosion, and this was studied to compare four selected soil erosion and sediment delivery models—WaTEM/SEDEM, USPED, InVEST, and TerrSet, with each working on several different algorithms. The models were compared based on the total volume of eroded and accumulated sediment within the catchment per unit time, and according to the spatial distribution of sites susceptible to soil loss or sediment accumulation. Despite the fact that the models are different in terms of calculation algorithms and data preprocessing requirements, comparable results in calculating the average annual soil loss and accumulation were obtained but they behaved differently in identifying the spatial distribution of specific locations prone to soil loss or accumulation processes.

6. Establishment of the Special Issue

Scope and coverage: When the Special Issue had been scheduled, it was anticipated to be a collection of contributions focusing on erosion and sediment transport issues in riverbeds or watersheds using analytical, numerical, and in situ experimental tools. However, laboratory experiments were also found to be an important consideration, due to their providing specific details in the transport of sediment.

Scale: Another point is the scale issue, that has varied greatly among the papers published in the Special Issue. The erosion and sediment transport issues have been investigated in channels and over watersheds. The channels have been either a prototype natural watercourse [45,54] or an artificial flume of a laboratory model [46,54]. As for the watersheds, the scale has changed from 1.25 km^2 subcatchments from the United Kingdom [53] to large, millions of km^2 watersheds in the Maghrib basin that extend over Morocco, Algeria, and Tunisia.

Geographical diversity: The Special Issue has not been a collection of studies concentrated on a specific region. The papers have large geographical variability, from the Maghrib area in North Africa to typhoon-dominated Taiwan and even Antarctica. Therefore, the erosion and sediment transport problem has not been limited to a particular region but extends over a geographical scale representing large diversity.

Large rivers vs. small rivers: Large rivers carry not only sand-type sediment but also gravel-dominated bedload which contradict with the uniform-size sediment assumption. However, in smaller rivers, the assumption that the sediment has a uniform grain size could be considered valid to a certain degree. In this sense, it is important that bedload sediment models should be dealt with differently when one is interested in a large river where the bedload is gravel-dominated and, thus, no uniformity exists in terms of sediment size.

In situ measurements and laboratory data: It is important that any model is validated with real data, which is a headache for all scientists and practitioners who deal with Earth sciences. Hydrology is no exception, and erosion and sediment transport are particular issues which experiences this problem at the highest level, due to the fact that sediment data are not collected as frequently as other hydrometeorological variables, such as precipitation, evaporation, humidity, streamflow, etc. Therefore, in situ problem-specific data collection efforts are quite valuable. Laboratory studies could also be as useful as fields studies. This is a fact that has been touched upon; the former could be replaced by the latter when it is hard to perform field studies and when the latter is possible.

Climate change and anthropogenic activities: It is an inevitable reality that hydrology is under change that is either linked to the climatology or Anthropocene. More frequent floods with higher peaks are reached in shorter time intervals than before. This is a challenge that can be associated with the stationary time series methodologies. The stationarity is now dead, and the past is no longer the mirror for the future as it was before.

7. Conclusions

The Special Issue has attracted the interest of hydrologists and water resources researchers. It is a common belief of the authors of this Editorial that all papers in this Special Issue could be considered a contribution to the erosion and sediment transport problem in hydrological watersheds at different scales of space and time and also under any kind of change.

Author Contributions: Conceptualization, G.M., H.A., M.M.; Writing—Original Draft Preparation, H.A.; Writing—Review & Editing, H.A., G.M., M.M.

Funding: This research received no external funding.

Acknowledgments: The authors thank editors for their help during the processes for the Special Issue. Yonca Cavus, PhD candidate from Istanbul Technical University, read the draft and edited it carefully. The authors are thankful to her for the time she dedicated to this Editorial.

Conflicts of Interest: The authors declare no conflict of interest.

References

1. Foster, G.R.; Meyer, L.D. A Closed-Form Soil Erosion Equation for Upland Areas. In *Sedimentation Symposium in Honor Prof. H.A. Einstein*; Shen, H.W., Ed.; Colorado State University: Fort Collins, CO, USA, 1972; pp. 12.1–12.19.
2. Prowse, T.D. Suspended sediment concentration during breakup. *Can. J. Civ. Eng.* **1993**, *20*, 872–875. [CrossRef]
3. Costar, F.; Dupeyrat, L.; Gautier, E.; Carey-Gailhardis, E. Fluvial thermal erosion investigations along a rapidly eroding river bank: Applications to the Lena River (Central Siberia). *Earth Surf. Process. Landf.* **2003**, *28*, 1349–1359. [CrossRef]
4. Knack, I.M.; Shen, H.T. A numerical model for sediment transport and bed change with river ice. *J. Hydraul. Res.* **2018**, *56*, 844–856. [CrossRef]
5. Burrell, B.C.; Beltaos, S. Effects and implications of river ice breakup on suspended-sediment concentrations: A synthesis. In Proceedings of the CGU HS Committee on River Ice Processes and the Environment 20th Workshop on the Hydraulics of Ice-Covered Rivers, Ottawa, ON, Canada, 14–16 May 2019.
6. ASCE Task Committee. Sediment sources and sediment yields. *ASCE J. Hydraul. Div.* **1970**, *96*, 1283–1329.
7. Kuznetsov, M.S.; Gendugov, V.M.; Khalilov, M.S.; Ivanuta, A.A. An equation of soil detachment by flow. *Soil Tillage Res.* **1998**, *46*, 97–102. [CrossRef]
8. Emmett, W.W. *The Hydraulics of Overland Flow on Hillslopes*; USGS (United States Geological Survey): Washington, DC, USA, 1970; Volume 662A, 68p.
9. Bennett, J.P. Concepts of mathematical modeling of sediment yield. *Water Resour. Res.* **1974**, *10*, 485–492. [CrossRef]
10. Croley, I.I. TE Unsteady overland sedimentation. *J. Hydrol.* **1982**, *56*, 325–346. [CrossRef]
11. Phien, H.N.; Arbhabhirama, A. A statistical analysis of the sediment accumulation in reservoirs. *J. Hydrol.* **1979**, *44*, 231–240. [CrossRef]
12. Skoklevski, Z.; Velickov, S. Suspended load transportation process within Vardar river basin in the Republic of Macedonia. In Proceedings of the XIXth conference Danube Countries on the Hydrological Forecasting and Hydrological Bases of Water Management, Osijek, Croatia, 15–19 June 1998; pp. 717–727.
13. Tingsanchali, T.; Lal, N.K. A combined deterministic-stochastic model of daily sediment concentrations in a river. In Proceedings of the Sixth IAHR International Symposium Stochastic Hydraulics, Taipei, Taiwan, 18–20 May 1992; pp. 221–228.
14. Rosen, T.; Xu, Y.J. A hydrograph-based sediment availability assessment: Implications for Mississippi River sediment diversion. *Water* **2014**, *6*, 564–583. [CrossRef]

15. Bogardi, J. *Sediment Transport in Alluvial Streams*; Akademiai Kiado: Budapest, Hungary, 1974.
16. Box, G.E.P.; Jenkins, G.M.; Reinsel, G.C. *Time Series Analysis, Forecasting and Control*; Prentice-Hall: Englewood Cliffs, NJ, USA, 1994.
17. Szidarovszky, F.; Yakowitz, S.; Krzysztofowicz, R. A Bayes approach for simulating sediment yield. *J. Hydrol. Sci.* **1976**, *3*, 33–44.
18. Phien, H.N. Reservoir sedimentation with correlated inflows. *J. Hydrol.* **1981**, *53*, 327–341. [CrossRef]
19. Aksoy, H.; Akar, T.; Unal, N.E. Wavelet analysis for modeling suspended sediment discharge. *Nord. Hydrol.* **2004**, *35*, 165–174. [CrossRef]
20. Hao, C.F.; Qiu, J.; Li, F.F. Methodology for analyzing and predicting the runoff and sediment into a reservoir. *Water* **2017**, *9*, 440. [CrossRef]
21. Garde, R.J.; Ranga Raju, K.G. *Mechanics of Sediment Transportation and Alluvial Stream Problems*; Wiley Eastern: New Delhi, India, 1977.
22. Tfwala, S.S.; Wang, Y.M. Estimating sediment discharge using sediment rating curves and artificial neural networks in the Shiwen River, Taiwan. *Water* **2016**, *8*, 53. [CrossRef]
23. Araujo, J.C.; Güntner, A.; Bronstert, A. Loss of reservoir volume by sediment deposition and its impact on water availability in semiarid Brazil. *Hydrol. Sci. J.* **2006**, *51*, 157–170. [CrossRef]
24. Baban, S.M.J.; Yusof, K.W. Modeling soil erosion in tropical environments using remote sensing and geographical information systems. *Hydrol. Sci. J.* **2001**, *46*, 191–198. [CrossRef]
25. Wicks, J.M. Physically-Based Mathematical Modelling of Catchment Sediment Yield. Ph.D. Thesis, Department of Civil Engineering, University of Newcastle Upon Tyne, Tyne, UK, 1988.
26. Nearing, M.A.; Foster, G.R.; Lane, L.J.; Finkner, S.C. A process-based soil erosion model for USDA-water erosion prediction project technology. *Trans. ASAE* **1989**, *32*, 1587–1593. [CrossRef]
27. Ayele, G.T.; Teshale, E.Z.; Yu, B.; Rutherfurd, I.D.; Jeong, J. Streamflow and sediment yield prediction for watershed prioritization in the Upper Blue Nile river basin, Ethiopia. *Water* **2017**, *9*, 782. [CrossRef]
28. Kavvas, M.L.; Chen, Z.Q.; Dogrul, C.; Yoon, J.Y.; Ohara, N.; Liang, L.; Aksoy, H.; Anderson, M.L.; Yoshitani, J.; Fukami, K.; et al. Watershed environmental hydrology (WEHY) model based on upscaled conservation equations: Hydrologic module. *ASCE J. Hydrol. Eng.* **2004**, *9*, 450–464. [CrossRef]
29. Kavvas, M.L.; Yoon, J.Y.; Chen, Z.Q.; Liang, L.; Dogrul, C.; Ohara, N.; Aksoy, H.; Anderson, M.L.; Reuter, J.; Hackley, S. Watershed environmental hydrology model: Environmental module and its application to a California watershed. *ASCE J. Hydrol. Eng.* **2006**, *11*, 261–272. [CrossRef]
30. Aksoy, H.; Kavvas, M.L. A review of hillslope and watershed scale erosion and sediment transport models. *Catena* **2005**, *64*, 247–271. [CrossRef]
31. Merritt, W.S.; Letcher, R.A.; Jakeman, A.J. A review of erosion and sediment transport models. *Environ. Model. Softw.* **2003**, *18*, 761–799. [CrossRef]
32. Milly, P.C.D.; Betancourt, J.; Falkenmark, M.; Hirsch, R.M.; Kundzewicz, Z.W.; Lettenmaier, D.P.; Stouffer, R.J. Stationarity is dead: Whither water management? *Science* **2008**, *319*, 573–574. [CrossRef]
33. Montanari, A.; Young, G.; Savenije, H.H.G.; Hughes, D.; Wagener, T.; Ren, L.L.; Koutsoyiannis, D.; Cudennec, C.; Toth, E.; Grimaldi, S.; et al. Panta Rhei-Everything Flows: Change in hydrology and society-The IAHS Scientific Decade 2013–2022. *Hydrol. Sci. J.* **2013**, *58*, 1256–1275. [CrossRef]
34. McMillan, H.; Montanari, A.; Cudennec, C.; Savenjie, H.; Kreibich, H.; Krüger, T.; Liu, J.; Meija, A.; van Loon, A.; Aksoy, H.; et al. PantaRhei 2013-2015: Global perspectives on hydrology, society and change. *Hydrol. Sci. J.* **2016**, *61*, 1174–1191. [CrossRef]
35. Ceola, S.; Montanari, A.; Krueger, T.; Dyer, F.; Kreibich, H.; Westerberg, I.; Carr, G.; Cudennec, C.; Elshorbagy, A.; Savenije, H.; et al. Adaptation of water resources systems to changing society and environment: A statement by the International Association of Hydrological Sciences. *Hydrol. Sci. J.* **2016**, *61*, 2803–2817. [CrossRef]
36. Bu, J.; Lu, C.; Niu, J.; Gao, Y. Attribution of runoff reduction in the Juma River basin to climate variation, direct human intervention, and land use change. *Water* **2018**, *10*, 1775. [CrossRef]
37. Koutsoyiannis, D. Hurst-Kolmogorov dynamics and uncertainty. *J. Am. Water Resour. Assoc.* **2011**, *47*, 481–495. [CrossRef]
38. Lins, H.F.; Cohn, T.A. Stationarity: Wanted dead or alive? *J. Am. Water Resour. Assoc.* **2011**, *47*, 475–480. [CrossRef]

39. Matalas, N.C. Comment on the announced death of stationarity. *J. Water Resour. Plann. Manag.* **2012**, *138*, 311–312. [CrossRef]
40. Koutsoyiannis, D.; Montanari, A. Negligent killing of scientific concepts: The stationarity case. *Hydrol. Sci. J.* **2015**, *60*, 1174–1183. [CrossRef]
41. Grimaldi, S.; Kao, S.-C.; Castellarin, A.; Papalexiou, S.-M.; Viglione, A.; Laio, F.; Aksoy, H.; Gedikli, A. *Statistical Hydrology, Treatise on Water Science*; Wilderer, P., Ed.; Academic Press: Oxford, UK, 2011; Volume 2, pp. 479–517.
42. Gedikli, A.; Aksoy, H.; Unal, N.E. Segmentation algorithm for long time series analysis. *Stoch. Environ. Res. Risk Assess.* **2007**, *22*, 291–302. [CrossRef]
43. Aksoy, H.; Gedikli, A.; Unal, N.E.; Kehagias, A. Fast segmentation algorithms for long hydrometeorological time series. *Hydrol. Process.* **2008**, *22*, 4600–4608. [CrossRef]
44. Gedikli, A.; Aksoy, H.; Unal, N.E.; Kehagias, A. Modified dynamic programming approach for offline segmentation of long hydrometeorological time series. *Stoch. Environ. Res. Risk Assess.* **2010**, *24*, 547–557. [CrossRef]
45. Szilo, J.; Bialik, R.J. Grain size distribution of bedload transport in a gliaciated catchment (Baranowski Glacier, King George Island, Western Antartctica). *Water* **2018**, *10*, 360. [CrossRef]
46. Song, Y.H.; Lee, E.H.; Lee, J.H. Functional relationship between soil slurry transfer and deposition in urban sewer conduits. *Water* **2018**, *10*, 825. [CrossRef]
47. Hallouz, F.; Meddi, M.; Mahe, G.; Toumi, S.; Rahmani, S.E.A. Erosion, suspended sediment transport and sedimentation on the Wadi Mina at the Sidi M'Hamed Ben Aouda Dam, Algeria. *Water* **2018**, *10*, 895. [CrossRef]
48. Sadaoui, M.; Ludwig, W.; Bourrin, F.; Bissonnais, Y.L.; Romero, E. Anthropogenic reservoirs of various sizes trap most of the sediment in the Mediterranean Maghreb Basin. *Water* **2018**, *10*, 927. [CrossRef]
49. Kotti, F.; Dezileau, L.; Mahe, G.; Habaieb, H.; Benabdallah, S.; Bentkaya, M.; Calvez, R.; Dieulin, C. The impact of dams and climate on the evolution of the sediment loads to the sea by the Medjerda River using a paleo-hydrological approach. *J. Afr. Earth. Sci.* **2018**, *142*, 226–233. [CrossRef]
50. Ben Moussa, T.; Amrouni, O.; Hzami, A.; Dezileau, L.; Mahe, G.; Condomines, M.; Saadi, A. Progradation and retrogradation of the Medjerda delta during the 20th century (Tunisia, Western Mediterranean). *Compte. Rendus Geosci.* **2019**, *351*, 340–350. [CrossRef]
51. Wang, Y.-C.; Lai, C.-C. Evaluating the erosion process from a single-stripe laser-scanned topography: A laboratory case study. *Water* **2018**, *10*, 956. [CrossRef]
52. Guo, S.; Zhu, Z.; Lyu, L. Effects of climate change and human activities on soil erosion in the Xihe River Basin, China. *Water* **2018**, *10*, 1085. [CrossRef]
53. Adams, R.; Quinn, P.; Barber, N.; Reaney, S. The role of attenuation and land management in small catchments to remove sediment and phosphorus: A modelling study of mitigation options and impacts. *Water* **2018**, *10*, 1227. [CrossRef]
54. Unal, N.E. Shear stress-based analysis of sediment incipient deposition in rigid boundary open channels. *Water* **2018**, *10*, 1399. [CrossRef]
55. Chao, Y.-C.; Chen, C.-W.; Li, H.-C.; Chen, Y.-M. Riverbed migrations in Western Taiwan under climate change. *Water* **2018**, *10*, 1631. [CrossRef]
56. Török, G.T.; Józsa, J.; Baranya, S. A shear Reynolds number-based classification method of the nonuniform bed load transport. *Water* **2019**, *11*, 73. [CrossRef]
57. Jakubínský, J.; Pechanec, V.; Procházka, J.; Cudlín, P. Modelling of soil erosion and accumulation in an agricultural landscape—A comparison of selected approaches applied at the small stream basin level in the Czech Republic. *Water* **2019**, *11*, 404. [CrossRef]

© 2019 by the authors. Licensee MDPI, Basel, Switzerland. This article is an open access article distributed under the terms and conditions of the Creative Commons Attribution (CC BY) license (http://creativecommons.org/licenses/by/4.0/).

Article

Grain Size Distribution of Bedload Transport in a Glaciated Catchment (Baranowski Glacier, King George Island, Western Antarctica)

Joanna Sziło [1,*] and Robert Józef Bialik [2]

1. Institute of Geophysics, Polish Academy of Sciences, 01-452 Warsaw, Poland
2. Institute of Biochemistry and Biophysics, Polish Academy of Sciences, 02-106 Warsaw, Poland; rbialik@ibb.waw.pl
* Correspondence: jszilo@igf.edu.pl; Tel.: +48-504-595-266

Received: 20 February 2018; Accepted: 20 March 2018; Published: 23 March 2018

Abstract: The relationships among grain size distribution (GSD), water discharge, and GSD parameters are investigated to identify regularities in the evolution of two gravel-bed proglacial troughs: Fosa Creek and Siodło Creek. In addition, the potential application of certain parameters obtained from the GSD analysis for the assessment of the formation stage of both creeks is comprehensively discussed. To achieve these goals, River Bedload Traps (RBTs) were used to collect the bedload, and a sieving method for dry material was applied to obtain the GSDs. Statistical comparisons between both streams showed significant differences in flow velocity; however, the lack of significant differences in bedload transport clearly indicated that meteorological conditions are among the most important factors in the erosive process for this catchment. In particular, the instability of flow conditions during high water discharge resulted in an increase in the proportion of medium and coarse gravels. The poorly sorted fine and very fine gravels observed in Siodło Creek suggest that this trough is more susceptible to erosion and less stabilized than Fosa Creek. The results suggest that GSD analyses can be used to define the stage of development of riverbeds relative to that of other riverbeds in polar regions.

Keywords: GSD; proglacial channels; bedload transport; field measurements; fluvial erosion

1. Introduction

One of the most important challenges in studies of bedload transport is investigating the differences between mobile and bottom particle sizes and taking these relations into account in new transport models [1–4]. This problem is mostly based on the lack of complex datasets for flow velocity, transport, and grain size distribution (GSD) of the bedload and bottom particles [2]. The lack of sufficient field observations and datasets collected during high flows when active transport is observed [5,6] is not the only problem associated with model development and geomorphological process assessments in Arctic or Antarctic catchments. The limited application of current models to natural gravel-bed rivers [7], and methods of predicting the sediment flux to the global ocean from small rivers [8] represent other problems that remain to be resolved.

Bedload transport is one of the main factors associated with changes in the morphology of troughs [9–14], and the difficulty describing the relationships among water flow conditions, GSD, and deposition processes was identified decades ago (e.g., [15–18]). Although a substantial amount of progress has been made in this field because of research conducted in laboratory channels (e.g., [19–21]) and field studies (e.g., [22–26]) and with the use of bedload transport models (e.g., [1,2,27–30]), the relationships among the factors associated with bedload transport are poorly understood in natural gravel-bed channels in general and in those located in polar regions in particular [31].

Ghoshal et al. [32] claimed that the bedload has a large influence on the GSD and determines the concentration of each size fraction of noncohesive particles under changing hydrological conditions. GSD analyses have frequently been used as indicators of the threshold for the initiation of particle motion, and shear stress dictates the specific role of particles in this process [10,33–37]. Johnson [38] suggested that the geomorphology of the bed as well as the sediment transport rate should be considered because of their relationship with grain size and shape. Moreover, the distribution of particles also depends on the mechanism of fluvial transport and can be used to define the influence of flow conditions on the geomorphology of the channel, as stated by Kociuba and Janicki [31] or Lisle [23]. Although the coarser fraction is entrained with greater discharge [39], because of protrusion and hiding effects, particles larger than the mean grain size are easier to move than smaller ones [40–42]. The above-mentioned situation may be understood by using the GSD for the analysis of selective bedload transport in natural gravel-bed channels [43] or by performing a direct analysis of certain parameters, such as the mean grain size, sorting, or skewness, which can be used to predict the direction of sediment transport [27,44]. These parameters allow us to obtain information about the predominant size of the transported particles, their dispersal during this process, and their symmetry or preferential dispersal to one side of the average. Ashworth et al. [45] concluded that sediment sorting can occur during entrainment, transport, or deposition. In this research, sediment sorting is calculated based on the sediment caught in traps during bedload transport.

The objective of this study is to answer the following questions: (1) How is GSD modified during the times of peak discharge in a polar catchment? (2) What is the relationship between the GSD and GSD parameters and the amount of transported material in gravel-bed rivers? (3) Does the modification of GSDs during efficient bedload transport events allow for the identification of the development stage of river troughs during changing meteorological conditions?

To answer these questions, insightful measurements have been performed in Fosa and Siodło creeks, two proglacial gravel-bed channels located on the forefield of the Baranowski Glacier on King George Island in Western Antarctica (Figure 1). This study is a continuation and development of the results presented by Szilo and Bialik [46], who identified the relationship between bedload transport and rapid outflow and high water discharge in the form of eight-loop hysteresis for these creeks.

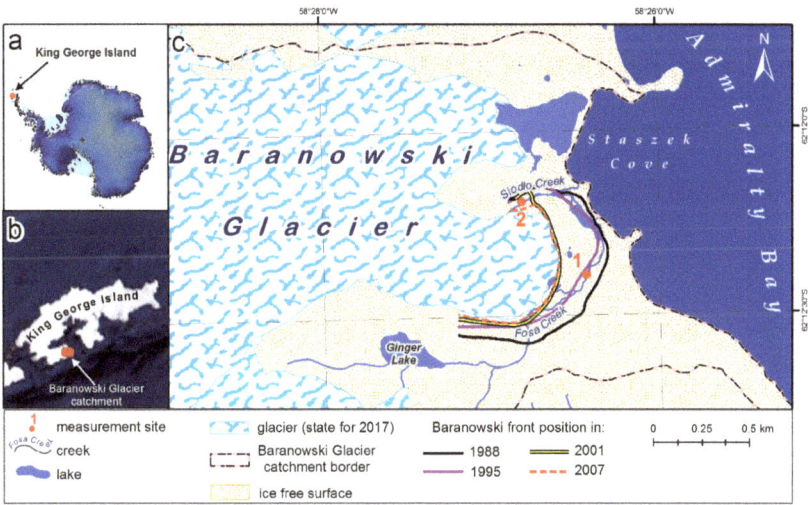

Figure 1. Location of the study site: (**a**) King George Island; (**b**) Baranowski Glacier catchment; (**c**) forefield of the glacier and measurement sites. Reference system: WGS 1984, UTM zone 21S, geoid EGM96.

2. Materials and Methods

The field campaign was conducted at two sites established in Fosa and Siodło creeks from 9 January to 11 February 2016 (Figure 1). Data on the bedload transport associated with water flow conditions measurements were collected at 24 h intervals over 35 days in both creeks. However, the measurements in Fosa Creek were extended to 2 March, and during those additional 14 days, bedload transport was not observed.

2.1. Study Site

The Antarctic Peninsula region, where the study site is located, is prone to climatic fluctuations [47–52]. The total catchment area of the creeks is 6.8 km² of which 69.1% (4.7 km²) is covered by the glacier (including Windy Glacier). The area of the forefield is 2.1 km² and is expanding yearly (status for 2017). The frontal part of the glacier is divided into two tongues: Northern and Southern. The Northern tongue is currently terminating on a narrow strip of land close to the shallow lagoon, and the Southern tongue presents several creeks flowing through the forefield. All measurements were initiated in two gravel-bed troughs of the Fosa and Siodło proglacial creeks, which were selected because of their continuous and dynamic water flows during the entire summer season. Fosa Creek receives water mainly from Ginger Lake and the Baranowski Glacier, whereas Siodło Creek is a subglacial outflow from the Baranowski Glacier. Both creeks are fed by meltwater from the glacier in the ablation season (which may represent the primary source of water) and intensive rainfalls. The measurement site in Fosa Creek was established in a location where ablation water cut the trough into moraine cover between 1988 and 1995, whereas in Siodło Creek, this trough was cut between 2001 and 2005. The total lengths (L) of Fosa and Siodło creeks are 1.40 km and 0.42 km, respectively; the longitudinal water surface slopes (S) are 0.07 mm^{-1} and 0.08 mm^{-1} for Fosa Creek and Siodło Creek, respectively (Table 1). Furthermore, Reynolds and Froude numbers are also presented for both creeks (Figure 2a,b).

Table 1. Basic hydraulic parameters of Fosa and Siodło creeks.

Parameter	Total Length (L) [km]	Average Bankfull Width (w) [m]	Average Bankfull Depth (d) [m]	Longitudinal Water Surface Slope (S) [mm^{-1}]	Maximum Discharge (Q_{max}) [m³ s^{-1}] [1]	Maximum Velocity (V_{max}) [ms^{-1}] [2]
Fosa Creek	1.40	2.82	0.14	0.07	0.75	0.91
Siodło Creek	0.42	2.15	0.09	0.08	0.26	1.25

Note: [1] Q_{max}: maximum water discharge measured at one day during field campaign; [2] V_{max}: maximum water velocity measured at one point during field campaign.

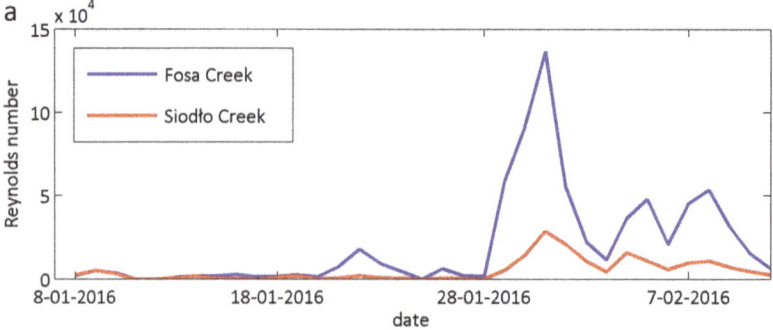

Figure 2. *Cont.*

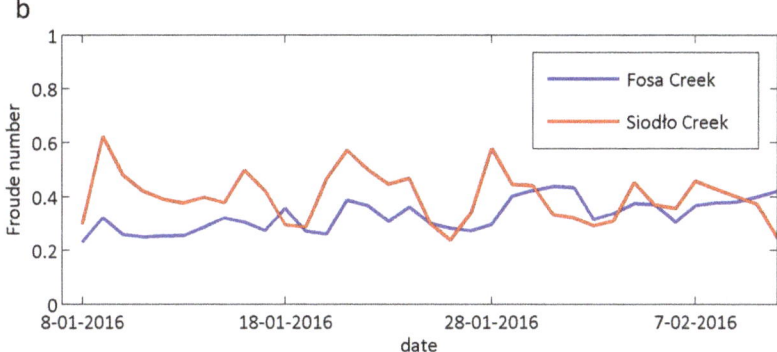

Figure 2. Temporal changes in (**a**) Reynolds number and (**b**) Froude number for both creeks.

The period of the field campaign was characterized by higher air temperatures relative to other summer months over the last decade [49,50,53]. The mean monthly air temperature was +1.0 °C in January and +1.1 °C in February for the period from 2007 to 2016 [54]. However, the mean air temperature during the field campaign was +1.8 °C, with a maximum of +5.9 °C on 29 January. On the same day, the highest precipitation of 8.8 mm was recorded. During the 35 days of continuous measurements, 16 days were without precipitation and 12 days had precipitation >1.0 mm. The total precipitation over the whole period was 47.7 mm. Meteorological conditions that contributed to greater water discharge were observed 2 days after intense rainfall (Figure 3) [46].

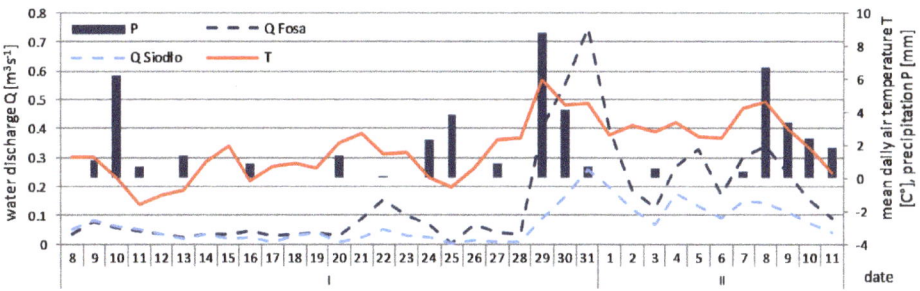

Figure 3. Mean daily air temperature, precipitation, and discharge during the field campaign in 2016 from Bellingshausen Station.

2.2. Bedload Transport Measurements

To collect the transported material, River Bedload Traps (RBT) [55,56] were anchored at the bottoms of the troughs in two cross sections (Figure 4). In Fosa Creek, a set of two RBT modules at a distance of 300 m from the forehead of the glacier was installed, and in Siodło Creek, a set of three RBT modules at a distance of 15 m from the front of the glacier was installed. One additional module was established in Siodło Creek because less rinsing of the bottom of the trough occurred in this creek compared with in Fosa Creek and a finer bedload was visible at the beginning of the measurements. The assumption was that the bedload can move more easily in a trough in an early stage of development (Siodło Creek) than in a later stage of development [46]. The locations of the modules in the profiles were chosen based on the most intense currents.

  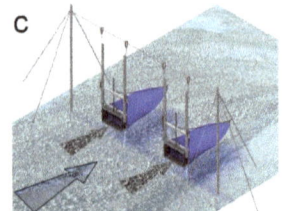

Figure 4. River Bedload Traps (RBT) in (**a**) Fosa Creek and (**b**) Siodło Creek (modified from Sziło and Bialik [46]); (**c**) schematic diagram for the bedload transport technique (Kociuba [56]).

2.3. Sieve Method and GSD Analysis

To obtain the GSD, each of the collected samples of the bedload was transported, drained, sieved, and weighed in the laboratory of the Henryk Arctowski Polish Antarctic Station. Furthermore, a comparison of the dry and wet sample weights was performed to determine the most reliable method of preparing the data for further analysis. Each sample that had a greater than 1% proportion of grain size <2 mm was truncated to exclude sediment that was predominantly suspended and because of limitations of the mesh diameter of the net. The upper limit to remove the samples because of the selective rejection of particles by the sample was set up at 32 mm [23]. Furthermore, for the GSD analysis, average weight values lower than 3% (chosen arbitrarily) for all collected data were rejected as not statistically important. To obtain information on the sedimentary environment, such as the sediment provenance, transport, and deposition conditions (e.g., [15,18,24]), the GSD parameters were evaluated. The Folk and Ward [15] method was employed to obtain the GSD parameters of mean grain size (Mz), standard variation (σ_ϕ), sorting (σ), skewness (Skl), and kurtosis (KG), which according to Blot and Pye [57] describe (a) the average size of grains (Mz); (b) the spread of the sizes around the average (σ); (c) the preferential spread to one side of the average (Skl); and (d) the degree of concentration of the grains relative to the average (KG). As suggested by Blot and Pye [57], the Folk and Ward [15] method *provides the most robust basis for routine comparisons of compositionally variable sediments* in the cases when the skewness or kurtosis of grain sizes are calculated. It should be noted that besides the presentation of grain diameters in metric units, according to standard procedure [15,57], data will be also shown and analyzed in phi (φ) units, which is the negative log to base 2 of the diameter in mm.

2.4. Bedload Transport Rate

The bedload transport rate (q_b) was calculated from the weighted arithmetic mean formula [46] given in Equation (1):

$$q_b = \frac{G_s}{S_w t} \cdot W_t \qquad (1)$$

where G_s represents the bedload material (kg), S_w represents the width of the module inlet (m), W_t represents the width of the trough (m), and t represents the measurement time (day). In addition, the water velocity was measured with an Electromagnetic Open Channel Flow Meter (manufactured by Valeport, model 801). For details, see Sziło and Bialik [46]. The average water velocity was measured at least two times [46] at each measurement section for 30 seconds [58]. If the standard deviation of the measurement values was higher than 10%, the measurements were repeated until an acceptable standard deviation was obtained. Based on the values of creek bathymetry and water velocities for each section, the water discharge was calculated following the standard procedure [58]. The accuracy of equipment employed for measurements was ±0.5% of reading values plus 5 mm/s.

2.5. Statistical Analysis

An analysis of variance (ANOVA) test was performed to determine whether or not the two investigated creeks had statistically significant differences in terms of bedload transport and hydrological conditions. The Anderson–Darling test was used to check for normality, and Bartlett's test was used to check for the homogeneity of variance to meet the assumptions of the ANOVA. To provide a better description of the changes that occurred in the channels of both creeks, sediment samples from the individual traps were analyzed independently even when collected from the same creek.

3. Results

3.1. Statistical Analysis: Water Discharge, Flow Velocity, and Bedload Transport

Figure 5 shows the bedload transport, water discharge, and flow velocity data for Fosa and Siodło creeks. Unfortunately, the assumptions of normality and homogeneity of variance were not observed for all presented data. When a variable does not fit the assumptions required for the ANOVA test, the data are usually transformed (log or square root) [59]. The square root transformation was used for water discharge and flow velocity, and the double square root transformation was used for bedload transport; the transformed variables met the above assumptions. A comparison of the results of all transformed data indicated only one of the variables (flow velocity) between the creeks presented statistically significant differences (p-value < 0.05).

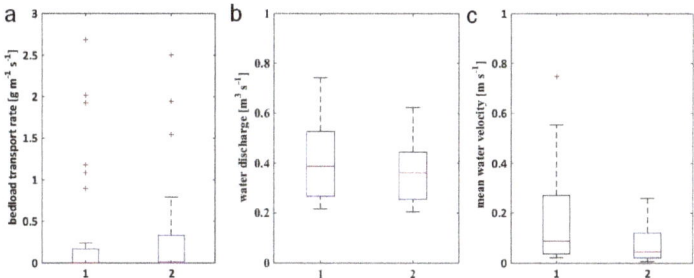

Figure 5. Bedload transport (**a**), water discharge (**b**), and flow velocity (**c**) data for Fosa Creek (1) and Siodło Creek (2).

3.2. Sieving Procedure Comparison (Wet and Dry)

Although standard procedures for wet sieving are acceptable for determining the GSD (e.g., [31]), the form of sieving (wet or dry) may influence the particle distribution [60,61]. To investigate how the drying process can influence the GSD, the weights of randomly chosen wet and dry bedloads not lower than 10 kg were compared (Figure 6). The results of the analysis revealed that the difference between the weights ranged from 15% to 28% (Figure 6). Hence, in this research, only the dried bedload was considered in further analyses.

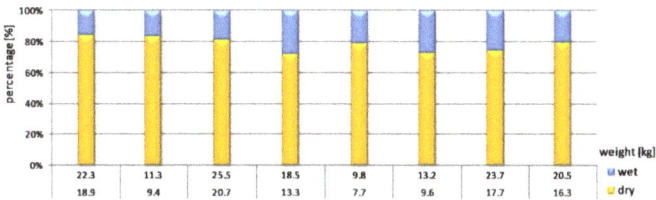

Figure 6. Weight fractions of the wet and dry samples.

3.3. GSD Parameters and Variation of Characteristic Diameters (D) of the Bedload Transport Rate

Figures 7 and 8 show the relationship between σ and Mz for all data, for both creeks separately, and for the individual traps. The analysis of the σ and Mz in the profiles established in the Fosa and the Siodło troughs suggested that linear correlations occurred with those indicators, and the correlation coefficients (R^2) were 0.82–0.92 (Figure 7). Moreover, an increase in Mz presents a corresponding increase in grain variability. The mean diameter of sediment from Fosa Creek for all measurement days varied from 2.9 mm to 6.2 mm and was higher than that in Siodło Creek, which varied from 1.5 mm to 3.6 mm. Furthermore, larger ranges of Mz values were observed for the individual traps in Siodło Creek than for those in Fosa Creek, and a maximum of 10.4 mm was observed on 6 February (Figure 8a,b). The strongest correlation for the material collected in the trap occurred in the Siodło middle (M) at $R^2 = 0.96$ (Figure 8b), for which 77% of the data ranged from 2.3 to 3.9 mm, whereas for the Siodło left (L), 80% of the data were in the interval from 4.1 to 8.9 mm.

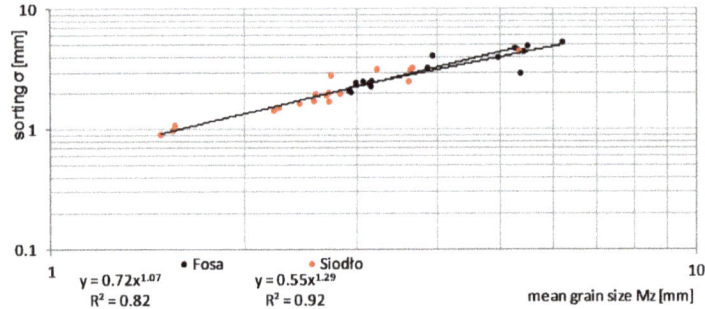

Figure 7. Relation between sorting (σ) and mean grain size (Mz) in Fosa and Siodło creeks in all profiles.

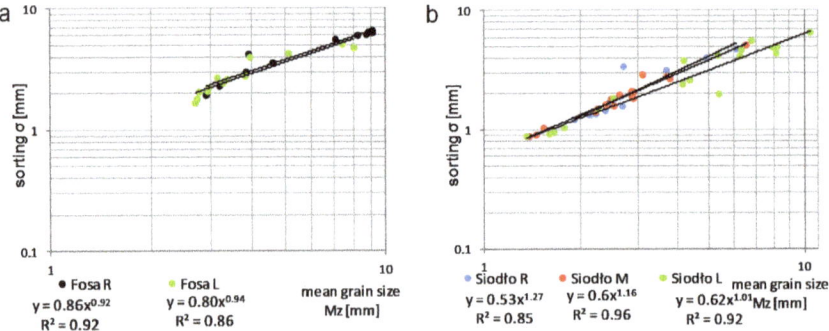

Figure 8. Relation between sorting (σ) and mean grain size (Mz) in Fosa (**a**) and Siodło (**b**) creeks in individual traps.

The temporal changes in bedload transport and σ for the studied creeks are presented in Figure 9. The σ in individual traps varied from 1 mm to 1.5 mm in 93% of the samples, which means that the collected sediments in both cross sections were poorly sorted (Figure 9a,b). Furthermore, after the peak bedload transport, which appeared from 29 to 30 January, an inversion of σ occurred from lower values in the Fosa right (R) section compared with in the Fosa left (L) section, where higher values were observed (Figure 9a). A similar situation occurred in the Siodło trough: after both of the bedload transport peaks (on 1 and 5 February), inversions of σ in the left trap from highest to lowest and then to highest values again were noticed (Figure 9b). Furthermore, the results suggest that better σ occurred in both troughs after each peak in bedload transport (Figure 9a,b).

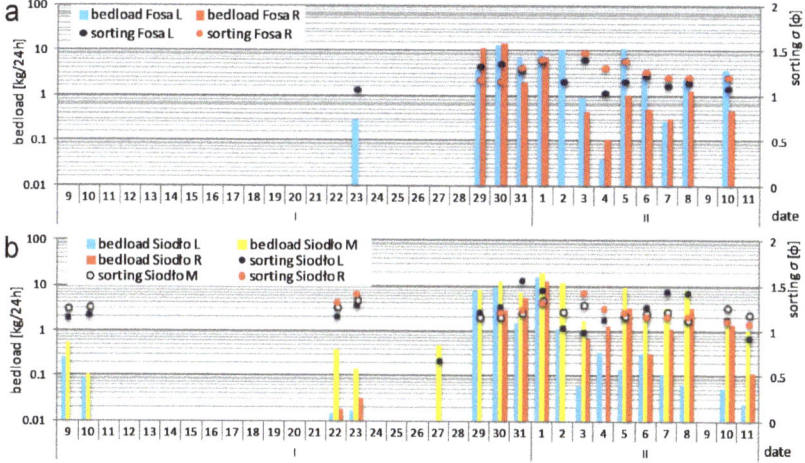

Figure 9. Bedload transport and sorting in individual traps in the Fosa (**a**) and Siodło (**b**) creeks.

The relation between bedload transport and σ was inversely proportional before and directly proportional after the peaks in bedload transport (Figure 10a,b). All bedload transport data from Siodło Creek were negatively skewed, which means that a large proportion of coarse sediment occurred (Figure 10c), and this finding is also confirmed based on the percentage of gravels in the collected material (Figure 11a,b). An increase in water discharge and a consequential increase in bedload transport resulted in symmetrical *Skl* and leptokurtic *KG* (Figure 10c,d). Before 31 January and after 4 February, when the maximum discharge was observed, strongly negative *Skl* was observed in Siodło (Figure 10c). Under low discharge conditions, leptokurtic *KG* was observed.

The bedload transport rate and grain size variation analyses indicate that an inverse correlation occurred for both creeks. In Fosa Creek, the grain size variations decrease when the bedload transport rate increases, whereas in Siodło Creek, the grain size variations are directly proportional to increases in bedload transport rate. Furthermore, in Fosa Creek, the D-value correlations increase from 0.15 for D_{50} to 0.26 for D_{95}, which is inconsistent with the data from Siodło Creek, where the correlation for D_{50} of 0.29 decreases to 0.11 for D_{95} (Figure 11a,b).

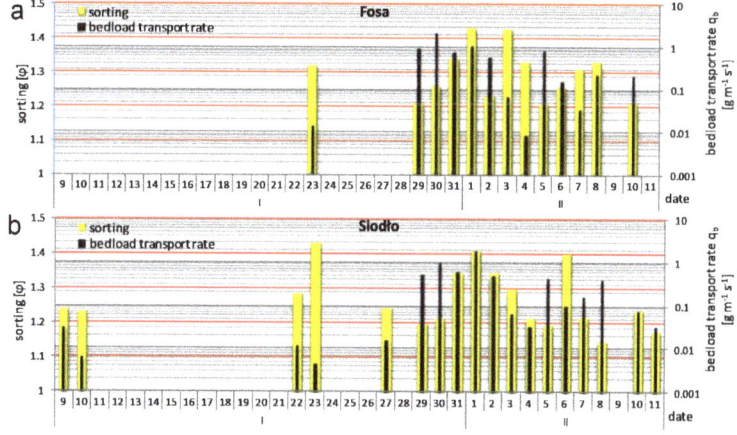

Figure 10. *Cont.*

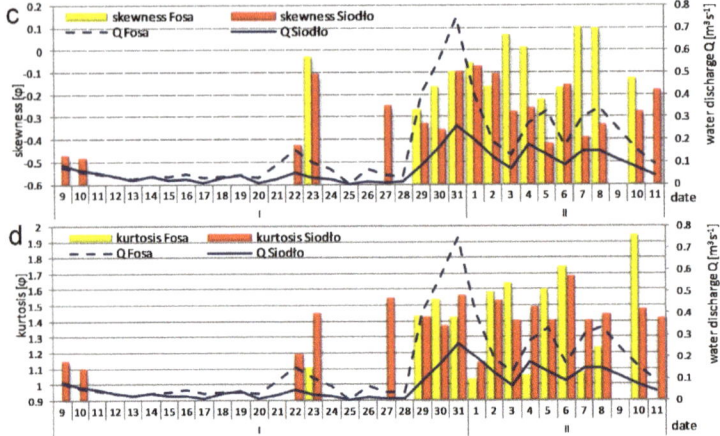

Figure 10. Grain size parameters (by Folk and Ward, 1957) for Fosa and Siodło Creeks: bedload transport rate and sorting for Fosa (**a**) and Siodło (**b**); skewness (**c**); and kurtosis (**d**).

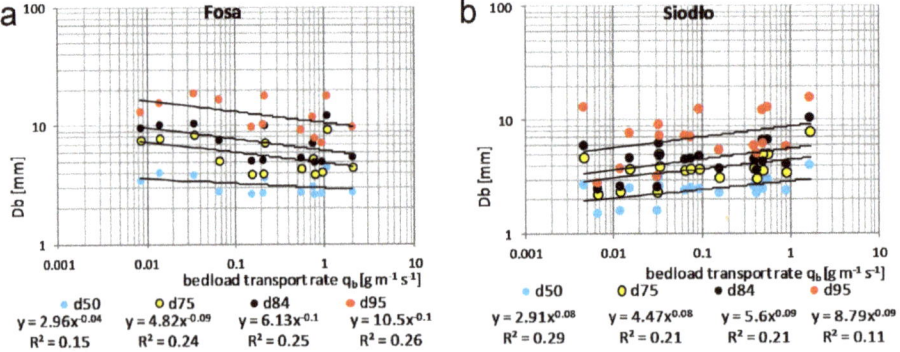

Figure 11. Correlation between bedload transport rate and characteristic diameters D_b for Fosa (**a**) and Siodło (**b**) creeks, where b stands for the percentage of the weight of the sample finer than this value.

3.4. Selective Transport and Bedload Transport versus Water Discharge and Water Velocity

Fine (4–8 mm) and very fine (2–4 mm) gravel accounted for 40 to 99% of the total sample weight in the troughs for each day of the measurement campaign. Nevertheless, a higher proportion of these gravels (2–8 mm) was observed in Siodło Creek, with the value reaching at least 63%. This increase was especially clear before the peaks of water discharge on 19 and 29 January, when the fine and very fine gravel constituted almost the entirety of the Siodło Creek samples. Furthermore, because of the increase in water discharge, the proportion of medium (8–6 mm) and coarse (16–22 mm) gravel also increased in both creeks, particularly on 29 to 31 January and 4 to 5 February (Figure 12a,b), when the highest water discharge occurred. In addition to the previously mentioned periods when the low water discharge was observed, certain coarse sediment samples were identified. Nonetheless, these samples had insignificant weight fractions and were disrupted by the presence of a few coarse pebbles. A similar situation was observed when the water discharge peaked. At that time, several boulders were caught in the Siodło trough, which contributed more than 50% of the sample in certain cases, although the water discharge was two times lower than that in Fosa Creek.

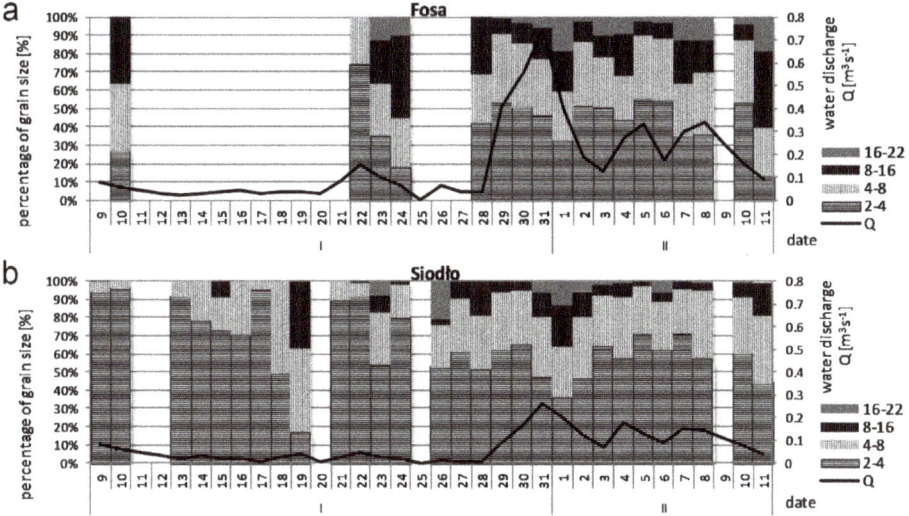

Figure 12. Percentage of bedload vs water discharge for Fosa (**a**) and Siodło (**b**) creeks.

Because of the high proportions of fine and very fine gravel in each sample, those grain sizes have been chosen for our examination of the correlation between bedload transport and water discharge or water velocity (Figures 13 and 14). The correlation between water discharge and bedload transport was equal to 0.65 in the Fosa and 0.55 in the Siodło, and these correlations were stronger than those between bedload transport and water velocity, which were 0.53 in Fosa and 0.28 in Siodło; this is likely due to the changes of the creek beds and shapes, due to the existence of scouring effects [56] and because only grains with diameter lower than 22 mm were taken into account for the analysis. Thus, the increase in water discharge was observable with the simultaneous increase in bedload transport, and the intensity or correlation in the relations between water velocity and bedload transport was lower.

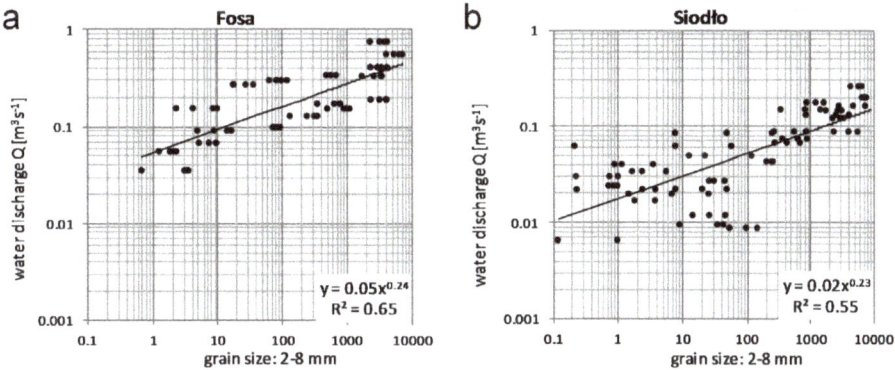

Figure 13. Relationship between water discharge and quantity of fine and very fine gravel in Fosa (**a**) and Siodło (**b**) creeks.

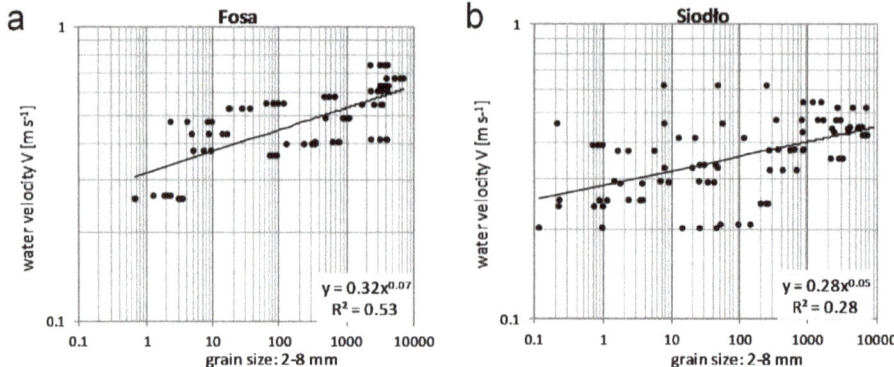

Figure 14. Relationship between water velocity and quantity of fine and very fine gravel in Fosa (**a**) and Siodło (**b**) creeks.

4. Discussion

The weights of the wet and dry samples differ significantly—by as much as 28% and by an average of 21.5%. These results are inconsistent with those of Kociuba and Janicki [31], who reported that the average water amount in a natural gravel-bed channel was 6%. In this situation, the error in the calculation can reach maximum values of 20.1–21.3 kg d^{-1} for the mean transport rate [31] and 33.3–67.8 kg d^{-1} for the mean daily bedload flux [62] based on exemplary data from the Scott River (Svalbard).

The flow velocity analysis for both creeks suggests that a statistically significant difference (p-value < 0.05) occurred in the hydraulic conditions. A stronger correlation with bedload transport was revealed for water discharge than for water velocity, which was estimated based on the measured values of the water velocity. This result explained the previously reported findings of Sziło and Bialik [46], who suggested that a similar relationship occurred between water discharge and bedload transport for Fosa and Siodło creeks. Moreover, the lack of significant differences in bedload transport clearly indicated that this catchment process was strongly associated with meteorological conditions, as presented in Figure 2. Certain differences noted by the authors in their previous work were likely caused by differences in the geomorphological features of the analyzed troughs, which could be explained based on the GSD analysis.

The detailed GSD analysis has shown that σ is inversely proportional to Mz, which is consistent with the results of studies in the Scott River performed by Kociuba and Janicki [31]. Kociuba and Janicki suggested that the bed material was poorly sorted, which is similar to the results from the Fosa and Siodło creeks (Figure 8a,b). Moreover, grains in the investigated troughs were finer (3 mm) than those in the Scott River (10 mm) [31]), which is likely because both creeks in this study are in the earlier stages of formation and more sensitive to morphological changes (especially Siodło Creek, as suggested by Sziło and Bialik [46]) in relation to the Scott River. These troughs were cut down into moraine cover later (in recent decades, Figure 1) than was the trough of the Scott River [63]. Poorly sorted bedload (Figure 9a,b) suggests high energetic differentiation of the particles and rapid changes in the geomorphology of both troughs. Significant bedload transport rates based on the definition of Batalla et al. [36] of 1 gm^{-1}s^{-1} were observed on 30 January and 1 February in the troughs. Nevertheless, the bedload transport peak was defined as q_b > 0.1 gm^{-1}s^{-1}. As a consequence, the sizes of movable grains are more variable and present a larger proportion of finer grains compared with that in the bed channels, where they could have already been rinsed. This finding could be related to the higher transport of the bedload and faster changes in the morphology of the troughs. Other GSD parameters, e.g., negative Skl, are typical for natural gravel-bed channels [23,31,64,65].

The investigation of the correlation between grain size variations and bedload transport rate revealed that both troughs were characterized by an inverse correlation. In Fosa Creek, this correlation increased proportionally to the D values and was strongest for D_{95}, which is more sensitive to water flow as concluded by Batalla et al. [36]. This result is similar to findings for the Salty River [36] and the Scott River [31]. Nevertheless, for Siodło Creek, a higher value was observed for D_{50} than for D_{95}.

For both creeks, finer bedload is transported before the peak of water discharge than that transported during and after this phenomenon, which is consistent with findings of the previous studies (e.g., [23]). Generally, two phases of selective transport can be distinguished in both creeks (Figure 11). In the first phase, very fine gravel was predominantly entrained; this was followed by the second phase, where gravel dominated. Similar observations of two-phase selective transport have been reported in previous studies of other natural gravel-bed channels [31,36,66,67]. Furthermore, the higher proportions of fine and very fine gravel in Siodło Creek compared with those in Fosa Creek over the entire field campaign confirm the previous hypothesis posed by Sziło and Bialik [46], who assumed that the Siodło trough may be in an earlier stage of formation than the Fosa trough and thus exhibit a more obvious erosive process. In addition, fine material collected from the Siodło, which constituted at least 63% in each sample, can also provide evidence that the channel is not yet armored and paved like the Fosa trough, and bottom erosion may occur faster [33]. Moreover, the presence of boulders (>64 mm) during the peaks of water discharge not exceeding even one-fourth bankfull, mostly in the Siodło trough [46], may suggest that the greater part of the sediment was entrained rather than laterally eroded, as has been observed in other gravel-bed channels during near-bankfull discharge [23,31].

5. Conclusions

According to the aims of this study, the analysis confirmed that the variability and modification of GSDs are strongly related to the daily variability of bedload transport dynamics, which was suggested for the first time for polar catchments by Kociuba and Janicki [31]. During the measurement period, i.e., the ablation period, the weather conditions played a significant role in the processes observed in the bed channels of streams, especially when low water discharge occurred. In addition, the increased proportion of medium and coarse gravel was strictly proportional to the increases in water discharge. The simultaneous measurements performed in two creeks that are located in the same catchment but present partial differences in the sources of the material (i.e., Fosa Creek was also fed with lake material) indicated the processes occurring in both troughs. The bedload material from Siodło Creek consisted mostly of fine and very fine gravels, suggesting an early stage of formation and an efficient erosive process, which is confirmed by the general conditions of both streams. Fosa Creek has been free of ice cover for more than 30 years and is at a more stable development stage in which water discharge is the predominant control mechanism for the channel geometry. However, the GSD analysis showed that bedload transport can also significantly influence this process.

Acknowledgments: This publication has been partially financed by funds from the Leading National Research Centre (KNOW) received by the Centre for Polar Studies for the period 2014–2018.

Author Contributions: J.S. and R.J.B. designed the study and conducted the field campaign and laboratory analysis. J.S. conducted the majority of the data analysis and wrote the majority of the paper. J.S. and R.J.B. both contributed to the final version of the manuscript.

Conflicts of Interest: The authors declare no conflicts of interest.

References

1. Wu, W.; Wang, S.S.; Jia, Y. Nonuniform sediment transport in alluvial rivers. *J. Hydraul. Res.* **2000**, *38*, 427–434. [CrossRef]
2. Wilcock, P.R.; Crowe, J.C. Surface-based transport model for mixed-size sediment. *J. Hydraul. Eng.* **2003**, *129*, 120–128. [CrossRef]
3. Lukerchenko, N.; Chara, Z.; Vlasak, P. 2D Numerical model of particle–bed collision in fluid-particle flows over bed. *J. Hydraul. Res.* **2006**, *44*, 70–78. [CrossRef]

4. Bialik, R.J. Numerical study of saltation of non-uniform grains. *J. Hydraul. Res.* **2011**, *49*, 697–701. [CrossRef]
5. Andrews, E.D.; Erman, D.C. Persistence in the size distribution of surficial bed material during an extreme snowmelt flood. *Water Resour. Res.* **1986**, *22*, 191–197. [CrossRef]
6. Powell, D.M.; Reid, I.; Laronne, J.B. Evolution of bed load grain size distribution with increasing flow strength and the effect of flow duration on the caliber of bed load sediment yield in ephemeral gravel bed rivers. *Water Resour. Res.* **2001**, *37*, 1463–1474. [CrossRef]
7. Török, G.T.; Baranya, S.; Rüther, N. 3D CFD modeling of local scouring, bed armoring and sediment deposition. *Water* **2017**, *9*, 56.
8. Syvitski, J.P.; Peckham, S.D.; Hilberman, R.; Mulder, T. Predicting the terrestrial flux of sediment to the global ocean: A planetary perspective. *Sediment. Geol.* **2003**, *162*, 5–24. [CrossRef]
9. Ashworth, P.J.; Ferguson, R.I. Interrelationships of channel processes, changes and sediments in a proglacial braided river. *Geogr. Ann. Ser. Phys. Geogr.* **1986**, *68*, 361–371. [CrossRef]
10. Carson, M.A.; Griffiths, G.A. Bedload transport in gravel channels. *J. Hydrol. N. Z.* **1987**, *26*, 1–151.
11. Ferguson, R.I.; Ashmore, P.E.; Ashworth, P.J.; Paola, C.; Prestegaard, K.L. Measurements in a braided river chute and lobe: 1. Flow pattern, sediment transport, and channel change. *Water Resour. Res.* **1992**, *28*, 1877–1886. [CrossRef]
12. Goff, J.R.; Ashmore, P. Gravel transport and morphological change in braided Sunwapta River, Alberta, Canada. *Earth Surf. Process. Landf.* **1994**, *19*, 195–212. [CrossRef]
13. Hassan, M.A.; Church, M.; Lisle, T.E.; Brardinoni, F.; Benda, L.; Grant, G.E. Sediment transport and channel morphology of small, forested streams. *JAWRA J. Am. Water Resour. Assoc.* **2005**, *41*, 853–876. [CrossRef]
14. Kociuba, W.; Janicki, G.; Siwek, K.; Gluza, A. Bedload transport as an indicator of contemporary transformations of arctic fluvial systems. In *Monitoring Simulation Prevention and Remediation of Dense and Debris Flows IV*; WIT Press: Boston, MA, USA, 2012; pp. 125–135.
15. Folk, R.L.; Ward, W.C. Brazos River bar: A study in the significance of grain size parameters. *J. Sediment. Petrol.* **1957**, *27*, 3–26. [CrossRef]
16. Visher, G.S. Grain size distributions and depositional processes. *J. Sediment. Res.* **1969**, *39*, 1074–1106.
17. Bagnold, R.A. Bed load transport by natural rivers. *Water Resour. Res.* **1977**, *13*, 303–312. [CrossRef]
18. Friedman, G.M. Differences in size distributions of populations of particles among sands of various origins: addendum to IAS Presidential Address. *Sedimentology* **1979**, *26*, 859–862. [CrossRef]
19. Proffitt, G.T.; Sutherland, A.J. Transport of non-uniform sediments. *J. Hydraul. Res.* **1983**, *21*, 33–43. [CrossRef]
20. Wilcock, P.R.; McArdell, B.W. Surface-based fractional transport rates: Mobilization thresholds and partial transport of a sand-gravel sediment. *Water Resour. Res.* **1993**, *29*, 1297–1312. [CrossRef]
21. Wilcock, P.R.; McArdell, B.W. Partial transport of a sand/gravel sediment. *Water Resour. Res.* **1997**, *33*, 235–245. [CrossRef]
22. Reid, I.; Frostick, L.E.; Layman, J.T. The incidence and nature of bedload transport during flood flows in coarse-grained alluvial channels. *Earth Surf. Process. Landf.* **1985**, *10*, 33–44. [CrossRef]
23. Lisle, T.E. Particle size variations between bed load and bed material in natural gravel bed channels. *Water Resour. Res.* **1995**, *31*, 1107–1118. [CrossRef]
24. Bui, E.N.; Mazzullo, J.M.; Wilding, L.P. Using quartz grain size and shape analysis to distinguish between aeolian and fluvial deposits in the Dallol Bosso of Niger (West Africa). *Earth Surf. Process. Landf.* **1989**, *14*, 157–166. [CrossRef]
25. Ashworth, P.J.; Ferguson, R.I.; Ashmore, P.E.; Paola, C.; Powell, D.M.; Prestegaards, K.L. Measurements in a braided river chute and lobe: 2. Sorting of bed load during entrainment, transport, and deposition. *Water Resour. Res.* **1992**, *28*, 1887–1896. [CrossRef]
26. Schneider, J.M.; Turowski, J.M.; Rickenmann, D.; Hegglin, R.; Arrigo, S.; Mao, L.; Kirchner, J.W. Scaling relationships between bed load volumes, transport distances, and stream power in steep mountain channels. *J. Geophys. Res. Earth Surf.* **2014**, *119*, 533–549. [CrossRef]
27. McLaren, P.; Bowles, D. The effects of sediment transport on grain-size distributions. *J. Sediment. Petrol.* **1985**, *55*, 457–470.
28. Bialik, R.J.; Nikora, V.I.; Rowiński, P.M. 3D Lagrangian modelling of saltating particles diffusion in turbulent water flow. *Acta Geophys.* **2012**, *60*, 1639–1660. [CrossRef]

29. Bialik, R.J.; Nikora, V.I.; Karpiński, M.; Rowiński, P.M. Diffusion of bedload particles in open-channel flows: Distribution of travel times and second-order statistics of particle trajectories. *Environ. Fluid Mech.* **2015**, *15*, 1281–1292. [CrossRef]
30. Bakke, P.D.; Sklar, L.S.; Dawdy, D.R.; Wang, W.C. The Design of a Site-Calibrated Parker–Klingeman Gravel Transport Model. *Water* **2017**, *9*, 441. [CrossRef]
31. Kociuba, W.; Janicki, G. Changeability of movable bed-surface particles in natural, gravel-bed channels and its relation to bedload grain size distribution (scott river, svalbard). *Geogr. Ann. Ser. Phys. Geogr.* **2015**, *97*, 507–521. [CrossRef]
32. Ghoshal, K.; Mazumder, B.S.; Purkait, B. Grain-size distributions of bed load: Inferences from flume experiments using heterogeneous sediment beds. *Sediment. Geol.* **2010**, *223*, 1–14. [CrossRef]
33. Parker, G.; Klingeman, P.C. On Why Gravel Bed Streams Are Paved. *Water Resour. Res.* **1982**, *18*, 1409–1423. [CrossRef]
34. Buffington, J.M.; Montgomery, D.R. A systematic analysis of eight decades of incipient motion studies, with special reference to gravel-bedded rivers. *Water Resour. Res.* **1997**, *33*, 1993–2029. [CrossRef]
35. Church, M.; Hassan, M.A. Mobility of bed material in Harris Creek. *Water Resour. Res.* **2002**, *38*. [CrossRef]
36. Batalla, R.J.; Vericat, D.; Gibbins, C.N.; Garcia, C. Incipient bed-material motion in a gravel-bed river: Field observations and measurements. *U. S. Geol. Surv. Sci. Investig. Rep.* **2010**, *5091*, 15.
37. Turowski, J.M.; Badoux, A.; Rickenmann, D. Start and end of bedload transport in gravel-bed streams. *Geophys. Res. Lett.* **2011**, *38*. [CrossRef]
38. Johnson, J.P. Gravel threshold of motion: A state function of sediment transport disequilibrium? *Earth Surf. Dyn.* **2016**, *4*, 685. [CrossRef]
39. Ferguson, R.I.; Prestegaard, K.L.; Ashworth, P.J. Influence of sand on hydraulics and gravel transport in a braided gravel bed river. *Water Resour. Res.* **1989**, *25*, 635–643. [CrossRef]
40. Komar, P.D.; Li, Z. Pivoting analyses of the selective entrainment of sediments by shape and size with application to gravel threshold. *Sedimentology* **1986**, *33*, 425–436. [CrossRef]
41. Naden, P. An erosion criterion for gravel rivers. *Earth Surf. Process. Landf.* **1987**, *12*, 83–93. [CrossRef]
42. Wiberg, P.L.; Smith, J.D. Calculations of the critical shear stress for motion of uniform and heterogeneous sediments. *Water Resour. Res.* **1987**, *23*, 1471–1480. [CrossRef]
43. Duan, J.G.; Scott, S. Selective bed-load transport in Las Vegas Wash, a gravel-bed stream. *J. Hydrol.* **2007**, *342*, 320–330. [CrossRef]
44. Ashworth, P.J.; Ferguson, R.I. Size-selective entrainment of bed load in gravel bed streams. *Water Resour. Res.* **1989**, *25*, 627–634. [CrossRef]
45. Ashworth, P.J.; Ferguson, R.I.; Powell, D.M. Bedload transport and sorting in braided channels. In *Dynamics of Gravel-Bed Rivers*; Bili, P., Hey, R.D., Thorne, C.R., Tacconi, P., Eds.; John Wiley: Hoboken, NJ, USA, 1992; pp. 497–513.
46. Szilo, J.; Bialik, R.J. Bedload transport in two creeks at the ice-free area of the Baranowski Glacier, King George Island, West Antarctica. *Pol. Polar Res.* **2017**, *38*, 21–39. [CrossRef]
47. Doran, P.T.; Priscu, J.C.; Lyons, W.B.; Walsh, J.E.; Fountain, A.G.; McKnight, D.M.; Moorhead, D.L.; Virginia, R.A.; Wall, D.H.; Clow, G.D. Antarctic climate cooling and terrestrial ecosystem response. *Nature* **2002**, *415*, 517–520. [CrossRef] [PubMed]
48. Ferron, F.A.; Simões, J.C.; Aquino, F.E.; Setzer, A.W. Air temperature time series for King George Island, Antarctica. *Pesqui. Antártica Bras.* **2004**, *4*, 155–169.
49. Kejna, M. Air temperature on King George Island, South Shetland Islands, Antarctica. *Pol. Polar Res.* **1999**, *20*, 183–201.
50. Kejna, M. Trends of air temperature of the Antarctic during the period 1958–2000. *Pol. Polar Res.* **2003**, *24*, 99–126.
51. Kejna, M.; Araźny, A.; Sobota, I. Climatic change on King George Island in the years 1948–2011. *Pol. Polar Res.* **2013**, *34*, 213–235. [CrossRef]
52. Van den Broeke, M.R. On the interpretation of Antarctic temperature trends. *J. Clim.* **2000**, *13*, 3885–3889. [CrossRef]
53. Pętlicki, M.; Sziło, J.; MacDonell, S.; Vivero, S.; Bialik, R. Recent Deceleration of the Ice Elevation Change of Ecology Glacier (King George Island, Antarctica). *Remote Sens.* **2017**, *9*, 520. [CrossRef]

54. Arctic and Antarctic Research Institute, St. Petersburg. Available online: http://www.aari.aq/data/data.php?lang=1&station=0 (accessed on 11 January 2018).
55. Kociuba, W.; Janicki, G. Continuous measurements of bedload transport rates in a small glacial river catchment in the summer season (Spitsbergen). *Geomorphology* **2014**, *212*, 58–71. [CrossRef]
56. Kociuba, W. Determination of the bedload transport rate in a small proglacial High Arctic stream using direct, semi-continuous measurement. *Geomorphology* **2017**, *287*, 101–115. [CrossRef]
57. Blott, S.J.; Pye, K. GRADISTAT: A grain size distribution and statistics package for the analysis of unconsolidated sediments. *Earth Surf. Process. Landf.* **2001**, *26*, 1237–1248. [CrossRef]
58. Bialik, R.J.; Karpiński, M.; Rajwa, A. Discharge measurements in lowland rivers: Field comparison between an electromagnetic open channel flow meter (EOCFM) and an acoustic Doppler current profiler (ADCP). In *Achievements, History and Challenges in Geophysics*; Springer: New York, NY, USA, 2014; pp. 213–222.
59. Feng, C.; Wang, H.; Lu, N.; Tu, X.M. Log transformation: Application and interpretation in biomedical research. *Stat. Med.* **2013**, *32*, 230–239. [CrossRef] [PubMed]
60. Robertson, J.; Thomas, C.J.; Caddy, B.; Lewis, A.J. Particle size analysis of soils—A comparison of dry and wet sieving techniques. *Forensic Sci. Int.* **1984**, *24*, 209–217. [CrossRef]
61. Kemper, W.D.; Rosenau, R.C. Aggregate stability and size distribution. In *Methods of Soil Analysis. Part 1. Physical and Mineralogical Methods—Agronomy Monography*; Soil Science Society of America: Madison, WI, USA, 1986; Volume 1, pp. 425–442.
62. Kociuba, W. Bedload transport in a High Arctic gravel-bed river (Scott River, Svalbard SW). In *New Perspectives in Polar Research*; Institute of Geography and Regional Development, University of Wrocław: Wrocław, Poland, 2014; pp. 231–246.
63. Zagórski, P.; Siwek, K.; Gluza, A.; Bartoszewski, S.A. Changes in the extent and geometry of the Scott Glacier, Spitsbergen. *Pol. Polar Res.* **2008**, *2*, 163–185.
64. Kondolf, G.M.; Wolman, M.G. The sizes of salmonid spawning gravels. *Water Resour. Res.* **1993**, *29*, 2275–2285. [CrossRef]
65. Komar, P.D.; Carling, P.A. Grain sorting in gravel-bed streams and the choice of particle sizes for flow-competence evaluations. *Sedimentology* **1991**, *38*, 489–502. [CrossRef]
66. Wathen, S.J.; Ferguson, R.I.; Hoey, T.B.; Werritty, A. Unequal mobility of gravel and sand in weakly bimodal river sediments. *Water Resour. Res.* **1995**, *31*, 2087–2096. [CrossRef]
67. Church, M.; Hassan, M.A. Upland gravel-bed rivers with low sediment transport. *Dev. Earth Surf. Process.* **2005**, *7*, 141–168. [CrossRef]

© 2018 by the authors. Licensee MDPI, Basel, Switzerland. This article is an open access article distributed under the terms and conditions of the Creative Commons Attribution (CC BY) license (http://creativecommons.org/licenses/by/4.0/).

Article

Functional Relationship between Soil Slurry Transfer and Deposition in Urban Sewer Conduits

Yang Ho Song [1], **Eui Hoon Lee** [2] **and Jung Ho Lee** [1,*]

[1] Department of Civil and Environmental Engineering, Hanbat National University, Daejeon 34158, Korea; syho@daum.net
[2] Research Center for Disaster Prevention Science and Technology, Korea University, Seoul 02841, Korea; hydrohydro@naver.com
* Correspondence: leejh@hanbat.ac.kr; Tel.: +82-042-821-1106

Received: 5 May 2018; Accepted: 21 June 2018; Published: 22 June 2018

Abstract: Soil slurry deposited on the surface of the Earth during rainfall mixes with fluids and flows into urban sewer conduits. Turbulent energy and energy dissipation in the conduits lead to separation, and sedimentation at the bottom lowers the discharge capacity of conduits. This study proposes a functional relationship between shear stress in urban sewer conduits and the physical properties of particles in a conduit bed containing less than 20 mm of soil. Several conditions were implemented for analyzing two-phase flow (soil slurry and fluid in urban sewer conduits) in terms of turbulent flow by considering soil slurry flowing into urban sewer conduits. The internal flows of fluid and soil slurry in urban sewer conduits were numerically analyzed and modeled by applying the Navier–Stokes equation and the k-ε turbulence model. The transfer deposition of the soil slurry in the conduits was reviewed and, based on the results, a limiting tractive force was calculated and used to propose criteria for transfer deposition occurring in urban sewer conduits.

Keywords: soil slurry; sedimentation; two-phase flow; transfer; deposition; limiting tractive force

1. Introduction

It is important to understand the flow characteristics of urban sewer conduits to maintain their flow control capacity and prevent flooding during rainfall. In the case of localized and heavy rainfall, soil slurry separates from the Earth's surface through outflows and flows into drainage. This lowers the discharge capacity of drains. Delaying the outflow leads to overflow that causes flooding. Under a balanced condition, sediments accumulated at the bottom of urban sewer conduits are directly related to a reduction in discharge capacity, and leads to system overload and flooding.

The purpose of urban sewer conduits is to prevent flooding in urban areas by discharging runoff generated during rainfall. Together with the design and construction of such urban sewer conduits, maintaining their discharge capacity is the most important factor influencing continual maintenance. If urban sewer conduits are not designed, constructed, and maintained properly, severe soil sedimentation can occur in them during or after rainfall events, which threatens their discharge capacity and ability to properly handle sediments. Reduced discharge capacity of urban sewer conduits can cause flooding during the rainy season or localized heavy rainfalls.

In general, soil slurry flowing into urban sewer conduits forms a layer of solid particles at the bottom of the conduit leading to deposition [1–4]. In terms of flow characteristics, when particles are transferred, the fluid and solid particles interact in a complex manner. Past research has relied on experiments [5], but it is challenging in experimental analyses to review the independent influence of each variable on transfer deposition. Moreover, soil slurry has a dynamic flow in the transfer process. In the case of particles, it is difficult to measure the internal flow velocity of a conduit because of their own viscosities, particle sizes, and fluid drag forces [6,7]. Using experimental results, many studies have analyzed the fundamental aspects of sediment transfer in turbulent flow conditions [8]. However, they propose representing sediment flow as an induction equation to relate it to flow velocity [9–12]. Knowledge of the characteristics of internal flow is critical. However, the internal flow of urban sewer conduits is completely different from that of a river, making it difficult to obtain accurate predictions. For this purpose, existing empirical equations based on flow along a river may be applicable if modified according to the hydraulic features of urban sewer conduits.

It is difficult to determine the initiation of particle movement because of various issues. The Shields diagram is a typical criterion for the incipient motion of sediments [13]. Shields suggested fundamental concepts for initiating the motion of a bed comprising non-cohesive particles resistant to erosion by flowing water. This is indicated by critical shear stress, which is commonly assumed to be a constant for a given sediment mixture and can be determined through laboratory experiments [14]. However, the proposed results cannot be directly applied owing to some limitations. Shields did not provide any method to obtain critical shear stress for transfer and deposition. Accordingly, Brownlie fitted a curve to the representative line using the Shields relationship [15]. In this study, the precise moment of incipient motion is suggested, mainly based on numerical modeling, and transport rate measurements are used for a more objective evaluation of the various conditions of the deposition environment.

To efficiently model the transfer deposition of soil in a conduit, it is necessary to determine the shear stress to smoothly model the transfer of particles without deposition in the conduit. To review the characteristics of the transfer deposition of particles in turbulent flow, some studies have applied obstacles in pipes or conduits, attended to the flow of particles in the relevant region, and applied the results to various technologies [16–18]. Some of these studies focused on one phase of the fluid to analyze the mechanism of its continuity and momentum equations, and applied the advection–diffusion equation to the soil slurry sediment. However, such methods are limited to the analysis of particle deposition and sedimentation.

Studies based on numerical analysis to reflect the conditions of turbulent flow have modeled the transfer deposition of soil and correlations among multiple parameters. The results have been compared with those of the above-mentioned empirical equation to suggest a proper method of application [19–22].

This study proposes a functional relationship between critical tractive force in urban sewer conduits and the physical properties of particles in a conduit bed containing less than 20 mm of soil. The inlet flow velocity (1.0, 2.0 and 3.0 m/s) of the soil slurry mixture, the volume concentration of the soil (10%, 30% and 50%), and its particle size were set as variables. Soil slurry flowing into urban sewer conduits was analyzed based on the two-phase flow (soil slurry and fluid in urban sewer conduits) as turbulent flow. In a relevant work, Song et al. studied the flow of soil slurry in urban sewer conduits [23]. In reviewing flow, the shear flow velocity and shear stress were found to be critical. Therefore, the results of Song et al. were applied in this study. The purpose was to review the criteria of deposition in the transfer process by assuming that soil slurry can flow in urban sewer conduits and by changing its flow conditions. It is noteworthy that the observation and analysis of the two-phase flow are difficult, and no relevant study has considered this for soil slurry. This study made the following assumptions:

1. The flow of a mixture of fluid and soil slurry in urban sewer conduits was considered, similar to simulating the mixing and flowing of a large amount of soil slurry and runoff.
2. In conditions of turbulent flow, it was assumed that flow in the conduit consisted of fluid and solid phases.
3. For accurate flow analysis, incidental simulations were excluded, and a method for reducing the time needed to analyze a short conduit was considered.
4. Soil flowing in a sewer conduit consists of particles of different sizes. For modeling, it is necessary to consider this distribution of particle size. In this study, it was assumed that the soil slurry had a uniform particle size distribution.

A commercial analysis tool ANSYS FLUENT 13.0 was used in this study [24]. This study analyzed the characteristics of flow in urban sewer conduits of 10 m length and 0.6 m diameter in light of the inlet flow velocity of the soil–fluid mixture (v), size of the soil particle (d), and the volume fraction in the soil mixture (v/f). The simulation results were used to investigate the limiting tractive force. Based on the results of a hydrodynamic analysis using 2D modeling, the limiting tractive force was calculated depending on the particle size of the soil slurry, and an equation for the functional relationship based on a different particle size was proposed. The limiting tractive force was found to be related to the physical properties of non-cohesive soil.

2. Numerical Method

In past studies on two-phase flow, the sensitivity of each phase was analyzed to propose the mechanism of their interaction [25–28]. These results provided a considerable amount of information concerning two-phase flow that was used to define the phenomenon for a variety of flows using a diversity of numerical analysis models. A few recent studies have proposed a functional equation based on particle density to define the criteria for sediment movement based on the flow velocity distribution of urban sewer conduits, in the context of two-phase flow containing a mixture of fluid and soil slurry [29–31].

2.1. Governing Equations and Mathematical Model

In this study, the models were constructed in FLUENT 13.0 to analyze two-phase flow using the ANSYS solver [32]. The control equations consisted of continuity and momentum equations and particle component conservation equations of mass and momentum [33]. The continuity and momentum equations can be implemented using Reynolds' transport theorem, and by applying Newton's second law [34,35].

The continuity equations are shown as Equation (1) and compressible fluid is presented as Equation (2). In Equation (3), the equations of momentum yield the volume fraction and density per unit time for each phase based on velocity. In the equation, $x_i (i = 1, 2, 3)$ represents the orthogonal coordinates, u_i the orthogonal (directional) element of the velocity vector that represents average flow velocity, ρ is fluid density, p is pressure, and F_i is the sum of influential forces, such as gravity, deflecting force, and centrifugal force.

$$\frac{\partial \rho}{\partial t} + \frac{\partial (\rho u_i)}{\partial x_i} = \frac{\partial \rho}{\partial t} + \nabla \cdot (\rho \bar{u}) = 0 \tag{1}$$

$$\nabla \cdot (\rho \bar{u}) = 0 \tag{2}$$

$$\frac{\partial (\rho u_i)}{\partial t} + \frac{\partial (\rho u_j u_i)}{\partial x_j} = \frac{\partial \tau_{ij}}{\partial x_j} - \frac{\partial p}{\partial x_i} + F_i \tag{3}$$

In Equation (4), τ_{ij} represents a viscous stress tensor representing Newtonian flow fluid, including the velocity of the deformed stress tensor; and ij represents structural correlation. In Equation (5), δ_{ij} represents the ratio of the strain (deformation) tensor described in Equation (4):

$$\tau_{ij} = 2\mu \delta_{ij} - \frac{2}{3} \mu \delta_{ij} \frac{\partial u_l}{\partial x_l} \tag{4}$$

$$\delta_{ij} = \frac{1}{2} \left(\frac{\partial u_i}{\partial x_j} + \frac{\partial u_j}{\partial x_i} \right) \tag{5}$$

2.2. Turbulence Model

To analyze the effect of turbulence in a conduit with soil slurry, a turbulence model is required. The shear force owing to the laminar viscous force of the neighboring fluid becomes its driving force and, consequently, a laminar flow with a certain layer appears. However, if the shear stress increases with increasing flow velocity, viscous force transmission becomes challenging to maintain. As a result, the shear stress breaks into very small turbulent eddies and ends up being transmitted inside the fluid.

In general, the Reynolds stress-based model is used to simulate turbulent flow [36–39]. To analyze turbulent flow, the authors used Launder and Spalding's standard k-ε model, the universally used turbulence model [40]. In the model used to calculate the Reynolds factors, two equations representing the turbulent kinetic energy and the rate of dissipation of energy were calculated to represent turbulent viscosity. It is well known that the prediction of rotational flow or vortex flow, generated when particles of the fluid rotate about a certain axis, and the analysis of an area near a wall with a low Reynolds' number is inaccurate. However, the model is excellent at convergence in basic turbulent analysis and shortening calculation time [32].

To analyze fluid flow in the standard k-ε model, two equations of turbulent energy and dissipation rate as well as a general transfer equation should be applied separately. These equations are called the turbulent kinetic energy equation and the turbulent kinetic energy dissipation equation, respectively, and are used as a two-equation model. C_μ is a model constant, which is generally 0.09. In this study, based on Equations (6) and (7) in the analysis of the transfer deposition of soil in a tube, k and ε are calculated in two separate transport equations (k-equation and ε-equation) to model turbulent stress [32].

$$\frac{\partial(\rho k)}{\partial t} + \frac{\partial(\rho \overline{u_j} k)}{\partial x_j} = \frac{\partial}{\partial x_j}\left[\left(\mu + \frac{\mu_t}{\delta_k}\right)\frac{\partial k}{\partial x_j}\right] + P_k - \rho\varepsilon \tag{6}$$

$$\frac{\partial(\rho\varepsilon)}{\partial t} + \frac{\partial(\rho \overline{u_j}\varepsilon)}{\partial x_j} = \frac{\partial}{\partial x_j}\left(\mu + \frac{\mu_t}{\delta_\varepsilon}\frac{\partial \varepsilon}{\partial x_j}\right) + C_{\varepsilon 1} P_k \frac{\varepsilon}{k} - C_{\varepsilon 2} P_k \frac{\varepsilon^2}{k} \tag{7}$$

$$P_k = \mu_t \left(\frac{\partial \overline{\mu_i}}{\partial x_j} + \frac{\partial \overline{\mu_j}}{\partial x_i}\right)\frac{\partial \overline{\mu_i}}{\partial x_j} \tag{8}$$

In Equation (8), P_k reflects the influence of viscous force, and each value represents the coefficient of the standard values of k and ε, which are obtained through experiments in a wide range of the turbulent area as proposed by Versteeg and Malalasekera (Table 1) [41]. The parameters presented in Table 1 were used to predict the shear flow of basic turbulence, including homogeneous and heterogeneous flows, and yield highly reliable results in terms of the boundary of the wall and free shear flow. C_μ, $C_{\varepsilon 1}$, and $C_{\varepsilon 2}$ are empirical constants determined through experiments, and δ_k and δ_ε are turbulent Prandtl numbers for turbulent kinetic energy and the rate of energy dissipation, respectively [42].

Table 1. Parameters of the standard k-ε model.

Parameter	C_μ	$C_{\varepsilon 1}$	$C_{\varepsilon 2}$	δ_k	δ_ε
Value	0.09	1.44	1.92	1.0	1.3

Data collected through experimental research on turbulent energy and the rate of dissipation were first applied as parameters in the model. Nevertheless, as this study focuses on analyzing phenomena based on modeling, the reliability of the parameters to be applied were not validated, and experimentally acquired values proposed in a manual were applied because the range of each parameter and ground for its application were required [32].

2.3. Setup and Boundary Conditions

In this study, the inlet, outlet, side wall, and symmetry of the entire conduit (volume) were defined as boundary conditions to consider the flow of particles and soil slurry in urban sewer conduits. The authors used ANSYS-FLUENT 13.0 for the analysis. In repeated numerical calculations, the relative error of a numerical solution was made to converge to 0.001. The relative error of the schemes is computed by comparing the result of the integration step. Reduce step size by half refine the step such that the change of outcome can be identified [24]. It took 17 h on average on an Intel(R) Xeon(R) CPU E5-2630, 2.40 GHz, with 16 GB RAM. The boundary conditions of the numerical model in this study are presented in Table 2.

Table 2. Boundary conditions for analysis.

Classification	Boundary Conditions
Multiphase flow	Fluid (water)-Solid (soil)
Applied models and flow conditions	Euler–Euler model Standard k-ε model Turbulent flow Unsteady and turbulent
Inlet conditions	Inlet velocity Inlet volume fraction Soil diameter
Outlet condition	Free fall
Wall condition	Non-slip
Convergence	0.001

It was assumed that the flow conditions did not change, and no deformed flow occurred. The authors simulated phenomena by analyzing various runoff flows in the conduit. This method helped reduce calculation time and implement an optimal solution through repeated calculations.

The inlet flow velocity of the soil slurry mixture, the volume concentration of the soil, and its particle size were set as variables. The flow velocity and particle distribution of the mixtures in all 63 conditions were compared and reviewed. Table 3 lists the values of parameters of the fluid and soil slurry mixture, which are similar to those used by Song et al. [23]. To simulate the soil slurry transfer and deposition in urban sewer conduits, short conduits were used, as shown in Figure 1. A mesh structure consisting of 140,000 rectangular meshes, each of 0.6 m in diameter and 10 m in length, was implemented. Wall boundary conditions were applied to confine fluid and soil regions. In viscous flows, the non-slip boundary condition is enforced at walls by default; no symmetrical boundaries were applied but the shape of the entire conduit was represented.

Table 3. Basic information concerning the parameters for analysis.

Parameter	Units	Value
Conduit specification	m	0.6 (D) × 10 (L)
Mesh specification	grid	140,000
Inlet velocity condition	m/s	1.0, 2.0, 3.0
Inlet volume fraction condition	%	10, 30, 50
Fluid density	kg/m^3	998.2
Fluid kinematic viscosity	Pa·s	0.001003
Soil density	kg/m^3	2,650
Soil diameter	mm	0.5, 1.0, 3.0, 5.0, 7.0, 15.0, 20.0

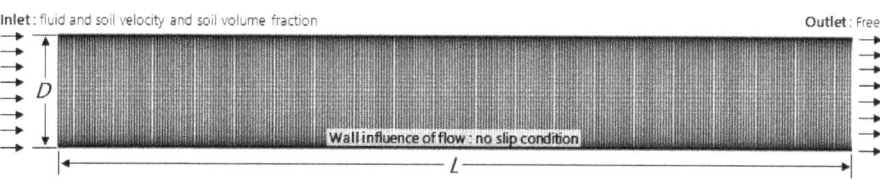

Figure 1. 2D mesh containing 140,000 cells.

As for the cross-section, the fluid and soil slurry flowed into the initial flow section (0 m), where the flow rate was constant for a certain period (500 s). The sectional volume fraction of the cross-section of the conduit was assumed to be the volume fraction of particles in the total flow. For the outlet (10 m), the results of the flow field and volume fraction in nine sections (1–9 m inside) were reviewed. In the case of the inlet, the flow velocity was constant, and thus flow field could be neglected. In the case of the outlet, the two-phase flow was such that the distribution of the flow field and volume fraction tended to be scattered. Therefore, it was excluded. Shields performed experiments in flumes with widths of 0.8 m and 0.4 m and beds composed of particles with diameters from 0.36 mm to 3.44 mm; mean velocity was increased in steps of 0.1 m/s to 0.6 m/s [13]. Brownlie conducted studies on the Colorado River, Mississippi River, and Red River. In his study, the flow velocity had minimum and maximum values of 0.37 m/s and 2.42 m/s, respectively, and beds were composed of particles ranging from 0.08 mm to 1.44 mm in diameter [15]. In this study, the same specification was applied to a sewer based on a numerical model. In the analysis, the particle size and inflow rate were changed to reflect the runoff condition in actual sewers. The transport and erosion phenomena in the conduit were reviewed. For discharge flowing into sewer pipes, most flow in as a mixture of fluid and soil slurry depending on the occurrence of rainfall, showing characteristics of turbulent flows. Accordingly, for soil slurry flowing into conduits, this study investigated the characteristics of flows inside conduits by determining the rate of flow corresponding to outflow and that to inflow occurred in mixed forms. The soil slurry flowing into the conduit contained particles of different sizes. The slurry was modeled by considering the dispersion of particle size. However, only particle size influenced velocity and particle distribution of the mixture. Therefore, we assumed particles of uniform size by considering a single particle size [23].

2.4. Model Validation

Song et al. [23] developed a model based on the experimental results obtained by Nabil et al. [43]. Based on the correction process, they suggested that the distribution of flow velocity and the changing pattern of the volume fraction of soil slurry were consistent in a conduit in their model. They reported experimental results for soil particles 1.4–2.0 mm in size, as used by Matousek [44].

Figure 2 shows the transfer of particles of size 1.4–2.0 mm in urban sewer conduits, with a volume concentration of 26% and flow velocities of 2.5 m/s and 3.0 m/s. A significant of the resulting flow was consistent. The internal distribution under the two conditions of flow velocity and volume concentration, which were used for examination, tended to be the same. Therefore, the analysis model was considered verified.

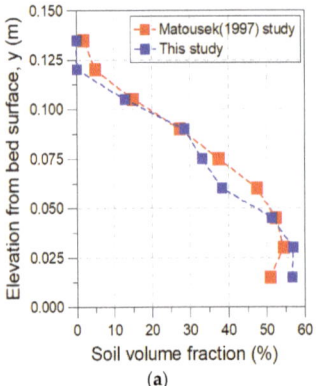
(a)

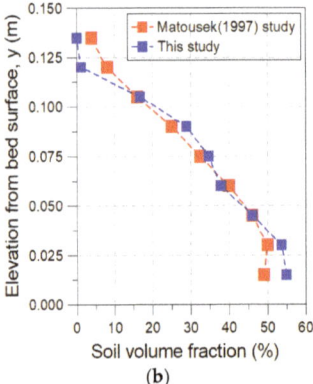
(b)

Figure 2. Results of examination using data obtained by Matousek. Results corresponding to flow velocities of: (**a**) 2.5 m/s; and (**b**) 3.0 m/s.

3. Numerical Modeling

3.1. Analysis of Flow Characteristics in Conduit Depending on Inlet Flow Velocity

The transfer of soil slurry in a conduit influences flow velocity under the action of inertia and gravity. Figures 3 and 4 show changes in the flow velocity in a conduit depending on the volume fraction at the same flow velocity for 1.0 mm and 20.0 mm of slurry. As shown in Figure 3, in the case of the flow velocity distribution of a single fluid, the flow velocity converged to zero at the bottom of the conduit, and then assumed the maximum value at the center with a rise in the diffusion coefficient through the volume fraction grade of soil slurry. However, this study analyzes two-phase flow rather than that of a single fluid, and revealed that the maximum value of flow velocity increased in the upper part of the conduit with changes in the volume fraction. The maximum flow velocity should have been observed at the center of the conduit because of the soil slurry deposited at its bottom.

In Figure 4, the flow velocity at the bottom changes slowly because of the soil slurry. In the upper part of the conduit, where the volume fraction had been reduced, the flow velocity tended to increase. With a rise in the volume fraction of the soil slurry in the lower part of the conduit, the damping effect of turbulence due to fluid friction largely worked and, consequently, the maximum value of flow velocity moved to the upper part of the conduit. Depending on deposition, the volume fraction of the soil slurry in the upper part of the conduit was smaller than that in the lower part. It was analyzed to be ideally consistent with the flow of a single fluid.

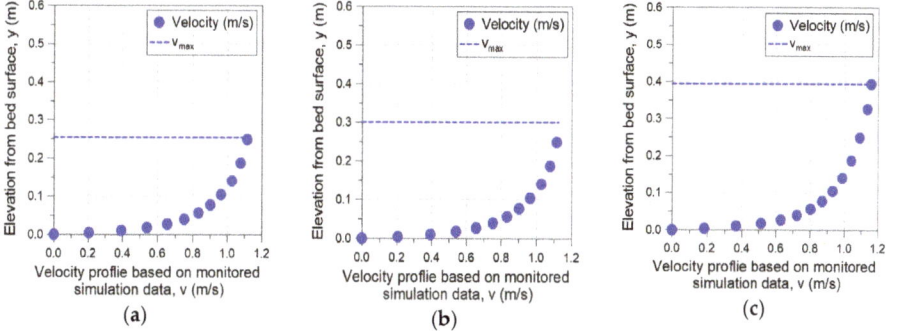

Figure 3. Vertical profile of longitudinal flow velocity of 1.0 mm of slurry in a conduit: (**a**) 1.0 m/s, 10% volume fraction; (**b**) 1.0 m/s, 30% volume fraction; and (**c**) 1.0 m/s, 50% volume fraction.

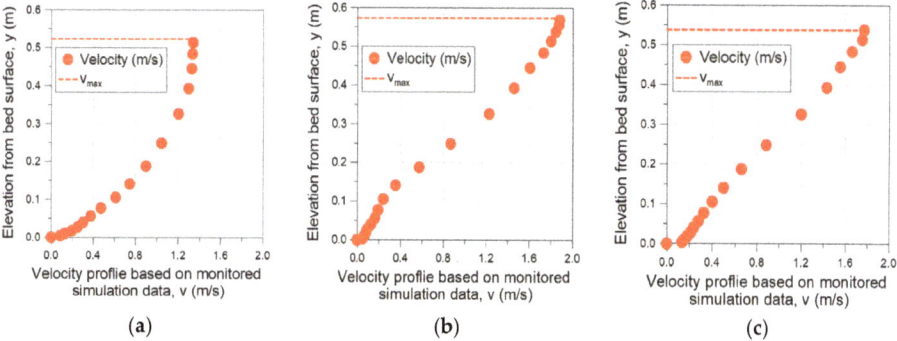

Figure 4. Vertical profile of longitudinal flow velocity of 20.0 mm of slurry in a conduit: (**a**) 1.0 m/s, 10% volume fraction; (**b**) 1.0 m/s, 30% volume fraction; and (**c**) 1.0 m/s, 50% volume fraction.

The distributions of the flow velocity of 1.0 mm and 20.0 mm of slurry were used in this study to verify that relatively smaller particles changed rapidly in the lower part of the conduit with a large volume fraction of soil slurry. It is fair to attribute this result to relative density. In general, the higher the volume fraction of soil slurry, the greater the interaction of particles in the mixture and the higher their viscosity. As a result, the flow velocity of the fluid in a relevant section decreases and, if the volume fraction decreases, flow velocity increases.

3.2. Analysis of Flow Characteristics of Conduit Depending on Inlet Volume Fraction

The most significant change with changing volume fraction of the inlet occurred in the distribution of flow velocity in the conduit. The volume fraction influenced the flow of each particle and changed the distribution of the flow velocity. The homogeneity of the distribution of flow velocity thus represented the homogeneity of the particle distribution. Figures 5 and 6 show the changes depending on volume fraction at velocities of 1.0 mm and 20.0 mm of the slurry. As shown in the results, the volume fraction in the upper part was close to zero, and a large volume fraction was observed only in the lower part of the conduit. The frictional force arising in the lower part did not significantly influence the center of the conduit with the maximum turbulence and flow velocity.

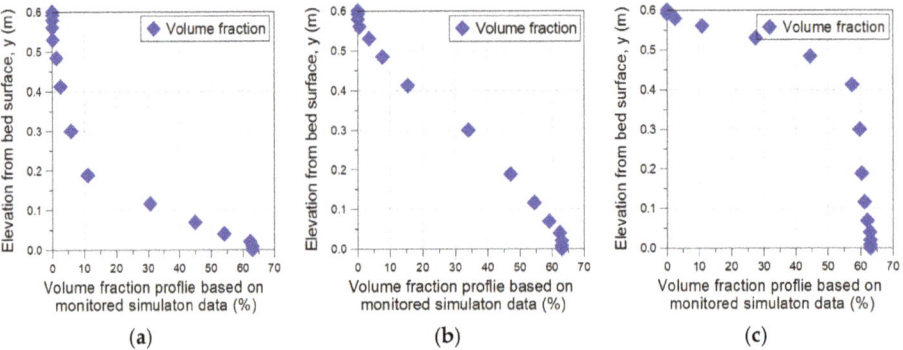

Figure 5. Vertical profile of the longitudinal volume fraction of 1.0 mm of slurry in the conduit: (**a**) 1.0 m/s, 10% volume fraction; (**b**) 1.0 m/s, 30% volume fraction; and (**c**) 1.0 m/s, 50% volume fraction.

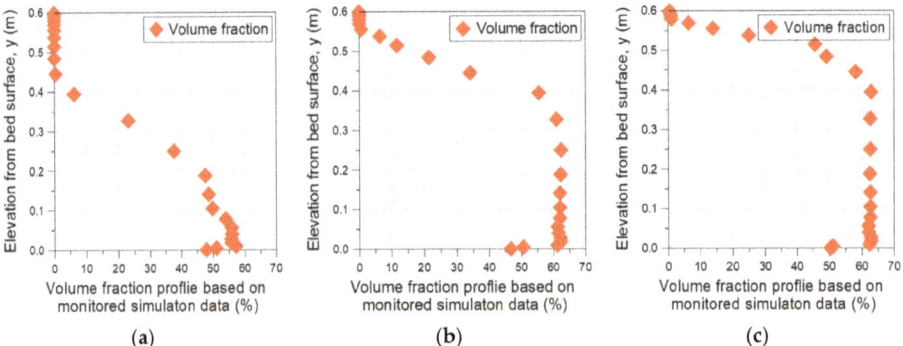

Figure 6. Vertical profile of the longitudinal volume fraction of 20.0 mm of slurry in the conduit: (**a**) 1.0 m/s, 10% volume fraction; (**b**) 1.0 m/s, 30% volume fraction; and (**c**) 1.0 m/s, 50% volume fraction.

Figure 5 shows the influence of the particle size of the soil slurry. As shown, when particle diameter was small, the distribution of volume fraction altered from heterogeneous to homogeneous along with a change in flow velocity. When the particles were small, they were easily moved by the eddy of the fluid. As shown in Figure 6, this was because particle movement was determined by gravity, which is based on unit particle weight. In general, with an increase in flow velocity, the fluid force became greater than the resistance of the hardened deposit and, thereby, moved the soil slurry. Friction increased with rising tractive force. Based on the change in flow velocity observed at the bottom of the conduit, the limiting tractive force could be estimated by calculating the shear flow velocity and shear stress in the conduit.

To sum up the numerical analysis, the larger the particle size of the soil slurry and the greater the increase in volume fraction, the greater the deposition. The higher the flow velocity, the smaller the deposition. An increase in the layer deposited led to increased flow velocity in the upper part of the conduit. With increased particle size of the soil slurry, deposition was evident.

4. Functional Relationship of Soil Slurry Transfer Deposition in Urban Sewer Conduits

4.1. Calculation and Review of Limiting Tractive Force

The major parameters needed to determine transfer deposition in the fluid transfer process are the vertical distribution of soil slurry in the conduit and the shear flow velocity (u_*) along the floor boundary. The shear flow velocity causes the soil slurry deposited at the bottom of the conduit to float in the re-floating process. At this time, the volume concentration can induce a smooth transfer or re-sedimentation. Accordingly, the shear flow velocity influenced by the layer of deposit was reviewed to measure the change in shear flow velocity at the bottom of the conduit. In general, this change in shear flow velocity represents a change in kinetic energy.

The characteristics of soil slurry include physical, such as particle size (d), density (ρ) and settling velocity (v_s), and volumetric constituents, such as the volume fraction of flow and the unit weight of mixed fluid flow. Raudkivi claimed that sediment transfer depends on shearing speed and the relative particle settling velocity [45]. Shearing speed can be calculated using Equations (9) and (10), and the settling velocity of the particle can be calculated using Equation (11). Shear stress is calculated through conversion. $(\rho_s - \rho_s)/\rho_s$ is the specific gravity of the sediment (submerged) in fluid, ρ_s is its density (kg/m^3), ρ is the density of water (kg/m^3), g is the acceleration due to gravity (m/s^2), d the particle diameter (m), and C_D is the coefficient of drag, which varies depending on the free motion of a particle but generally converges to 0.44 in turbulent flow. It is known that the drag coefficient of particles in a turbulent fluid may be significantly different, depending on concentrations and particle characteristics. The viscosity of the mixture, however, increases rapidly with volumetric sand concentrations beyond 20% [46,47]. For mixture flow, pressure drops could not be predicted accurately. The amount of error increased rapidly with the slurry concentration. These problems will be addressed through further research. The objective of this study is to provide a better description of the physical properties in agreement with numerical data. Accordingly, the average drag coefficient was considered to be 0.44 of a perfect sphere with the same volume and density depending on the particle diameter. Therefore, it was assigned a fixed value without any separate calculation. The particle settling velocity to be applied in this study was calculated using Equation (11). The weight of a submerged particle was calculated by $4/3(\rho_s - \rho_s)/\rho_s = 2.2$. Subsequently, calculations using particle size could be performed with (gd/C_D).

$$\tau_b = \rho g R_h S_f \tag{9}$$

$$\tau_c^* = \frac{\tau_b}{\rho\left(\frac{\rho_s-\rho}{\rho}\right)gd} = \frac{u_*^2}{\left(\frac{\rho_s-\rho}{\rho}\right)gd} \tag{10}$$

$$v_s = \sqrt{\frac{4}{3}\left(\frac{\rho_s-\rho}{\rho}\right)\left(\frac{gd}{C_D}\right)} \tag{11}$$

$$\frac{u}{u_*} = \frac{1}{\kappa} \ln\left(\frac{z}{z_0}\right) \qquad (12)$$

Given the distribution of flow velocity in a certain cross-section, such as flow in the conduit, Shear flow velocity in Equation (12) could be calculated, where u is the average flow velocity at the upper distance z relative to the bottom, z_0 is the distance from the bottom, where the flow velocity is zero, u_* is the flow shear speed, and κ is the von Kármán constant, which is approximately 0.41 [48,49]. In the case of shear flow velocity using Equation (12), assuming a certain flow velocity distribution in a certain cross-section, an approximate value of velocity can be estimated using the equation below. It can also be calculated using the law of the wall, which was used to explain the relatively thin layer (z/H <0.2) at the bottom [29].

Therefore, by applying the change owing to an increase in continuous velocity occurring along the boundary of the bottom of the conduit, the shear flow velocity and shear stress in transfer deposition can be calculated. If y increases discontinuously at a certain calculation point, $\partial u/\partial y$ in Equation (13) increases infinitely. Therefore, the shear stress at the point where there is no change in flow velocity can be defined as the limiting shear stress. In Equation (13), μ is the dynamic viscosity of the fluid. Shear force is the value of friction generated along the side wall due to the velocity of fluid flow in a cross-section of the conduit. Therefore, this friction is proportional to flow velocity and inversely proportional to the cross-sectional area of a conduit, and is used as shear rate or shear stress. In Equation (13), τ_b represents shear stress at the bottom. In Equation (14), τ_c^* is the critical shear stress at the bottom to initiate motion in the sedimentation layer (z) in the conduit, and is used to calculate the limiting tractive force that forms the basis of transfer deposition.

$$\tau_b = \mu \frac{\partial u}{\partial y} = \rho g \sin\theta (h - y) \qquad (13)$$

$$\tau_c^* = \frac{\tau_b}{(\rho_s - \rho) g d} \qquad (14)$$

Table 4 lists the results of shear stress based on shear flow velocity in 63 cases. Soil slurry of 0.5 mm diameter had a shear stress of 0.030–0.060 at a velocity of 1–2 m/s. Although there were slight deviations, it generally remained stable at 0.043. At a velocity of 3 m/s, the soil slurry had a value of over 0.284, which means that at relatively high flow velocity, the soil slurry was deposited more rapidly at the bottom of the conduit. Given that the amount of soil slurry (flow amount defined as the mass of soil slurry per total volume in unit flow) flowed in proportion to the inlet flow velocity, a larger deposit was expected when a large amount of slurry flowed in. This means that small particles could be deposited in the conduit. Soil slurry of 20.0 mm diameter had an average value of 0.053, which was maintained even with changes in flow velocity. The probability of sedimentation based on deposition was 0.056, the value suggested by Shields [13]. Given these results, if particle size increased, the density and specific gravity of the soil slurry increased. Deposition thus easily occurred and the settling velocity increased.

Table 4. Calculation of shear stress based on results of numerical analysis.

d (mm)	τ_c^*								
	1.0 m/s			2.0 m/s			3.0 m/s		
	10% v/f	30% v/f	50% v/f	10% v/f	30% v/f	50% v/f	10% v/f	30% v/f	50% v/f
0.5	0.030	0.060	0.056	0.035	0.040	0.041	0.284	0.308	0.307
1.0	0.036	0.036	0.029	0.074	0.730	0.068	0.224	0.259	0.243
3.0	0.061	0.037	0.041	0.033	0.029	0.028	0.086	0.098	0.102
5.0	0.046	0.051	0.056	0.056	0.032	0.056	0.027	0.036	0.034
7.0	0.047	0.047	0.036	0.045	0.047	0.041	0.035	0.055	0.056
15.0	0.052	0.052	0.060	0.038	0.045	0.048	0.042	0.042	0.051
20.0	0.048	0.060	0.054	0.053	0.048	0.051	0.049	0.055	0.056

Shear flow velocity and shear stress at the bottom represent the force accompanying and transferring the deposits. The shear stress along the critical boundary or the critical erosion speed is the limiting condition of deposit transfer and shear stress (limiting point of motion) at the start of the yield. The value of roughness at the bottom depends on the size of sediment particles there, and influences changes in the flow velocity distribution and the transfer capability of the soil slurry.

In general, the settling velocity influences the movement of the particles once the dynamic force of the fluid is imposed on them, and thus creates a boundary layer. In turbulent flow, if the flow velocity increases, the amount of force on the particles exceeds a threshold and acts on the sediment particles at the bottom of the conduit under a constant velocity. This means that, when the threshold is reached, shear stress in the conduit dominates the transfer velocity of the deposits, and significantly depends on flow velocity.

4.2. Functional Relationship of Transfer Deposition Due to Soil Slurry Particles

The method proposed by Shields [13] is commonly used to calculate the limiting tractive force as it can be used to directly compare the results obtained at different densities and viscosities, and thus makes it possible to easily judge the conditions of flow. However, in this method, the velocity of shear flow and the size of particles belong to both the longitudinal and transverse axes, and thus, the relationship between them is unclear. Therefore, the limiting tractive force for a given particle size could not be directly calculated. Problems and improvements regarding Shields' methodology have been discussed by Maa [50], Yalin and Karahan [51], and Smith and Cheung [52]. To solve these problems, Julien introduced a dimensionless parameter given by $(0.1Re^2/\tau_c^*)^2$ and proposed the following modified Equation (15) [14]. d_* is defined as exactly analogous to Rouse's auxiliary parameter or Rouse Reynolds number, eliminating the critical shear stress from the abscissa definition (in Reynolds number coordinates) [53]. Here, "SPG" represents the specific gravity of the particle. The Reynolds number for a multiparticle system, given in Equation (16), is defined in a similar manner as a function of the void fraction [54,55].

$$\theta_c = \tau_c^* = f(d_*) = f\left[\frac{(SPG-1)gd}{v^2}\right]^{1/3} \tag{15}$$

$$Re = \frac{u_* d}{v} \tag{16}$$

The works by Shields, Brownlie, and Parker et al. have improved and applied the limited diameter-based equation. To review the case where the limited diameter is small, this study used the results of a numerical analysis [15,56].

The results for particles of sizes 0.5–20.0 mm are presented without dimensions and compared with the results reported by Shields' limiting tractive force as shown in Figure 7. Shields used 15 data items, whereas this study used 63 for comparison.

As shown in Figure 7, the limiting tractive force for relatively smaller particles (0.5 mm and 1.0 mm) was close to the criteria suggested by Shields' curve. A velocity of 1.0 m/s was obtained for consistent conditions. The pattern of the increasing amount of soil slurry flowing with increased flow velocity was different from that in Shields' diagram. The range of limiting tractive force was 0.224–0.307, which means that with increasing flow velocity, the inflow of a large amount of soil slurry led to higher sedimentation. In a turbulent state, high flow velocity leads to a more active deposition of particles, which means that small particles can be deposited in the conduit.

With increasing particle size, a pattern similar to that in Shields' diagram was observed. In the case of a relatively larger particle of diameter of 20.0 mm, the variable for transfer deposition was calculated to be 0.053, identical to the result suggested by Shields' curve even with changes in flow velocity. This means that, if the particle size increased, the density and specific gravity of the soil slurry also increase, and particles are rapidly deposited at the bottom of the conduit regardless of the values of other variables. Therefore, particle size is the basic criterion for soil slurry transfer in calculating the limiting tractive force, but its influence is considered to be little to negligible if particle size is greater or less than a particular value. Small particles flowing into an urban sewer conduit constitute an important variable, and, thus, the conduit must be designed considering its discharge capacity.

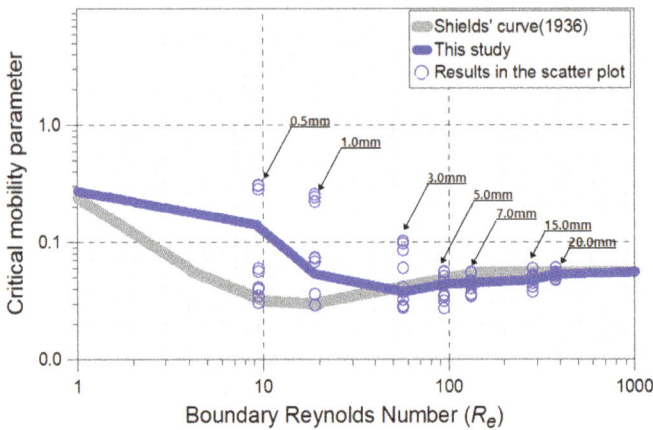

Figure 7. Shields' diagram depending on the boundary Reynolds number and comparison results of the particles.

To account for measurement errors in Table 5, the evaluation accuracy was examined by calculating two different evaluation statistics: SSQ (the sum of the squares) and RMSE (root mean squared error). These are expressed as Equations (17) and (18).

$$\text{SSQ}(x_1, x_2) = \sum (x_1 - x_2)^2 \tag{17}$$

$$\text{RMSE}(x_1, x_2) = \sqrt{\frac{\sum_{i=1}^{n}(x_1 - x_2)^2}{n}} \tag{18}$$

where x_1 is the amount of observed values, x_2 is the amount of calculated values (Table 5), subscript i indicates i_{ts} data in the set, and n is the number of data [57–60]. The results of the numerical model were compared with those of Shields and Brownlie [13,15]. To measure errors, we compared the values obtained in Table 5 with those of Shields and Brownlie, as shown in Table 6.

Table 5. Functional relationship of transfer deposition in a conduit using particle size.

d (mm)	d_* (non-dimensional)	Relationship Equation
$d \leq 0.5$	$d_* \leq 9.4$	$\theta_c = 0.280 \times d_*^{-0.288}$
$0.5 < d \leq 1.0$	$9.4 < d_* \leq 18.8$	$\theta_c = 2.336 \times d_*^{-1.248}$
$1.0 < d \leq 3.0$	$18.8 < d_* \leq 56.4$	$\theta_c = 0.135 \times d_*^{-0.306}$
$3.0 < d \leq 5.0$	$56.4 < d_* \leq 94.0$	$\theta_c = 0.012 \times d_*^{0.288}$
$5.0 < d \leq 7.0$	$94.0 < d_* \leq 131.6$	$\theta_c = 0.032 \times d_*^{0.067}$
$7.0 < d \leq 15.0$	$131.6 < d_* \leq 282.1$	$\theta_c = 0.029 \times d_*^{0.086}$
$15.0 < d \leq 20.0$	$282.1 < d_* \leq 376.1$	$\theta_c = 0.007 \times d_*^{0.345}$
$20.0 < d$	$376.1 < d_*$	$\theta_c = 0.056$

Table 6. Comparison between the results of this study and those of Shields and Brownlie.

d (mm)	d_* (non-dimensional)	Shields and Brownlie (x_1)	This Study (x_2)
0.5	9.4	0.0334	0.1468
1.0	18.8	0.0298	0.0600
3.0	56.4	0.0419	0.0395
5.0	94.0	0.0485	0.0438
7.0	131.6	0.0535	0.0450
15.0	282.1	0.0556	0.0478
20.0	376.1	0.0556	0.0530
	SSQ		0.0020
	RMSE		0.0446

The overall SSQ and RMSE values were found to be 0.0020 and 0.0446, respectively. Theoretically, a functional relationship equation is accepted as excellent when RMSE and SSQ values are equal to zero. The RMSE and SSQ values in this study indicate that the assessed results were highly correlated. Accordingly, this study proposed a functional relationship of the limiting tractive force for particles of different sizes, as shown in Table 5. The functional relationship can be used as the criterion to assess the transfer deposition of soil slurry in a conduit. The modified equation proposed in this study expresses this criterion and can be used to calculate the amount of sedimentation in a conduit.

Nevertheless, further studies should focus on analyzing errors considering various diameters and mathematical representation of the curve of critical tractive force results, such as flow conditions and other derived particle parameters.

5. Conclusions

In this study, a numerical analysis was performed to investigate the influence of the transfer deposition of soil slurry in urban sewer conduits. The results were analyzed based on various flow conditions. Particle size was set as the main parameter and settling velocity was applied to analyze deposition behavior in the conduit. By considering the turbulence model, the shear flow velocity was investigated according to flow velocity and volumetric concentration. Accordingly, a functional relationship used to identify the boundary of the transfer deposition of soil slurry in urban sewer conduits. Based on the entire process of the numerical analysis, the following conclusions can be drawn:

1. Particle size is the basic criterion for soil slurry transfer in calculating the limiting tractive force, but its influence is little to negligible if particle size is greater or less than a particular value. Nevertheless, small particles flowing into urban sewer conduits are an important variable, and thus conduits must be designed considering their discharge capacity.
2. A turbulence model was applied to calculate average flow velocity, shear flow velocity, and shear stress. In a conduit with sedimentation, the distribution of flow velocity was weakened overall by a similar drag-based volume fraction, and a slant in the flow velocity in the conduit increased due to deposition. The shear flow velocity and turbulent stress were large when a large value was calculated around the boundary of the bottom of the conduit. Therefore, it was estimated by applying an overall inclination.
3. Based on the results, the authors proposed a functional relationship between the limiting tractive and particle size. This relationship can be used as criterion to judge the transfer deposition of soil slurry in a conduit, and can be applied to urban sewer conduits, unlike previous studies. If the phenomena in urban sewer conduits are measured for comparison in an improved study, a more reasonable research method can be devised.
4. The results of this study can help overcome inaccuracy in simulating particles of small diameters in one-dimensional models, which are used to estimate the flow of sediment in urban sewer conduits. In future work, it will be necessary to further investigate the deposition rate of soil in conduits and the pattern of cohesion of each particle.

The findings of this study may be helpful in designing conduits with the aim of preventing conduit clogging or conduit damage that may occur during heavy rainfall events. To analyze drainage flow under the influence of various soil shapes and floating particles, it is necessary to consider different conditions through model experiments and numerical modeling. This study considered only the inflow of a large amount of soil slurry at the beginning of rainfall events. Therefore, the proposed method must be further investigated by introducing various types of rainfall events. Moreover, to validate the results of this study, data from real conduit systems must be collected and analyzed using a rainfall–runoff model. Overall, further research is required to determine the relationship considering the volume fraction of small particles.

Author Contributions: Y.H.S. carried out the literature survey and wrote the draft of the manuscript. Y.H.S. and E.H.L. worked on subsequent drafts of the manuscript. Y.H.S. and J.H.L. performed the simulations. Y.H.S., E.H. and J.H.L. conceived the original idea of the proposed method.

Funding: This research was supported by a grant (MOIS-DP-2015-03) from the Disaster and Safety Management Institute funded by the Ministry of Interior and Safety of Korea.

Acknowledgments: This research was supported by a grant (MOIS-DP-2015-03) from the Disaster and Safety Management Institute funded by the Ministry of Interior and Safety of Korea

Conflicts of Interest: The authors have no conflict of interest to declare.

References

1. Chebbo, G.; Gromaire, M.C.; Ahyerre, M.; Garnaud, S. Production and transport of urban wet weather pollution in combined sewer systems: The "Marais" experimental urban catchment in Paris. *Urban Water* **2001**, *3*, 3–15. [CrossRef]
2. Gromaire, M.C.; Garnaud, S.; Saad, M.; Chebbo, G. Contribution of different sources to the pollution of wet weather flows in combined sewers. *Water Res.* **2001**, *35*, 521–533. [CrossRef]
3. Ab. Ghani, A.; Md. Azamathulla, H. Gene-expression programming for sediment transport in sewer pipe systems. *J. Pipeline Syst. Eng. Pract.* **2010**, *2*, 102–106. [CrossRef]
4. Azamathulla, H.M.; Ghani, A.A.; Fei, S.Y. ANFIS-based approach for predicting sediment transport in clean sewer. *Appl. Soft Comput.* **2012**, *12*, 1227–1230. [CrossRef] [PubMed]
5. Shu, A.P.; Wang, L.; Zhang, X.; Ou, G.Q.; Wang, S. Study on the formation and initial transport for non-homogeneous debris flow. *Water* **2017**, *9*, 253. [CrossRef]
6. Lee, H.Y.; Lin, Y.T.; Yunyou, J.; Wenwang, H. On three-dimensional continuous saltating process of sediment particles near the channel bed. *J. Hydraul. Res.* **2006**, *44*, 374–389. [CrossRef]
7. Li, X.; Wei, X. Analysis of the relationship between soil erosion risk and surplus floodwater during flood season. *J. Hydrol. Eng.* **2013**, *19*, 1294–1311. [CrossRef]
8. Meyer-Peter, E.; Müller, R. Formulas for bed-load transport; appendix 2. In Proceedings of the 2nd Meeting of the International Association for Hydraulic Structures Research (IAHSR), Delft, The Netherlands, 7 June 1948; IAHR: Stockholm, Sweden, 1948.
9. Song, Z.; Liu, Q.; Hu, Z.; Li, H.; Xiong, J. Assessment of sediment impact on the risk of river diversion during dam construction: A simulation-based project study on the Jing River, China. *Water* **2018**, *10*, 217. [CrossRef]
10. Ribberink, J.S. Bed-load transport for steady flows and unsteady oscillatory flows. *Coast. Eng.* **1998**, *34*, 59–82. [CrossRef]
11. Ungar, J.E.; Haff, P.K. Steady-state saltation in air. *Sedimentology* **1987**, *34*, 289–299. [CrossRef]
12. Joshi, S.; Xu, Y.J. Bedload and suspended load transport in the 140-km reach downstream of the Mississippi River avulsion to the Atchafalaya River. *Water* **2017**, *9*, 716. [CrossRef]
13. Shields, A. *Application of Similarity Principles, and Turbulence Research to Bed-Load Movement*; California Institute of Technology: Pasadena, CA, USA, 1936.
14. Julien, P.Y. *Erosion and Sedimentation*; Cambridge University Press: New York, NY, USA, 1995.
15. Brownlie, W.R. *Prediction of Flow Depth and Sediment Discharge in Open Channels*; W. M. Keck Laboratory of Hydraulics and Water Resources, California Institute of Technology: Pasadena, CA, USA, 1981.
16. Eaton, J.K.; Johnston, J.P. A review of research on subsonic turbulent flow reattachment. *AIAA J.* **1981**, *19*, 1093–1100. [CrossRef]

17. Agelinchaab, M.; Tachie, M.F. PIV study of separated and reattached open channel flow over surface mounted blocks. *J. Fluids Eng.* **2008**, *130*, 1–9. [CrossRef]
18. Brevis, W.; García-Villalba, M.; Niño, Y. Experimental and large eddy simulation study of the flow developed by a sequence of lateral obstacles. *Environ. Fluid Mech.* **2014**, *14*, 873–893. [CrossRef]
19. Georgoulas, A.N.; Kopasakis, K.I.; Angelidis, P.B.; Kotsovinos, N.E. Numerical investigation of continuous, high-density turbidity currents response, in the variation of fundamental flow controlling parameters. *Comput. Fluids* **2012**, *60*, 21–35. [CrossRef]
20. Chuanjian, M.A.N.; Jungsun, O.H. Stochastic particle based models for suspended particle movement in surface flows. *Int. J. Sediment Res.* **2014**, *29*, 195–207.
21. Pang, A.L.J.; Skote, M.; Lim, S.Y.; Gullman-Strand, J.; Morgan, N. A numerical approach for determining equilibrium scour depth around a mono-pile due to steady currents. *Appl. Ocean Res.* **2016**, *57*, 114–124. [CrossRef]
22. Coker, E.H.; Van Peursem, D. The Erosion of horizontal sand slurry pipelines resulting from inter-particle collision. *Wear* **2017**, *400–401*, 74–81. [CrossRef]
23. Song, Y.H.; Yun, R.; Lee, E.H.; Lee, J.H. Predicting sedimentation in urban sewer conduits. *Water* **2018**, *10*, 462. [CrossRef]
24. ANSYS. *ANSYS Fluent 12.1 Theory Guide*; ANSYS Inc.: Canonsburg, PA, USA, 2010.
25. Newton, C.H.; Behnia, M. Numerical calculation of turbulent stratified gas–liquid pipe flows. *Int. J. Multiph. Flow* **2000**, *26*, 327–337. [CrossRef]
26. Ghorai, S.; Nigam, K.D.P. CFD modeling of flow profiles and interfacial phenomena in two-phase flow in pipes. *Chem. Eng. Process. Process Intensif.* **2006**, *45*, 55–65. [CrossRef]
27. De Schepper, S.C.; Heynderickx, G.J.; Marin, G.B. CFD modeling of all gas-liquid and vapor-liquid flow regimes predicted by the Baker chart. *Chem. Eng. J.* **2008**, *138*, 349–357. [CrossRef]
28. Bhramara, P.; Rao, V.D.; Sharma, K.V.; Reddy, T.K.K. CFD analysis of two phase flow in a horizontal pipe-prediction of pressure drop. *Momentum* **2009**, *10*, 476–482.
29. Nezu, I.; Nakagawa, H.; Jirka, G.H. Turbulence in open-channel flows. *J. Hydraul. Eng.* **1994**, *120*, 1235–1237. [CrossRef]
30. Durán, O.; Andreotti, B.; Claudin, P. Turbulent and viscous sediment transport—A numerical study. *Adv. Geosci.* **2014**, *37*, 73–80. [CrossRef]
31. Guo, J.; Mohebbi, A.; Zhai, Y.; Clark, S.P. Turbulent velocity distribution with dip phenomenon in conic open channels. *J. Hydraul. Res.* **2015**, *53*, 73–82. [CrossRef]
32. Park, C.W.; Hong, C.H. *User Guide of ANSYS Workbench*; Intervision: Seoul, Korea, 2008.
33. Anjum, N.; Ghani, U.; Ahmed Pasha, G.; Latif, A.; Sultan, T.; Ali, S. To investigate the flow structure of discontinuous vegetation patches of two vertically different layers in an open channel. *Water* **2018**, *10*, 75. [CrossRef]
34. Török, G.T.; Baranya, S.; Rüther, N. 3D CFD modeling of local scouring, bed armoring and sediment deposition. *Water* **2017**, *9*, 56. [CrossRef]
35. Fan, F.; Liang, B.; Li, Y.; Bai, Y.; Zhu, Y.; Zhu, Z. Numerical investigation of the influence of water jumping on the local scour beneath a pipeline under steady flow. *Water* **2017**, *9*, 642. [CrossRef]
36. Loth, E.; Kailasanath, K.; Löhner, R. Supersonic flow over an axisymmetric backward-facing step. *J. Spacecr. Rocket.* **1992**, *29*, 352–359. [CrossRef]
37. Sahu, J.; Heavey, K.R. Numerical investigation of supersonic base flow with base bleed. *J. Spacecr. Rocket.* **1997**, *34*, 62–69. [CrossRef]
38. Jalali, P. *Theory and Modelling of Multiphase Flows*; Lecture Material; Lappeenranta University of Technology: Lappeenranta, Finland, 2010.
39. Wallis, G.B. *One-Dimensional Two-Phase Flow*; McGraw-Hill Book Company: New York, NY, USA, 1969.
40. Launder, B.E.; Spalding, D.B. The numerical computation of turbulent flows. In *Numerical Prediction of Flow, Heat Transfer, Turbulence and Combustion*; Pergamon Press: New York, NY, USA, 1983; pp. 96–116.
41. Versteeg, H.K.; Malalasekera, W. *An Introduction to Computational Fluid Dynamics: The Finite Volume Method*; Pearson Education: London, UK, 2007.
42. Prandtl, L. Bericht uber Untersuchungen zur ausgebildeten Turbulenz. *Z. Angew. Math. Mech.* **1925**, *5*, 136–139.

43. Nabil, T.; El-Sawaf, I.; El-Nahhas, K. Sand-water slurry flow modelling in a horizontal pipeline by computational fluid dynamics technique. *Int. Water Technol. J.* **2014**, *4*, 13.
44. Matousek, V. Flow Mechanism of Soil–Water Mixtures in Pipelines. Ph.D. Thesis, Delft University, Delft, The Netherlands, 15 December 1997.
45. Raudkivi, A.J. *Loose Boundary Hydraulics*, 4th ed.; CRC Press: Rotterdam, The Netherlands, 1998.
46. O'Brien, J.S.; Julien, P.Y. Laboratory analysis of mudflow properties. *J. Hydraul. Eng.* **1988**, *114*, 877–887. [CrossRef]
47. Wichtmann, T.; Niemunis, A.; Triantafyllidis, T. Strain accumulation in sand due to drained uniaxial cyclic loading. In *Cyclic Behaviour of Soils and Liquefaction Phenomena*; Taylor & Francis Group: London, UK, 2004; pp. 233–246.
48. Nezu, I. Experimental Study on Secondary Currents in Open Channel Flows. In Proceedings of the 21st IAHR Congress, Melbourne, Australia, 13–18 August 1985.
49. Schlichting, H.; Gersten, K.; Krause, E.; Oertel, H.; Mayes, K. *Boundary-Layer Theory*; McGraw-Hill Book Company: New York, NY, USA, 1955.
50. Maa, J.P.Y. The bed shear stress of an annular sea-bed flume. In *Estuarine Water Quality Management*; Springer: Berlin/Heidelberg, Germany, 1990.
51. Yalin, M.S.; Karahan, E. Inception of sediment transport. *J. Hydraul. Div.* **1979**, *105*, 1433–1443.
52. Smith, D.A.; Cheung, K.F. Initiation of motion of calcareous sand. *J. Hydraul. Eng.* **2004**, *130*, 467–472. [CrossRef]
53. Guo, J. Hunter rouse and shields diagram. In *Advances in Hydraulics and Water Engineering*; World Scientific Publishing Co.: Singapore, 2002; Volumes I & II, pp. 1096–1098.
54. Syamlal, M.; O'Brien, T.J. *The Derivation of a Drag Coefficient Formula from Velocity-Voidage Correlations*; Technical Note; US Department of Energy, Office of Fossil Energy, NETL: Morgantown, WV, USA, 1987.
55. Wen, C.Y.; Yu, Y.H. A generalized method for predicting the minimum fluidization velocity. *AIChE J.* **1966**, *12*, 610–612. [CrossRef]
56. Parker, G.; Toro-Escobar, C.M.; Ramey, M.; Beck, S. Effect of floodwater extraction on mountain stream morphology. *J. Hydraul. Eng.* **2003**, *129*, 885–895. [CrossRef]
57. Karahan, H.; Gurarslan, G.; Geem, Z.W. Parameter estimation of the nonlinear Muskingum flood-routing model using a hybrid harmony search algorithm. *J. Hydrol. Eng.* **2012**, *18*, 352–360. [CrossRef]
58. Hyndman, R.J.; Koehler, A.B. Another look at measures of forecast accuracy. *Int. J. Forecast.* **2006**, *22*, 679–688. [CrossRef]
59. Willmott, C.J.; Matsuura, K. On the use of dimensioned measures of error to evaluate the performance of spatial interpolators. *Int. J. Geogr. Inf. Sci.* **2006**, *20*, 89–102. [CrossRef]
60. Pontius, R.G.; Thontteh, O.; Chen, H. Components of information for multiple resolution comparison between maps that share a real variable. *Environ. Ecol. Stat.* **2008**, *15*, 111–142. [CrossRef]

© 2018 by the authors. Licensee MDPI, Basel, Switzerland. This article is an open access article distributed under the terms and conditions of the Creative Commons Attribution (CC BY) license (http://creativecommons.org/licenses/by/4.0/).

Article

Erosion, Suspended Sediment Transport and Sedimentation on the Wadi Mina at the Sidi M'Hamed Ben Aouda Dam, Algeria

Faiza Hallouz [1,2,*], Mohamed Meddi [1], Gil Mahé [3], Samir Toumi [1] and Salah Eddine Ali Rahmani [4]

1   Laboratoire Génie de l'Eau et de l'Environnement, Higher National School of Hydraulic, 09000 Blida, Algeria; m.meddi@ensh.dz (M.M.); s.toumi@ensh.dz (S.T.)
2   University of Khemis Miliana, 44225 Ain Defla, Algeria
3   IRD Hydrosciences, 34090 Montpellier, France; gilmahe@hotmail.com
4   Geo-Environment Laboratory, Faculty of Earth Sciences, Geography and Spatial Planning, University of Sciences and Technology Houari Boumediene, 16111 Bab Ezzouar Algiers, Algeria; alirahmani101990@gmail.com
*   Correspondence: hallouzfaiza@gmail.com; Tel.: +213-660-319-854

Received: 28 March 2018; Accepted: 26 June 2018; Published: 4 July 2018

Abstract: The objective of this study was to follow-up on the evolution of the hydro-pluviometric schemes and particular elements of Wadi Mina (6048 km^2) to the Sidi M'Hamed Ben Aouda Dam to evaluate the silting origin and status of this dam situated in the northwest of Algeria. The pluviometric study targeted a series of rains during 77 years (1930–2007), the liquid discharge data cover a period of 41 years (1969–2010) and the solids and suspended sediment concentrations data cover very variable periods, starting from 22 to 40 years for the entire catchment area. The statistical tests for ruptures detection on the chronological series of rains and discharges indicate a net reduction of rains of more than 20% on the entire basin since 1970. The evolution of solids inputs was quantified: the maximum values are registered on autumn start and at the end of spring. The Wadi Mina basin brings annually 38×10^6 m^3 of water with a specific degradation of 860 t·km^{-2}·year^{-1}. By comparing the results found, we thus observe that the basin upstream of SMBA (1B) Dam is the greatest sediment producer towards the dam because it shows a specific degradation equal to 13.36 t·ha^{-1}·year^{-1}.

Keywords: Wadi Mina; Algeria; sediment; ruptures; SMBA Dam; specific degradation

1. Introduction

Algeria has always been confronted with extreme phenomena: periodic droughts which were sometimes severe and lasting and floods that are more often than not catastrophic. These events constitute a major constraint for the economic and social development and have been the subject of numerous studies [1–12].

Following a rain over a catchment area, a complex mechanism of interactions between hydrological and erosive phenomena is started. The streaming related to strong precipitations entails the soil slide by erosion in quantities that are sometimes very abundant. This has an impact on the environment and one of the results is the soils fertility loss and the early filling of water impoundments (dams, ponds, lakes), thus reducing their storing capacity and possibly entailing as well as degradation of the water quality and usage conflicts [13].

Since the Independence of Algeria, the total number of dams in exploitation at national level increased from 14 in 1962 to 65 in 2014 and, if we add the inter-basin water transfer major works, there are 7 billion m³ of water that are mobilized to be used for drinking water supply, irrigation and industry [14].

The scale of aggradations and the raise of the bottom of these dam impoundments by successive deposits of sediments brought by the watercourses and the wind are a serious problem whose negative consequences are considerably felt in the agriculture, farming, fishing, electricity and navigation fields [15–17]. The study on the silting of lakes, catchment areas or dams is of considerable interest for attempting to explain the complex mechanisms of solid transport and to quantify the volumes of transported sediments [18–21] and for a potential dredging of the impoundment [22].

The sedimentation in North Africa dams is very high in relation to what is noted at international level. According to Lahlou [23], the 23 major dams that are being exploited in Morocco lose annually 50 Mm³ of their storage capacity. In Tunisia, the annual loss of the storage capacity is estimated between 0.5% [24] and 1% for a total initial capacity of about 1430 Mm³ [25]. Algeria's rivers transport a large quantity of sediments [26,27]. The sediment deposited in Algerian dams is estimated to be 20 million m³·year^{-1} [28]. Competition for water among agriculture, industry, and drinking water supply—accentuated by a drought in Algeria—has shown the need for greater attention to be paid to water [29] and for it to be managed at the large basin scale [30]. Surface water resources in Algeria are evaluated to be approximately 8376 billion m³ for an average year [31]. These water resources in Algeria are characterized by wide variability—the resources for the last nine years have been significantly below this average [32]. Under these circumstances, several dams were built in Algeria to ensure water resources for the supply of drinking water to all its cities and allowed approximately 12,350 km² of irrigated land to be developed [31–33]. However, the sediments deposits in the dam reservoirs are estimated, on average, at 20 million m³ per year [34]. The high efficiency of sediments transport of the Algerian rivers [27,35] encouraged numerous authors to study the transfer of sediments in suspension in this area [36–41]. Considering these problems and to establish the lifespan of dams and to implement a better strategy of water resources management, research on the dynamics of sediments underlines the need to better understand the variations of erosion, deposits and dynamics of storage of sediments in a variety of catchment areas and rivers [42]. In this context, a certain number of studies used the relation of concentration of suspended particles matters (SPM) for acquiring a more complete understanding of the drainage basins processes [19,43,44].

Among the factors that favor erosion (slope, nature of rocks, hilly area, climate, and human activities), climate is recognized as the main factor in the semiarid Mediterranean regions of Algeria that go through short and intense rain episodes, a strong power of evaporation of the wind, prolonged droughts and freezing cycles [45,46]. The erosion is extremely active and the average concentration is at least higher than the global average [42].

Indeed, the main works for surface waters mobilization (big dams, dams and small lakes) were built on the watercourses of basins situated in the northern part of the country. These basins are characterized by various hilly areas dominated, on the one side, by marly lands and soils vulnerable to erosion and, on the other side, by a forest cover limited in the most watered parts and a seasonal vegetation cover. To these natural conditions is added the torrential nature of precipitations that fall at the end of the dry season and that would be responsible for a major part of the erosion and the solid load of water [21].

This article presents a study of erosion, sedimentation and climate change on Wadi Mina, at the Sidi M'Hamed Ben Aouda Dam in the northeast of Algeria, in a semiarid area. It is essential to know the evolution of water resources available for this dam in a variable climatic environment. In addition, since 1976 the rains reduction starting [47] and the increase in extreme events [48], the erosion and solids transport conditions were modified. The first results of the SIGMED (Approche Spatialisée de l'Impact des activités aGricoles au Maghreb sur les transports solides et les ressources en Eau De grands bassins versants) program showed as well the impact of human activities on environment and erosion [49,50], but it is still difficult to quantify the anthropogenic and climatic parts responsible for the recent modifications of the dynamic erosion on the basin [51].

The watershed of Wadi Mina (6048 km², northwest Algeria) is the subject of numerous works in the field of water erosion and siltation of dams [52–67].

The hydrologic part was a study on the rain and water flows in the basin towards available stations supplied by ANRH (National Agency for Hydraulic Resources). The solid transport of suspended particles matters (SPM) was estimated based on specific measures taken by the hydrological services of ANRH during the period 1969–2010. The water volume of the SMBA (Sidi M'Hamed Ben Aouda) Dam was estimated based on the results of the bathymetric measures supplied by ANBT (National Agency for Dams and Transfers) in the period 1978–2004.

In the last part, we studied the sedimentation study on the SMBA Dam and the origin of sediments, comparing the soil loss measured at the hydrometric stations with the dam silting data (bathymetry) with the results of a study of soils erosion by the USLE (Universal Soil Loss Equation) method carried out by Toumi et al. [66]. This is the first comparative study in Algeria which compares the erosion and sedimentation rates on a river basin and the associated reservoir, from several different approaches; this is the originality of this article in relation to the literature.

2. Materials and Methods

2.1. Study Area

Situated in a semiarid area, the catchment area of Wadi Mina presents strong erodibility and the good availability of pluviometric and hydrometric data. Situated at around 300 km west of Algiers, the catchment area Wadi Mina drains, at the level of the Sidi M'Hamed Ben Aouda (SMBA) Dam, a surface of 6048 km² (Figure 1). Wadi Mina is among the main tributaries of Wadi Cheliff. There are two dams on Wadi Mina, the SMBA Dam, subject of our study, and the Bakhadda Dam, situated at 90 km. The topographic study [68] allowed classifying the studied sub-basins in the hard hilly landscape class, in accordance with the IRD (French National Institute for Research for Sustainable Development) classification [69]. The basin of Wadi Mina spreads on dissimilar natural units:

- In the south, the Tellian plateaus also include the mountains of Frenda and Saida, with an altitude ranging from 900 to 1300 m, where limestone shows on the surface.
- In the north, a unit of mountain ranges and basins, generally of inferior altitude of 900 m, within a complex geological structure, individualized in blocks, where Ouarsenis is one of the most remarkable. They are characterized by an abundance of outcrops of marls, very sensible at hydric erosion.

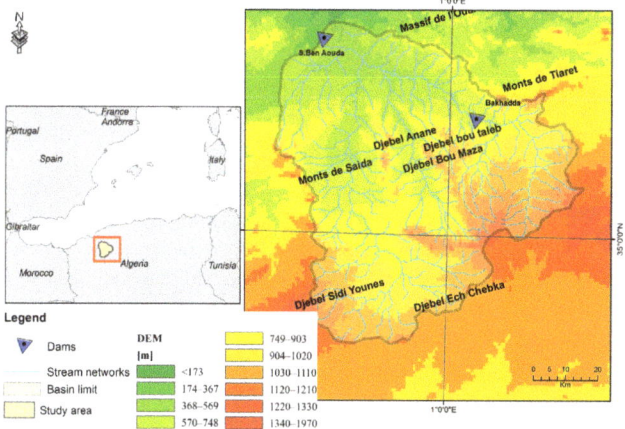

Figure 1. Localization of the catchment area of Wadi Mina.

The climate of the region is Mediterranean semi-arid with an annual average precipitation of 305 mm, marked by an irregularity which is both seasonal and inter-annual [68]. Moreover, the analysis of geographical repartition and the diversity of vegetal formations on the basin showed the existence of two parts clearly distinct [59,70]: the north area which is purely marlacious, strongly eroded and lacking vegetation, except for a few small islands of reforestation and plantations of fruit trees in the valley and the south area which is less eroded and where about 50% of the surface is covered by scattered vegetation with very variable density, ranging from the localized forest (pine of Alep) to the very scattered covering scrub. It must be noted that subsistence agriculture reigns with excessive exploitation of soils, a permanent clearing and an intensive overgrazing.

2.2. Acquisition and Preparation of Data

The pluviometric data came from the National Agency of Hydraulic Resources (ANRH). They are registered at 26 pluviometric posts (Table 1) distributed on the catchment area and the related observation duration is variable. The longest chronicles (1930–2007) are registered at the pluviometric stations in the south of the basin (Figure 2).

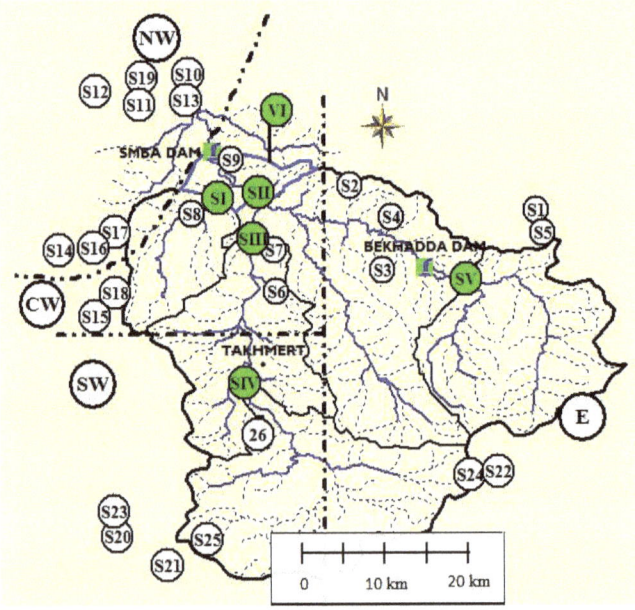

Figure 2. Localization of pluviometric and hydrometric stations (group of regions: E: east; NW: northwest; CW: center-west; and SW: southwest).

For the entire 26 stations of the basin of Wadi Mina over a period of 77 years (1930–2007), it has been estimated at around 4% the total number of gaps on all the time series of monthly and annual precipitation. Gaps in rainfall series were filled using the method of the regional vector to facilitate the use of rupture tests [47].

Table 1. Inventory of pluviometric stations (latitudes and longitudes in decimal degrees, altitude in m).

N°	ANRH Station Code	Name	Latitude	Longitude	Altitude	Period
1	012507	Wadi Lili	35.50417	1.288056	588	1969–2007
2	012703	Djdiouia amont	35.6525	0.8277778	562	1930–2003
3	012901	Bakhadda Bge	35.35444	1.035	572	1972–2007
4	012909	Sidi Ali Benamar	35.32444	1.129722	630	1971–1994
5	012917	Tiaret-ANRH	35.38528	1.306944	1106	1990–2007
6	013302	Ain Hamara	35.38028	0.6797222	288	1968–2007
7	013306	Wadi El-Abtal	35.46056	0.6972222	354	1952–1968
8	013401	Sidi AEK Djilali	35.48555	0.5877778	236	1968–2007
9	013410	SMBA	35.58528	0.5944445	195	1968–2007
10	013505	Relizane DEMRH	35.76278	0.5347222	58	1966–2005
11	013506	El Matmar	35.72333	0.4847222	70	1990–2007
12	013507	L'Hillil	35.72556	0.3480555	120	1968–2006
13	013511	Relizane aval	35.73695	0.5336111	69	1938–1969
14	111401	Mascara	35.38278	0.2408333	487	1930–1961
15	111404	Aouf	35.18639	0.3688889	968	1930–1959
16	111405	Matmor	35.33333	0.2	470	1930–2003
17	111407	Tighenif	35.42417	0.3252778	540	1938–1969
18	111418	Nesmoth	35.25945	0.3913889	906	1990–2007
19	111609	Bouguirat	35.76472	0.2291667	65	1930–2003
20	080504	Bled Bel Hammar	34.73583	0.22	1170	1970–2006
21	080606	Maamora	34.68389	0.5269445	1148	1974–2007
22	080701	Medrissa	34.89611	1.233333	1110	1930–2006
23	111216	Med El Habib	34.8	0.2541667	1106	1930–2007
24	013201	Ain Kermès	34.90861	1.1025	1112	1976–2007
25	013204	Sidi Yousef	34.80056	0.5944445	1091	1930–2007
26	013304	Takhmert	35.115	0.6913889	663	1930–2007

The longest possible series of hydrometric data (1969–2010) were used (Table 2). We have at our disposal five hydrological series on the Wadi Mina basin, with little or no gaps (Figure 3).

The solid discharges and concentrations of SPM data cover 40 years periods from 1970/1971 to 2009/2010 for the Wadi El Abtal and Sidi Abdel Kader Djilali stations, from 1975/1976 to 2009/2010 for the Ain Hamara station (35 years old), from 1985/1986 to 2005/2006 for the Sidi Ali Ben Amar station, and from 1972/1973 to 1993/1994 for the Takhmert station (22 years old).

The measurement of suspended sediment concentrations (C) consists of a systematic sampling of water samples using a 500 cL bottle. These samples are taken at a single point, either on the edge or in the middle of the Wadi. Samples are more numerous during floods, while in low water or when the liquid discharge (Q) is constant during the day, only one sample is taken. The difficulty of the measurement is the non-uniformity of the sediment concentration in the measurement section. The turbidity measurement procedure used by ANRH is developed in the work of Bourouba [35] and Meddi [68].

C and Q measurements were used to define rating curves that estimate C from measured values of Q on the Wadi Mina, according to a common approach [71–76]. The curves most commonly used to estimate sediment transport is a power function [77–79]: $Q = aC^b$, where a and b are regression coefficients determined empirically that account for the effectiveness of erosion and transport [42].

Table 2. Inventory of hydrometric stations used for the study.

Basin	S/Basin	Stations	Surface (km²)	Latitude	Longitude	Altitude	Period	Number of Instantaneous Solid Flow Measurements
Wadi Haddad	2B	Sidi Aek Djilali: I	499	35.48555	0.5877778	236	1969/2010	6262
Wadi Mina	1B	Wadi El Abtal: II	5365	35.58528	0.5944445	195	1969/2010	17,692
Wadi Abd Aval	3B 4B	Ain Hamara: III Takhmert: IV	2474 1488	35.38028 35.115	0.6797222 0.6913889	288 663	1969/2010 1969/2007	10,222 737
Upstream Wadi Mina	5B	Sidi Ali Ben Amar: V	1163	35.32444	1.129722	630	1969/2007	848
Wadi Mina	6B	Intermediary zone: VI	192	/	/	/	/	/

2.3. Methods Used

The main objective of the analysis of a chronological series, in accordance with Kendall and Stuart [80], is to bring clarity on the statistical mechanisms generating this series of observation.

Discontinuities in the studied datasets were analyzed using the rupture tests of Pettitt, Buishand, Lee, and Heghinian and the segmentation procedure of Hubert [81]. The null hypothesis tested is that there is no rupture in the series and a rupture is considered as very likely when it is statistically significant and is detected by two tests or two stations. These procedures are well defined by Maftei et al. [82] and in Khomsi et al. [83]. The choice of retained methods is based on the robustness of their foundation and the conclusions of a simulation study of artificially disturbed random series [84]. The study was performed with the software KHRONOSTAT [85], developed by IRD [86].

These tests were applied on the precipitations series in Algeria [87], Morocco [88] and Sub-Saharan Africa [89] as well as on the hydro-pluviometric studies series of the Sudan-Sahelian area [89,90], in West and Central Africa [91], and Tunisia [92]. These methods are thus largely validated in many regions of the world, on rains, discharges or temperatures series.

2.3.1. Flood Frequency Analysis (FFA)

The HYFRAN-PLUS software (version-V2.1) developed by the Canadian Institute (INRS-ETE, Eau Terre Environnement Research Centre of Institut national de la recherche scientifique), is designed for Hydrological Frequency Analysis (FFA) especially for extreme value. In the FFA, data corresponds to observations $X_1, \ldots, X_N$ which are independent and identically distributed (IID), i.e., verify the hypothesis of homogeneity, stationarity and independence. Several models are available to fulfill this task and the selection of the most robust approach to fit the model to data is a great challenge [93]. The prediction procedure is mainly based on the selection of the best fit for a given dataset. Conventional estimates of flood exceedance quantiles are highly dependent on the form of the underlying flood frequency distribution [94]. Several frequency distributions have been developed to fit different probability shapes, especially for extreme values. The first models related to extreme value theory (EVT) are the Generalized Extreme Value distribution [95], Gumbel [96] and then other models such as Lognormal distribution, Pearson family, Halphen family, Logistic distribution related models are proposed to give more flexibility when the EVT hypotheses are not fulfilled [97].

A decision support system (DSS) has been developed in the HYFRAN-PLUS software [98] to discriminate the different classes of the statistical distributions, and especially to estimate the quantiles for high return period T. Note that, for a random variable Q, the quantile Q_T with a return period T is the solution of the equation: $\Pr[Q > Q_T] = \frac{1}{T}$. DSS is mainly based on the classification of statistical distributions according to their asymptotic behavior. DSS is mainly based on the classification of statistical distributions according to their asymptotic behavior [99].

There are three nested classes of distributions that are commonly used for extremes [94,100] and implemented in the HYFRAN-PLUS software: Class C (regularly varying distributions) containing Fréchet (EV2), Halphen Inverse B (HIB), Log-Pearson type 3 (LP3) and Inverse Gamma (IG) distributions, class D (the sub-exponential distributions) involving Halphen type A (HA) and B (HB), Gumbel (EV1), Pearson type 3 (P3) and Gamma (G) distributions and the class E of distributions with inexistent exponential moments (Figures 3 and 4). The Log-Normal (LN) distribution has an asymptotic behavior between the C and D classes [94]. However, the use of the Generalized Extreme Value (GEV) distribution is currently a standard, thanks to the simplicity of its quantile function and the availability of software for parameter estimation [94].

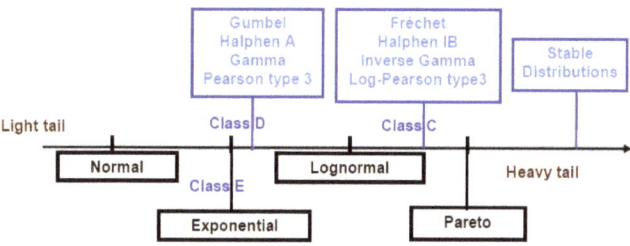

Figure 3. Distributions ordered with respect to their right tails [94].

The methods developed in the DSS allow identifying the most adequate class for adjusting a given sample. These methods are (Figure 4) [94]:

- the test Jarque–Bera (JB test): considered to test the log-normality with selected a priori based on the diagram (represented by coefficients of variation Cv and skewness Cs);
- the graphic log-log: used to discriminate firstly the class C and secondly D and E classes;

Mean Excess Function (MEF) used to discriminate the classes D and E; and two statistics: the ratio of Hill and Jackson statistics that can be used to perform a confirmatory analysis of suggested conclusions from the two previous methods (log-log graph and MEF).

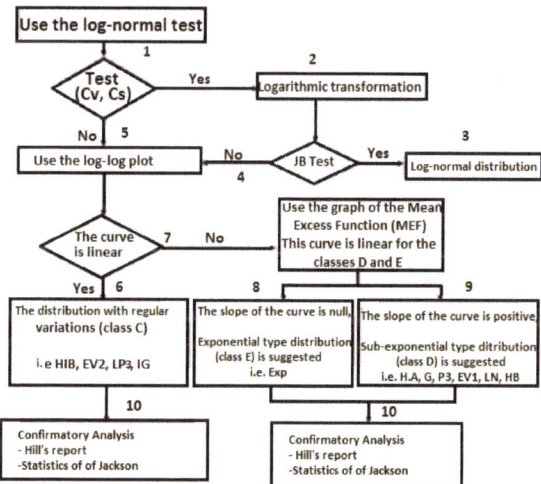

Figure 4. Diagram for class discrimination used in the DSS (according to [94]).

2.3.2. Generalized Extreme Value (GEV) Distribution and the Extreme Value Theory (EVT)

The generalized extreme value (GEV) distribution appears to be a universal model for hydrological extremes in several regions of the world (in particular in the United Kingdom), the same as the Log-Pearson type 3 distribution in the USA and Australia or Lognormal distribution in China [94].

When modeling the maxima of random variables, extreme value theory plays the same fundamental role as the central limit theorem when modeling sums of variables. In both cases, the theory tells us what are the limiting distributions. Generally, there are two main ways to identify extremes in real data. Let $Y_1, \ldots, Y_N$, be a sequence of N random variables, and X their maximum: $X = \max\{Y_1, \ldots, Y_N\}$ taken in successive periods. If the variables $Y_1, \ldots, Y_N$ are independent and identically distributed and F_0 is their distribution function (referred here as parent distribution), the distribution of the maximum X is given by:

$$\begin{aligned} H_N(x) &= P(X = \max(Y_1, \ldots, Y_N) < x) \\ &= P(Y_1 < x, \ldots, Y_N < x \\ &= \prod_{i=1}^{N} P(Y_1 < x) \\ &= [F_0(x)]^N \end{aligned} \quad (1)$$

Let $X_1, \ldots, X_N$ constitute the extreme events, also called block of maxima, taken in n successive periods, for example n years. The limiting distribution for the block of maxima is given by Fisher and Tippett [101]. The Fisher–Tippett theorem states that if the sequence $Y_1, \ldots, Y_N$ are Independent and Identically Distributed (IID) and if there exist constants $C_N > 0, d_N \in R$ and some non-degenerate function H such that H_N converges, when N tends to infinity, towards H in distribution ($H_N \to H$), then H corresponds to one of the three standard extreme value distributions:

$$\text{Gumbel (EV1)} : A(x) = \exp(-e^{-x}), \ x \in R \quad (2)$$

$$\text{Fréchet (EV2)} : \Phi_k(x) = \begin{cases} 0 & x \leq 0 \\ \exp(-x^k) & x > 0 \end{cases} \quad k < 0 \quad (3)$$

$$\text{Reverse Weibull (EV3)} : \Psi_k(x) = \begin{cases} \exp(-(-x^k)) & x \leq 0 \\ 0 & x > 0 \end{cases} \quad k > 0 \quad (4)$$

Depending on each case, the distribution of the original observation F_0 belongs to the Max-Domain of Attraction (MDA) of A, Φ_k or Ψ_k:

- The Gumbel MDA contains a great variety of distributions such as Normal, Gamma, Exponential and Gumbel.
- The Fréchet MDA contains distributions with power decay tail as the Pareto, Student, Cauchy, Inverse Gamma and Fréchet.
- The Weibull MDA distributions are light tailed with finite right endpoint. Such domain of attraction encloses Uniform and Beta distributions and is not of interest in FFA. The Weibull distribution has an upper bound and is not considered for high extremes.

Von Mises [101] proposed the generalized extreme value (GEV) distribution, which includes the three limit distributions EV1, EV2 and EV3. Another re-parameterization given by Fisher and Tippett [102] is generally used in hydrology:

$$F_{GEV}(x) = \begin{cases} \exp\left[-\left(1 - \frac{k}{\alpha}(x - \mu)\right)^{\frac{1}{k}}\right] & k \neq 0 \ (\text{EV2 and EV3}) \\ \exp\left[-\exp\left(-\frac{(x-\mu)}{\alpha}\right)\right] & k = 0 \ (\text{EV1}) \end{cases} \quad (5)$$

whenever $1 - k\frac{(x-\mu)}{\alpha} \geq 0$. When the shape parameter k is positive (respectively negative), the GEV corresponds to the inverse Weibull (EV3) (respectively Fréchet, EV2) distribution. The limiting case, $k = 0$, corresponds to the Gumbel (EV1) distribution. The simplicity of the GEV distribution function lead to an explicit expression of the quantile function to compute quantile for a given non-exceedance probability p:

$$x_p = \begin{cases} \mu + \frac{\alpha}{k}(1 - (-\ln p)^k) & k < 0 \text{ (EV2)} \\ \mu - \alpha \ln(-\ln p) & k = 0 \text{ (EV1)} \\ \mu + \frac{\alpha}{k}(1 - (-\ln p)^k) & k > 0 \text{ (EV3)} \end{cases} \quad (6)$$

This explicit quantile formula is also useful to generate a GEV samples by inversion method based on the generation of uniform distribution.

2.3.3. USLE Method

A classic example is the Universal Soil Loss Equation (USLE) [103]. This method expresses the long-term average annual soil loss as a product of rainfall erosivity factor (R factor), soil erodibility factor, slope length factor, slope steepness factor, cover-management factor, and support practice factor. USLE only predicts the amount of soil loss that results from sheet or rill erosion on a single slope and does not account for additional soil losses that might occur from gully, wind or tillage erosion. This erosion model was created for use in selected cropping and management systems, but is also applicable to non-agricultural conditions such as construction sites. The USLE can be used to compare soil losses from a particular field with a specific crop and management system to "tolerable soil loss" rates. Alternative management and crop systems may also be evaluated to determine the adequacy of conservation measures in farm planning. However, erosion is seen as a multiplier of rainfall erosivity (the R factor, which equals the potential energy); this multiplies the resistance of the environment, which comprises K (soil erodibility), SL (the topographical factor), C (plant cover and farming techniques) and P (erosion control practices). Since it is a multiplier, if one factor tends toward zero, erosion will tend toward zero.

3. Results

3.1. Analysis of Precipitations

At the monthly scale, some significant patterns arise. There are two periods of maximum occurrence of rupture in the rainfall time series of the stations studied, Autumn until January, which concentrate more than 40% of the annual amount of precipitation, with a peak in December and January, and Spring between April and June with more than 20% of the annual amount and two peaks in June and April. Intermediate seasons show much less occurrences of ruptures during the 1970s. However, during the 1980s the maximum of occurrence of ruptures is in August, and similarly during the 2000s in August and September, while during the 1990s the maximum is in March (Table 3).

The pluviometric posts for which a rupture could be detected are more numerous in the north than in the south. The pluviometric deficits noted are frequently close to 19–20%. This pluviometric deficit is thus felt for more than three decades and seems to be accentuated during the 1970–1980 decade. The areas in the west center and northwest have higher totals ruptures than the other areas on a monthly and yearly scales.

Table 3. Ruptures detected at the level of each pluviometric station. Ruptures dates detected by three statistical tests (red: strong rupture), by at least two statistical tests (dark gray: probable ruptures), by at least one statistical test (light gray: weak rupture) on the catchment areas (in bold: the ruptures dates detected for more than one catchment area the same year or the same month) (empty box = no probable rupture detected).

Station	Regional Groups		N	D	J	F	M	A	M	J	J	A	Annual
Oued lili			72		75	77	80	75	85	71	82	70	74
Djdiouia amont			75	01	81	72	86	96		76		01	76
Bekhadda bge		75	77		74	76	77	76	70		74	69	06
Sidi Ali Ben Amar	E		74		78		81			78	78		81
Tiaret—ANRH		84		73		74			79		78	76	02	93
Medrissa		93				01		99		96	96	01	98
Ain kermes		92	80	71		72		75	79		76		84	
Total ruptures by region		04	04	04	03	07	02	06	04	02	06	05	06	06
Relizane demrh		71	83	00	87	72	95	74	75	70	76	03	93	75
El Matmar		02	80	71	03	72	06	05	79	06	76	06	87	05
Hillil		02	70	00	81	72	70	74	75	71	76	03	75	75
Relizane aval	NW	81		82	78	70	81		73	77	80	02	82	65
Mascara			72		74	81	77	82	73	70		75	01	60
Motmor			75	99	74			96			76		01	76
Tighenif		82	80	71		72				79	02	76	80	87	67
Bouguirat				01	74	01		96				76		01	76
Total ruptures by region		05	06	07	07	07	05	06	06	06	07	06	08	08
Ain Hamara		84	80	06	70	06	68	74	81	70	76			75
Oued El Abtal		02			74	71		80	74	92	75	78		64
Sid Aek Djilali		02	06	02	03	72	68	74	02	71	03	05	87	75
Sidi M'Hamed Ben Aouda	CW	02	80	71	03	72	06	05	79	06	76	06	87	69
Aouf		81		71		77	91		94	79		82	81	71
Nesmoth			80	74	83	72			79		76		80	04
Total ruptures by region		05	04	05	05	06	04	04	06	05	05	04	04	06
Bled Bel Hammar		78	74		73	75	76	76	76	80	75	76	81	77
Maamora		92			74	82	71		72	74	80	79	70	89
Med El Habib	SW		75	99	81	72		96	98		96	96		76
Sidi Youcef		06			74			96			96	96		76
Takhmert			75		74			96	98		76			76
Total ruptures by region		03	03	01	05	03	02	04	04	02	05	04	0	05
% of annual rain		06	08	11	13	11	13	12	13	08	02	01	02	
Number of ruptures in the decades	1960						02						01	05
	1970	03	09	08	12	18	05	07	15	09	17	07	03	13
	1980	05	07	01	05	02	02	04	01	02	02	03	09	02
	1990	03		02			02	07	03	01	03	03	01	02
	2000	06	01	05	03	03	02	02		03	01	05	06	03

Indeed, 13 stations present a considerable rupture during the 1970–1980 decade for the entire catchment area of Wadi Mina, which represents 50% of the pluviometric stations used in this study (Figure 5).

It is equally interesting to note that the stations that present ruptures are localized differently during each decade: in the northwest of the basin before the 1970 decade, on the entire basin, except for the highlands of south and southeast, between the years 1970 and 1980, and on the entire basin, except for the center, after 1980, but in proportion representing less than half.

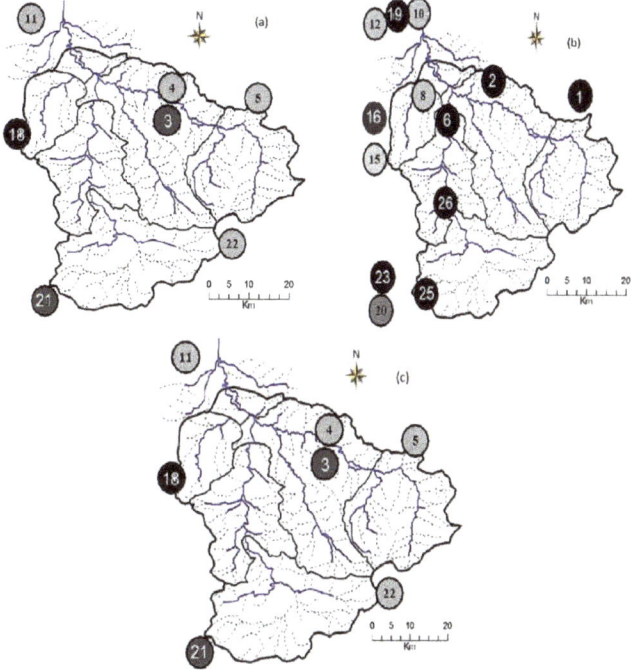

Figure 5. Ruptures detected at rainfall stations: (**a**) before 1970; (**b**) 1970–1980; and (**c**) after 1980. Rupture dates were detected: by three statistical tests (black: strong ruptures); by two statistical tests (dark grey: probable rupture); or by at least one statistical test (light grey: weak rupture).

Indeed, we have noted a rupture corresponding to a decrease of precipitations during the mid-1970s over the whole Wadi Mina basin (58% series present a rupture in 1976) (Table 3), date presented by numerous researchers [6,104], who already indicated the tendency towards drought starting with 1970 in north Algeria. These results are compliant with the recent works in Algeria [8,11] and which concluded to a rains decrease noted during the last thirty years which hit the entire country of Algeria, but especially its western part, adding up to: (a) 30–40% at annual level and more than 40% during winter and spring in the Mascara region; (b) 20–30% on the littoral, starting from the center to the western part of the country; and (c) 20% at annual level and 25% during winter in the center of the country. They confirm as well the results of works carried out on the basins of the extreme west in Algeria [6,11] which concluded to a decrease of rainfall. A rupture in the rainfall time series is frequently observed in 1979–1980, except 1972 for the Wadi Mina basin, at the Wadi El Abtal and Ain Hamara stations, and 1976 for the west-central part of Algeria. These results confirm as well the occurrence in Morocco of a pluviometric deficit starting with 1970 and its continuation during the 1980s, an extremely severe and long drought episode [105]. At the level of North Africa, the results obtained by Meddi et al. [11] in Algeria and by Bouzaiane and Laforgue [106] in Tunisia indicate the same rupture period and show the spatial extension of the drought, accompanied by a net pluviometric reduction. The conclusions of the Intergovernmental group on the climate evolution from 2001 and 2007 [107] and those of the regional report of the United Nations on the changes in North Africa [108], go in the same direction.

3.2. Analysis of Hydrological Data

The discharges register a global decrease since 1970 (Table 4), or the probable ruptures are noted as well during rainy season and dry season (Table 4), which means that even the low water levels are slow, for a sudden drop, it indicates that the level of the groundwater has also abruptly fall.

Table 4. Detection of ruptures in the annual and monthly flow series and annual maximum (Q_{maxd}) and minimum (Q_{mind}) daily discharge (1968/1969–2006/2007). Dates of ruptures detected by three statistical tests (red: strong ruptures), by at least two statistical tests (dark gray: probable ruptures), by at least one statistical test (light gray: weak rupture) (on the catchment areas (in bold: the ruptures dates detected for more than one catchment area the same year or the same month) (empty box = no probable rupture detected).

Station	S	O	N	D	J	F	M	A	M	J	J	A	Annual Discharge	Q_{maxd} Annual Daily	Q_{mind} Annual Daily
Ain Hamara	72		99	71	74	06	81	84	80	70	00	79	72	92	72
Takhmert	93	75	96	89		95				02	97		94	85	94
Sidi Ali Ben Amar		74	82		83	88	87	87	87	89	90	90	81	70	81
Wadi El Abtal	73	73		70		95	72	81	80	76	79	75	74	84	72
Sidi Aek Djilali	70	75	90	70	90	87	91	75	76	76	75	74	75	95	74
% annual discharge														Total	
Number of ruptures in the decades 1960														0	
1970	3		4		3	1		1	1	3	2	3	3	22	
1980		1		2	1		2	2	3	3	1	2	1	15	
1990	1		3		1	2	1				2	1	1	11	
2000						1					1	1		3	

At annual level, the probable rupture dates are 1974, 1975, 1980 and 1994, which corresponds generally to the ruptures detected at month level. In addition, the ruptures on the series of annual daily maximum were for the most part noted in 1970, 1984 and 1995 and the ruptures dates for the annual daily minimum series are detected in 1972, 1981 and 1994. The rupture in the daily maximum flows series is thus never noted the same year as in the average annual flows series, while there are the same ruptures dates for the average annual flows and the minimum daily flows.

In Sahelian and non-Sahelian Africa, many authors have revealed, based on hydro-pluviometric series, a dry phase which started around 1970 [109,110] and which lasts until today [111,112], it was noted as well a decrease of rivers flows [109,113,114], lakes levels [113], as well as a shortening of the rainy season [115,116].

These results indicate a satisfactory concordance with the dates detected on the annual and monthly rainfall of late 1970s/early 1980s. We thus assume that the decrease of runoff would be caused by a global decrease in precipitations on the catchment area of Wadi Mina [47]. However, we must not rule out a possible impact of storages and uses on the drastic decline of low water levels compared to groundwater.

In addition, several authors, among which [117], have as well reported rain decreases for the Mediterranean perimeter: in Spain, Italy, Turkey and Cyprus from 1951 to 1995 [118], between 1950 and 2000 in October in Mediterranean Iberia, March in Atlantic Iberia, January and winter in Greece and winter in the Middle East. Hulme et al. [119] showed a strong decrease of rains for the Mediterranean basin between the start and the end of 20th century. Likewise, subsequent to an analysis of the evolution of pluviometric inputs in the plains of Ghriss (northwest Algeria) [10], it was shown a net decrease starting with 1973, estimated at more than 25%. We can quote as well other results that have indicated that rains simulation over north Algeria, with an annual decrease of 15%, would entail a reduction of the annual runoff of 40% [120]. In Morocco, a climatic rupture that starts around 1976 in certain regions and which extends largely around 1979–1980, to nearly the entire territory, was noted [69,121].

As a consequence of the decrease of rainfall after the 1970s rupture which was observed in most basins of western Algeria, river discharges generally decreased as well. Meddi and Hubert [6] showed that the decrease in river discharge varied between −37% and −70% from eastern Algeria to western

Algeria. In the Macta basin in northwest Algeria, runoff was estimated to be 28–36% lower in 1976–2002 as compared to 1949–1976 [11]. In the Tafna basin, also in northwest Algeria, research [122,123] showed that the decrease in precipitation after the rupture date was, on average, 29% over the whole basin and was accompanied by a decrease of 60% in river flow [42]. In Wadi Sebdou in west Algeria, Norrant and Douguédroit [118] showed that rainfall and discharge began to decrease in the 1970s. The diminution of rainfall by 24% between the period 1939–1975 (526 mm) and the period 1975–2004 (401 mm) induced a decrease by 55% of the yearly volume discharged in 1975–2004 as compared to 1939–1975. A similar variability was observed on Wadi Abd basin (Wadi Mina) [43].

Finally, we note that annual discharges have been decreasing overall since the 1970s, which generally corresponds to ruptures detected at the monthly time step. There is therefore a good agreement with the rupture dates in the annual rainfall series, which suggests a climatic cause, in the first place, the observed decline in discharges in the Wadi Mina basin [47].

The study of floods is a precondition for the choice of an adjustment distribution and thus the floods flows on the entire Wadi Mina basin are classified in category C and follow a GEV distribution (Table 5).

The results of stationary nature and independence hypotheses tests of the studied data concerning the annual floods of the stations of the sub-basins of Wadi Mina indicate the acceptance of the independence tests of Wald–Wolfowitz and the stationary nature tests of Kendall for all the stations (Table 5).

Table 5. Distribution of floods for each station of the Wadi Mina basin.

Basin	Stations	Independence (Wald–Wolfowitz)	Stationary (Kendall)	Class Distributions
Wadi El Abd Aval	Ain Hamara	Accepted to the 5% threshold	Accepted to the 1% threshold	C (GEV distribution)
	Takhmert	Accepted to the 5% threshold	Accepted to the 5% threshold	C (GEV distribution)
Wadi Haddad	Sidi Abdel Kader Djilali	Accepted to the 1% threshold	Accepted to the 1% threshold	C (GEV distribution)
Upstream Wadi Mina	Wadi El Abtal	Accepted to the 5% threshold	Accepted to the 5% threshold	C (GEV distribution)
	Sidi Ali Ben Amar	Rejected to the 1% threshold	Accepted to the 5% threshold	C (GEV distribution)

The study of the floods recurrence variability before and after the ruptures was performed on all the stations of the basin of Wadi Mina (Figure 6). For all the stations, the maximum discharges values decreased for the same return period with about 30 to 40%. For example, at the Ain Hamara station and before the rupture of 1992, the floods with intensity lower than 80 m$^3 \cdot$s^{-1} only return once in five years (Figure 6a). However, beyond this value, precisely starting with 99 m$^3 \cdot$s^{-1}, the floods present longer return periods (20–100 years). However, after the same rupture date and for the same return periods, the intensities of the floods considerably decreased. For example, with a return period of 20 years before 1992, the flood with an intensity of 99 m$^3 \cdot$s^{-1} became centenary.

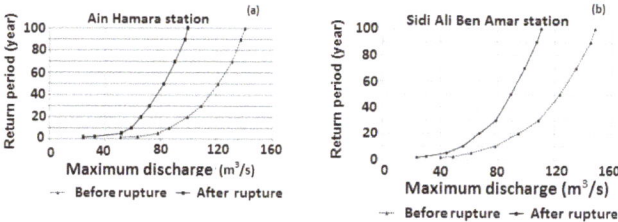

Figure 6. Cont.

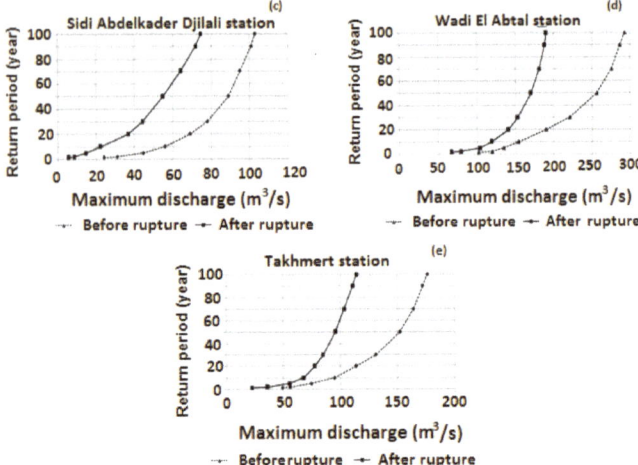

Figure 6. Impact of ruptures in the hydrometric series on the floods recurrence. (**a**) Ain Hamara station; (**b**) Sidi Ali Ben Amar station; (**c**) Sidi Abdelkader Djilali station; (**d**) Wadi El Abtal station; (**e**) Takhmert station.

The change of the pluviometric regime of the 1970s had thus as an impact a variance of recurrence of floods on all the sub-basins of the catchment area of Wadi Mina.

The pluviometry reduction has, of course, consequences on the watercourses regimes and thus on the availability of water resources which is the key to the success of many development projects. Nevertheless, the consequences of this reduction of flown volumes are already obvious with regard to the water and environment resources exploitation. Agriculture and the alimentation of impoundments are largely sanctioned by this resource decrease. The consequences of this phenomenon are thus very worrying with regard to the good functioning and the cost-effectiveness of projects already performed or planned.

3.3. Solids and Liquids Annual Transports

The annual average concentrations of SPM are compared with annual average discharge (Figure 7). Despite year-to-year irregularities at the Sidi Aek Djilali station, concentrations and discharges have changed in opposite directions. Discharges increased from 14 $m^3 \cdot s^{-1}$ in the late 1980s to less than 3 $m^3 \cdot s^{-1}$ in the early 2000s, while concentrations increased from 44 $g \cdot L^{-1}$ in 1974 to nearly 78 g. In 2005, flows at the El Abtal Wadi station increased from more than 40 $m^3 \cdot s^{-1}$ in 1974 to 0.48 $m^3 \cdot s^{-1}$ in 1997, whereas the concentrations evolved by 12 $g \cdot L^{-1}$, in 1971, to almost more than 67 $g \cdot L^{-1}$ in 2005. However, ruptures were detected on the rainfall and discharges from the mid-1970s, but this result show that is no effect of ruptures in precipitation and discharge on the solid load response since the discharge and rainfall decrease in the 1970s and increase in the early 2000s while the sediment load is still increasing. Geomorphologists have noted the difficulty of predicting the impacts of climate change on sediment yield due to nonlinear effects [124–126]. In addition, Achite and Ouillon [42] in their study on Wadi Abd basin, found that the change in flow regime induced a fully nonlinear effect between river discharge and sediment yield. This must be considered in the forecasts, especially in small river basins in semi-arid areas.

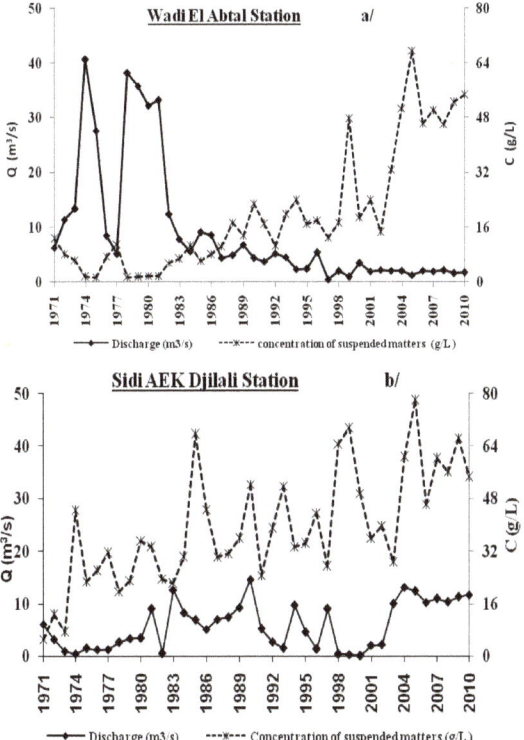

Figure 7. Evolution of annual average of discharges and concentrations of SPM in the Wadi Mina basin: (**a**) Wadi El Abtal station; (**b**) Sidi Aek Djilali station.

Regression analysis was performed between the concentration in instant suspended sediment (c) and the instant discharge (Q) at Wadi Mina. Generally, the power model $Q = aC^b$, where a and b are regression coefficients, is the most used for the estimation of the sediment concentration (Figure 8) [77–79].

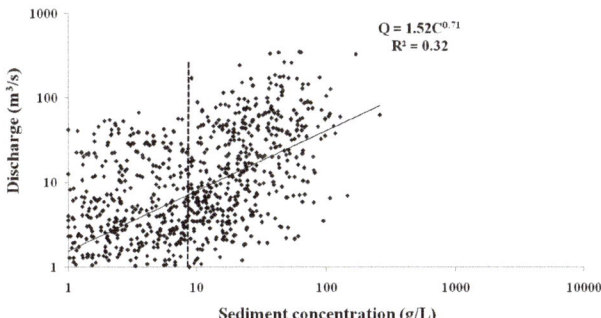

Figure 8. Relation between the liquids instantaneous flows and the concentrations of matters in suspension (Station of Wadi Abtal).

As for the majority of watercourses, we found a power relation [127] for Wadi Mina, connecting the concentration of SPM in g·L^{-1} to the liquid discharge in m^3·s^{-1} (Figure 8). A first graphic analysis of this figure shows a fairly strong dispersion around the regression line, with a very broad concentration range for low discharges below 10 m^3·s^{-1} and concentrations that become significantly higher, above 10 g·L^{-1} for discharges greater than 50 m^3·s^{-1}.

For studying the basin responses in terms of discharges and SPM during the hydrological year, it was considered useful to regroup the instantaneous values pursuant to the different seasons (Table 6).

Table 6. Seasonal models connecting the instantaneous flows with the concentration of matters in instant suspension.

STATION	Correlation	Autumn	Winter	Spring	Summer
W. El Abtal	Model	$Q = 0.506C^{1.092}$	$Q = 1.940C^{0.719}$	$Q = 0.889C^{0.969}$	$Q = 0.483 C^{0.891}$
	Coefficient R^2	0.60	0.20	0.59	0.37
S.A.E.K. Djilali	Model	$Q = 0.031C^{1.137}$	$Q = 0.125C^{0.561}$	$Q = 0.024C^{1.12}$	$Q = 0.161C^{0.584}$
	Coefficient R^2	0.60	0.20	0.63	0.30
A. Hamara	Model	$Q = 1.038C^{0.854}$	$Q = 1.7158C^{0.458}$	$Q = 0.893C^{0.867}$	$Q = 1.292C^{0.655}$
	Coefficient R^2	0.51	0.34	0.55	0.38
Takhmert	Model	$Q = 0.8C^{1.022}$	$Q = 1.788C^{0.621}$	$Q = 1.13C^{0.934}$	$Q = 0.842C^{0.845}$
	Coefficient R^2	0.63	0.43	0.66	0.44
S.A.B. Amar	Model	$Q = 0.799C^{0.733}$	$Q = 2.437C^{0.167}$	$Q = 0.67C^{0.712}$	$Q = 0.277C^{0.896}$
	Coefficient R^2	0.56	0.21	0.64	0.44

The analysis in Figure 9 shows that autumn and spring are distinguished by their strong liquid discharges generating an important flow of SPM. The maximum solid flow is of 42,365 kg·s^{-1}, achieved in September 1994, resulting from a liquid discharge of more than 351 m^3·s^{-1}.

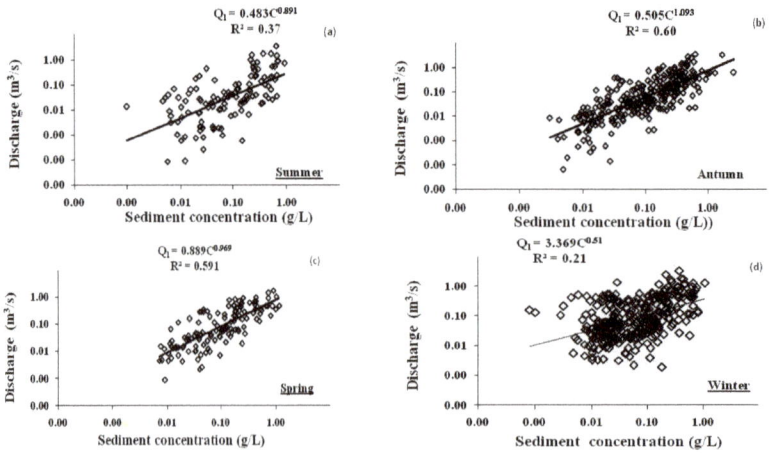

Figure 9. Seasonal variations (summer: June, July and August; autumn: September, October, November; winter: December, January, February; and spring: March, April, May) of the instantaneous values of discharges depending on the concentrations of suspended particles matters (SPM) (Station Wadi El Abtal). (**a**) Summer; (**b**) Autumn; (**c**) Spring; (**d**) Winter.

In contrast, during winter and summer, we note a net regression of liquid discharge that do no longer exceed 84 m^3·s^{-1} in winter and 38 m^3·s^{-1} in summer. However, we note a few specificities for each season.

For autumn, the heavy charge is explained by the fact that rain often falls on a dried up soil and badly protected by vegetation [68].

The basin of Wadi Mina is bare in autumn and, after a long warm period, the autumn floods imply systematically suspended solids inputs in maximum number and this phenomenon is relatively reduced for the following season, in winter or when the lands forming the basin are saturated and the vegetation cover is developed [68].

On the one hand, the torrential and aggressive rains are the ones occurring in October and November and they tear off major quantities of SPM and, on the other hand, the weak vegetation cover of the preceding season (summer) favors the destruction of the soil aggregates that shall be subsequently transported in suspension by the watercourse.

In winter, the erosion of the bed and the banks along the watercourse becomes important because of the increase of discharges in Wadi, and as well runoffs are generated by rainy sequences which are relatively abundant but of weak intensity (0.4 mm) in Takhmert and (29.1 mm) in Ain Hamara. They find a loose soil with vegetation cover that confers soil roughness and favors water retention [128].

In winter, the runoffs remain important, although this season is a little rainy and cold transition period with a succession of freezing and thawing which is caused essentially by the important water reserves, stored in the soil, following the autumn rains.

In spring, the erosion of the bed and the banks along the watercourse becomes important as this is essentially related to the increase discharges in the Wadi and the herbaceous cover and cultures that considerably reduce the mobilization of thin materials on the slopes. We can add as well the reduction of rains of heavy intensity compared with the winter season.

At the end of spring, the soil is humid and the vegetation cover is well developed. On the slopes, the thin soils saturate rapidly and develop important erosion in the gully and mass movement [129,130]. In addition, in this period, the floods have an important role in the production of sediments in the drainage networks. In addition, Hallouz et al. [67], in their study on the Wadi Mina basin, found that the maximum values of the solid inputs, for all the sub-basins, are observed at the beginning of autumn and at the end of the spring, this variability is explained on the one hand by the variation of the vegetal cover (bare soils) during the year and the erosive nature of the autumn rains (high intensities) and on the other hand by the releases made by the Bakhadda Dam, so these two factors allow the first autumn floods to transport large quantities of sediment after a long dry season characterized by high temperatures and the destruction of soil aggregates. Rainfall erosivity is the potential ability of rainfall to cause soil loss [131]. The rainfall erosivity index represents the climate influence on water related soil erosion [132]. Rainfall erosivity is the impact of the kinetic energy of raindrops on soil. Higher velocity and larger size of the raindrops results in higher kinetic energy and higher soil loss. According to Yu [132], most soil erosion researchers and soil conservationists recognize the positive correlation between erosivity and rainfall intensity [133].

3.4. Annual Inputs

The calculation of the sediment average specific load (Ds) in tons is given by the product sum of three variables Q (m$^3 \cdot$s^{-1}), C (g·L^{-1}) and T (duration in seconds). Table 7 resumes the average specific exportations in tons per km^2 and per year for each sub-basin.

Table 7. Results of the calculation of sediment charge in tons.

Basins	W. Mina (I + II)	W. Haddad	W. Mina	W. Abd		W. Mina Amont
Station name	(W. Abtal + S.A. Djilali)	S. Aek Djilali	W. El Abtal	A. Hamara	Takhmert	S. A. B. Amar
Average specific load (t·km^{-2}·year^{-1})	860	965	762	502	220	239
Period of observation	1969/2010	1969/2010	1969/2010	1969/2010	1969/2007	1969/2007

The average specific sediment discharge of Wadi Mina at the Sidi M'Hamed Ben Aouda Dam is 860 t·km^{-2}·year^{-1} (the sum of degradations measured at the stations Wadi El Abtal and Sidi Abdelkader El Djilali to which is added the load produced on the surface of the transitional basin VI). We note that this value is very moderated in relation to those published for other basins in the region, such as the basin of Wadi Mazafran (Algerian coastal basins), Wadi Isser (Lakhdaria), Wadi Sebdou (basin Tafna), Wadi Agrioum (east Algerian), whose specific degradations amount to 1610, 2300 and 1120 t·km^{-2}·year^{-1} [134,135] and 5000 t·km^{-2}·year^{-1} [136]. Meddi [68] already reported that Wadi Haddad à Sidi Abdelkader El Djilali has a high average specific load in relation to other catchment areas (Ds = 965 t·km^{-2}·year^{-1}), which means that this sub-basin continues to be the busiest supplier of sediments towards the dam. This catchment area is the most favorable for the runoff from the point of view of topography and vegetation cover [68]. It is characterized by a discontinuous and poor vegetation cover during warm season. The concentration of floods after the warm season in October and sometimes in May, with strong intensities, would generate strong concentrations in sediments during these months. This context favorable to the runoff and, consequently, to the solid transport explains this serious degradation of soils for this catchment area, in addition to its geological nature which is very favorable as well to this event (marl) [68].

The lowest degradation values are observed for Wadi Abd at Takhmert (220 t·km^{-2}·year^{-1}) and Wadi Mina upstream of Sidi Ali Ben Amar (239 t·km^{-2}·year^{-1}). These values can well be explained by the less favorable nature of the runoff of basins which are characterized by floods of average or light power [68].

3.5. Where from Does Originate the Sediments of the Dam?

Thus, the results of bathymetric surveys performed at the level of the SMBA Dam indicated that the capacity loss of the dam since its commissioning in 1978 and until 2004 is evaluated at 82 × 10^6 m^3 (Figure 10), equivalent to a third (1/3) of its initial capacity. Remini and Bensafia [64] estimated that, in 2010, the SMBA Dam registers a sediment volume equal to 109 × 10^6 m^3, namely an annual average loss of about 3.5 × 10^6 m^3·year^{-1}, equivalent to a sediment load equal to 5.6 × 10^6 t·year^{-1} that deposits in the impoundment of the dam (considering an average density of the sediment of 1.6 t·m^{-3}).

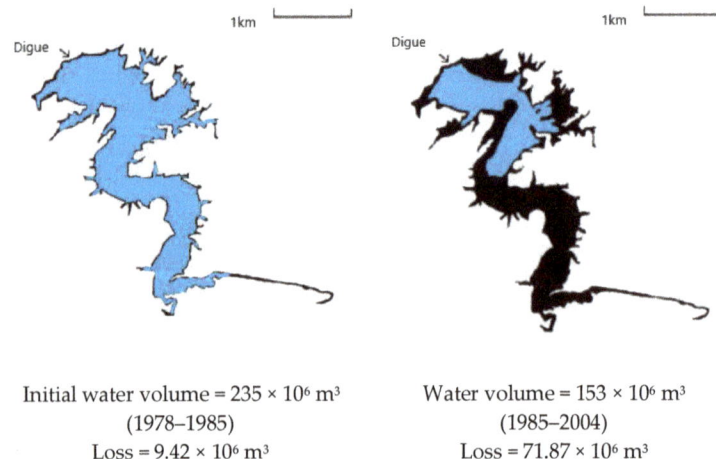

Initial water volume = 235 × 10^6 m^3
(1978–1985)
Loss = 9.42 × 10^6 m^3

Water volume = 153 × 10^6 m^3
(1985–2004)
Loss = 71.87 × 10^6 m^3

Figure 10. Evolution of the basin morphology and the surface in free water of the Sidi M'Hamed Benaouda Dam [14]. (**a**) Initial water volume (1978–1985); (**b**) Water volume (1985–2004).

Indeed, according to bathymetric surveys conducted in 1978, 1985, 2000 and 2004, the SMBA Dam has lost much of its capacity (Figure 11), from 1978 to 1985 annual loss estimated to 1.35 million m^3, from 1985 to 2000, an annual loss of about 4.41 million m^3 and from 2000 to 2004 an annual loss was estimated to 1.42×10^6 m^3, over all these periods of bathymetric measurements, the highest loss rate was observed between 1985 and 2000, which is probably due to exceptional rains that hit the region during this period, thus in October 1993: "Twenty minutes of rain do 16 dead and a dozen of missing persons in Relizane" [137].

It is important to note that at the start of the dam exploitation, the reservoir filling with fine particles is carried out linearly according to weather conditions. Starting with 2000, there was a decrease in siltation, mainly caused by the dam exploitation mode (periodic shelving of fine particles by the outlet gates) [14].

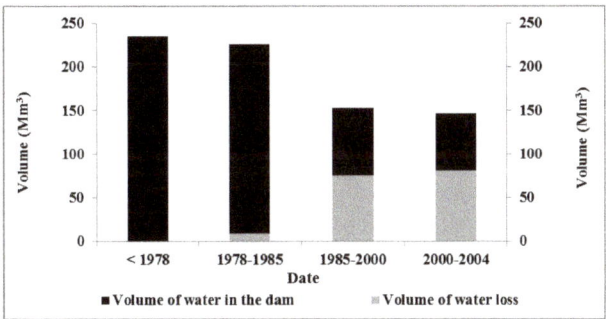

Figure 11. Evolution of the accumulated loss of water from the Sidi M'Hamed Benaouda Dam [14].

Conversely, the calculations made by the concentration measures for the basin of Wadi Mina at the level of the station of Wadi El Abtal and Wadi Haddad, at the Sidi AEK Djilali station level, and which control a surface of 5856 km^2, generated a load equal to 5.2×10^6 t·year^{-1} (Table 8). The difference of 400,000 t·year^{-1} corresponds to the sedimentary inputs of the micro-basin surrounding the impoundment (192 km^2) and the autochthon sedimentation in the impoundment, without being able to establish the parts of each source. To elaborate further on this topic, we should deploy specific measures as those presented by Maleval [138]. This difference of 400,000 t·year^{-1} corresponds to a specific degradation of 21 t·ha^{-1}·year^{-1}, namely 2.4 times more than the specific degradation for the entire basin of 5856 km^2 (Table 8). This difference can be explained, on the one hand, by potential autochthon sedimentation, but which is not quantifiable for the moment, and, on the other side, by a specific degradation intrinsically higher because of geomorphology reasons, as the eroded products on the slopes arrive very rapidly at the impoundment, without the existence of an intermediary decantation area. Thus, this specific degradation value of 21 t·ha^{-1}·year^{-1} is even higher than the "potential" erosion value of about 12 t·year^{-1}, calculated by Toumi et al. [66] for the entire basin, based on the USLE equation.

Table 8. Sedimentary load and specific degradation in the basin of Wadi Mina.

Designation	SMBA Dam
Sedimentary load deposited in the dam impoundment (bathymetry).	5.6×10^6 t·year^{-1}
Sedimentary load by concentration measure on Wadi Haddad and Wadi Mina upstream.	5.2×10^6 t·year^{-1}
Sedimentary load coming from the dam slopes.	4×10^5 t·year^{-1}
Specific degradation (dam slopes).	21 t·ha^{-1}·year^{-1}
Specific degradation (Stations: W.El Abtal + Sidi Abdelkader Djilali).	8.6 t·ha^{-1}·year^{-1}

For concluding this part, we can note that, in accordance with the digits in Table 8, over 90% of sediments of the SMBA Dam appears to come from the contributions of the Wadi Haddad and Wadi Mina upstream of the dam.

Thus, Touaibia [45] showed that the USLE equation allowed to map each parameter of this equation for the entire basin of Wadi Mina and superposition of layers (for each parameter) in a GIS and resulted in values of degradation in $t \cdot ha^{-1} \cdot year^{-1}$ (Table 9).

Indeed, the results of the concentration measures as well as those found by the erosive potentials equation (USLE) show that the basin upstream of the SMBA (1B) Dam (Figure 12) presents an important erosive risk, because it produces a specific degradation equal to 13.36 $t \cdot ha^{-1} \cdot year^{-1}$ and 19.64 $t \cdot ha^{-1} \cdot year^{-1}$, respectively, this because of the predominance of tertiary marls in this part of the basin and which are covered in northeast by calcareous sandstone and dolomites and the slopes, which are from 12% to 30%, generating important corrosive risks with occurrence of the signs caused by diffuse streaming, without forgetting the sparse vegetation cover [139]. In addition, the specific degradations vary from 11.54 to 12.64 $t \cdot ha^{-1} \cdot year^{-1}$ at the basin level (5B) upstream of the Bakhadda Dam and, respectively, the basin of Wadi Haddad (2B), calculated by the USLE equation [66] against 2.39 to 9.64 $t \cdot ha^{-1} \cdot year^{-1}$ on the same sub-basins, these values remaining high, considering that the basin of Wadi Haddad presents a lithology constituted by the Continental Quaternary (this formation is represented by alluvia, regs and terraces), the Continental Pliocene (represented by lacustrine limestone) and the marine Lower Miocene (represented by a marly formation). The basin vegetation is formed essentially of scrubland, representing almost 30% from the entire basin (degraded forest). The vegetation is overgrazed and is discontinuous in the space [68] and the basin (5B) upstream of the Bakhadda Dam is dominated by limestone and dolomites alternating with marls and an average dense vegetation and slopes varying from 5% to 12% that present weak erosion risks, but can be strong with the presence of agricultural activities, mainly grain farming that dominates soil occupation [139].

The values of specific degradation are low in the southwest part of the basin (basin of Wadi El Abd) and thus the fields in this regional are quasi-flat (slope lower than 10%) [139], and thus the specific degradations are equal to 3.86 $t \cdot ha^{-1} \cdot year^{-1}$ at the level of the Takhmert station and 8.82 $t \cdot ha^{-1} \cdot year^{-1}$ at the Ain Hamara station [66], in accordance with the calculations generated using the USLE equation, and varying from 2.20 $t \cdot ha^{-1} \cdot year^{-1}$ at the level of the Takhmert station and 5.02 $t \cdot ha^{-1} \cdot year^{-1}$ at the Ain Hamara station (Table 9), this basin marking the existence of marls which are friable rocks, thus a favorable factor for degradation and an insufficiency of forest vegetation and the surfaces are often used for pasture [128]. The climatic conditions semiarid are associated with a high degree of rainfall irregularity with average precipitations of 266 mm in Ain Hamara and 238 mm in Takhmert.

The maxim value of losses in soil is registered at the level of the dam basin (400,000 $t \cdot ha^{-1}$), this value representing the sum of the solid transport by suspension and that of the bed load which is estimated between 15% and 25% of the suspension [140], namely a loss in soil equal to 392,157 $t \cdot ha^{-1}$ and by bed load equal to 78,431 $t \cdot ha^{-1}$, estimating a load transport equal to 20% of the suspension.

Compared to the found values, we observe that the basin upstream of the SMBA Dam (1B) is the greatest producer of sediments towards the dam because it shows a specific degradation equal to 13.36 $t \cdot ha^{-1} \cdot year^{-1}$, value found after the elimination of release of water from the Bekhadda Dam, estimated at 1.32 $t \cdot ha^{-1} \cdot year^{-1}$ on average, considering that this dam registers an average loss amounting to 0.27×10^6 m^3 since its launching in 1936; this result corroborates with that found by Touaibia [45] as well as with the works of GTZ [141], because this area is dominated by a marly lithological formation situated often on the piedmonts of a slope of 10 to 20% (Table 9), favouring soil erosion.

Table 9. Values of specific degradation per sub-basin.

Sub-Basin	Wadi Name	Specific Degradation (t·ha^{-1}·year^{-1})		Difference (USLE, Concentrations Measures)
		USLE Equation	Concentrations Measure	
1B		19.64	13.36	6.28/32%
5B	Wadi Mina	11.54	2.39	9.15/79%
6B		/	21	/
2B	Wadi Haddad	12.64	9.64	3.00/24%
3B		8.82	5.02	3.80/43%
4B	Wadi El Abd	3.86	2.20	1.66/43%

In addition, the other sub-basins record specific degradation varying from 2.2 to 9.64 t·ha^{-1}·year^{-1}, and these values are close to those found by [66] using the USLE equation, except for the basin of Wadi Mina upstream at the level of the station of Sidi Ali Ben Amar, where the calculated value using the USLE equation is 4.83 times higher than the one calculated by the direct concentration measures (Figure 12). The value of the basin 5B using the concentrations measure considers the solids quantities sedimented in the dam impoundment in contrast to that generated by the USLE equation which does not consider the volumes of sediments caught in the Bakhadda dam but it only models potential soil erosion taking place. This calculation approach generated the difference between the estimated values by the two techniques (Table 9).

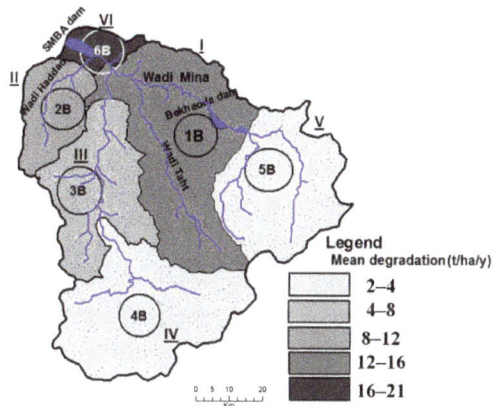

Figure 12. Specific degradation per sub-basin.

In fact, we only measured the sediment that actually reached the stream system, so the difference between what is measured in the current and what is predicted by USLE is largely due to the deposits of halfway that do not reach the watercourse. This does not prevent lake like Bakhadda dam also playing a role of intermediate trapping on the watercourse.

Soil erosion is such a highly variable process that one should avoid establishing rules for larger catchments without measuring at least a few of the processes taking place directly. Sediment concentration may point towards the real processes, but they are still an indirect value for material eroded. In this study, the purpose of the comparison of the different sources of information that we use is to evaluate the validity of the observed and modelled values: are the fluxes measured in the rivers (SPM) compatible with the rates of erosivity provided by USLE? Does the sedimentation rate in the dam reflect the orders of magnitude of SPM observed in streams? Thus, sub-basins can be used to locate a potential source of sediment in small sub-basins close to the reservoir, for which the erosivity rate could be higher, and/or the proximity to the reservoir allows for greater fraction of SPM carried to reach the reservoir.

The analysis of the photo of the slopes of the dam basin (Figure 13) indicates a favorable environment (inclined slopes and formed of slightly resistant soil, little or no vegetation cover, aggressive climate and human activities), for causing important inputs in sediments that shall deposit directly in the lake. This would explain largely the very serious specific degradation calculated for the small catchment area downstream with 192 km^2 that surrounds the dam (6B) after the last station of hydrologic measurement, obtained by the difference between the equivalent in SPM, deducted from the dam silt volume and the concentration measures in SPM at the upstream hydrological station.

Indeed, the average annual streaming can reach 15–30%, varying with the years and depending on the fallow land silting by weeds between two weedings, but as well with the distribution of showers depending on the agricultural works and the prior humidity of the soil [142]. Average showers falling on the bare, saturated and crusted surface soil can cause a streaming exceeding 60–80%. Here lies the greatest risk for the dams silting because, gathering in gullies and Wadis, the floods peaks move very important quantities of materials [142].

We consider that the matters in suspension, measured at the catchment area outlet, do not reflect the current dynamics of all the slopes, but they reflect the transport capacity of streaming and runoff waters in the gullies. There exists a stock of fine sediments which is progressively evacuated towards the SMBA Dam and in a rhythm that depends on the available force: that of floods, related in particular to the quantity of rains received and the streaming coefficient [143].

Figure 13. Bare banks of the SMBA Dam (Shot Hallouz, August 2013).

Thus, depending on the gullied lands and the type of gully, the input to silting of an impoundment can reach and even exceed 30% [144].

4. Discussion

In this study, we could use the maximum of information available which allowed us, on the one hand, to calculate the quantities of solid transport on the whole of the watershed, and, on the other hand, to propose the hypotheses on the origin of the material sedimented in the SMBA Dam.

The rain, the discharge, and concentrations of SPM are fairly well known on the watershed of the Wadi Mina and its main tributaries.

Rainfall series for which a rupture could be detected are more to the north than to the south. The recorded rainfall deficits frequently are around 19–20%. This rainfall deficit is thus felt for over three decades and seems to have spread over the basin during the 1970s. Indeed, 13 stations time series show a significant rupture during the 1970s across the entire watershed of Wadi Mina which represents 50% of rainfall stations used in this study.

The largest number of monthly ruptures was detected during the 1970s especially during the months of January and April with, respectively, 18 and 15 ruptures whose average monthly rainfall accounts for about 12% of annual rainfall.

This decline of rains has resulted in a decrease in discharge, the latter, know an overall decrease since the 1970s, or the probable ruptures are observed both in the rainy season and in the dry season. At the annual time step, probable rupture dates are 1974, 1975, 1980 and 1994 which generally correspond to years where most of the ruptures detected at the monthly time step took place.

These results show a satisfactory agreement with the dates detected on the series of annual and monthly rainfall of late 1970s/early 1980s. Thus, we suppose that reduced flows would be due to an overall decrease in precipitation over the watershed of Wadi Mina [47]. Small changes in precipitation can have significant impact on the flow of surface water [145]. In fact, climate is recognized to be the main factor in semi-arid Mediterranean areas of Algeria which experience short and intense rain episodes, high evaporating power of wind, prolonged droughts and freezing and thawing cycles [42,45,46].

The recent increase of rainfall represented by a significant occurrence of positive ruptures in time series during the 2000s is not followed by an increase in discharges, which might be explained by an increasing impact of dams and small hydraulic structures over the basins that mitigate the effect of rainfall on the hydrological regime of sub basins.

With the aid of HYFRAN-PLUS software used in the second part, the assumptions of stationary and independence are verified on all the floods of the Wadi Mina basin and these are classified in Category C, following the GEV distribution.

In addition, the dates of rupture detected on each sub basin showed a decrease of floods from 1970 onward. The flood values change around the ruptures date, and the study concluded that, on the entire study area, a flood, for a twenty-year return period before the rupture date, became centennial after.

The two parts of this study have shown that the feeling of local population of a decrease in rainfall regime in the region is absolutely founded.

The last part of the study concerns the relationship between the suspended sediment transport on the Wadi Mina and siltation of the SMBA Dam.

Thus, the annual average concentrations of *SPM* were compared with annual average discharge. Despite year-to-year irregularities at the Sidi Aek Djilali station, concentrations and discharges have changed in opposite directions [68]. We found that ruptures were detected on the rainfall and discharges from the mid-1970s, this result show that is no effect of ruptures in precipitation and discharge on the solid load response since the discharge and rainfall decrease in the 1970s and increase in the early 2000s while the sediment load is still increasing.

The watershed of Wadi Haddad in Sidi Abdelkader El Djilali has a specific charge very high compared to other basins (Ds = 965 $t \cdot km^{-2} \cdot year^{-1}$) which means that even after the rupture in the rainfall time series, and the development of agriculture and dams, this sub-basin is still the provider of the largest amount of sediment to the reservoir. The lowest degradation values are observed on the Wadi in Abd Takhmert (220 $t \cdot km^{-2} \cdot year^{-1}$) and the Wadi Mina upstream in Sidi Ali Ben Amar (239 $t \cdot km^{-2} \cdot year^{-1}$). Wadi Mina basin annually provides 38 million m^3 of water with a flow of 2.6 million tons of suspended sediment, which therefore corresponds to a relatively low specific degradation of about 860 $t \cdot km^{-2} \cdot year^{-1}$, a value significantly below published estimates for neighboring basins in Algeria and North Africa, This value indicates an increase in recent decades, which could be related to an increased susceptibility to soil erosion in the basin, under the combined effect of climatic and anthropogenic forcing [51]. However, the quantities of sediment exported by the Wadi are highly variable from one year to another. Note that the year 1994/95 alone represents a contribution of 8.15 million tons from the Wadi Abd, i.e., a specific degradation of more than 1500 $t \cdot km^{-2} \cdot year^{-1}$, a value that corresponds to estimates found for Maghreb watersheds as a whole.

We found that the basin upstream of the SMBA Dam (1B) is the greatest producer of sediments towards the dam because it shows a specific degradation equal to 13.36 $t \cdot ha^{-1} \cdot year^{-1}$, a value found after the elimination of release of water from the Bekhadda Dam, considering that this dam registers an average loss amounting to 0.27×10^6 m^3 since its launching in 1936. In addition, the other sub-basins record specific values varying from 2.2 to 9.64 t·ha^{-1}·year^{-1}, and these values are close to those found by Toumi et al. [66] using the USLE equation, except for the basin of Wadi Mina upstream at the level of the station of Sidi Ali Ben Amar, where the calculated value using the USLE equation is 4.83 times higher than the one calculated by the direct concentration measures. The value of the basin 5B using the concentrations measure considers the solids quantities sedimented in the dam impoundment in contrast to that generated by the USLE equation which does not consider the volumes of sediments caught in the Bakhadda Dam but only models potential soil erosion taking place. In fact, we only measured the sediment that actually reached the stream system, so the difference between what is measured in the current and what is predicted by USLE is largely due to the deposits of halfway that do not reach the watercourse. This does not prevent lake-like Bakhadda Dam also playing a role of intermediate trapping on the watercourse.

Soil erosion is such a highly variable process that one should avoid establishing rules for larger catchments without measuring at least a few of the processes taking place directly. Sediment concentration may point towards the real processes, but they are still an indirect value for material eroded. In this study, the purpose of the comparison of the different sources of information that we use is to evaluate the validity of the observed and modeled values. Thus, sub-basins can be used to locate a potential source of sediment in small sub-basins close to the reservoir, for which the erosivity rate could be higher, and/or the proximity to the reservoir allows for greater fraction of SPM carried to reach the reservoir.

According to the bathymetric surveys carried out at the SMBA Dam, loss in water capacity of this dam since its commissioning in 1978 until 2004 is estimated at 82 million m^3, equivalent to a third of its original capacity. According to Remini and Bensafia [63], in 2010 the SMBA Dam contains a volume of sediments equal to 109×10^6 m^3 or an annual average loss of about 3.5×10^6 m^3·year^{-1}.

Based on these results, we concluded that over 90% of the sediments of the SMBA Dam come from the contributions of the Wadi Haddad and Wadi Mina upstream of the dam, but the specific degradation around the dam is higher than over the rest of the basin. Currently the dam filling by sediments is about 48%, so it should be completely silted up by 2048, without dynamic change.

Despite all the efforts made in recent years, the dam silting phenomenon has increased due to the increase in anthropogenic influence and erosion.

Definitively, the decrease in rainfall, which hit the region since the mid-1970s, induced a progressive degradation of the vegetation cover combined with the increase of summer thunderstorms, and induces changes in the functioning of the watershed with a greater susceptibility to erosion. For specific years, moderate rainfall and runoff generated considerable sedimentary flow.

5. Conclusions

This article presents a study of erosion, sedimentation and climate change on Wadi Mina, at the Sidi M'Hamed Ben Aouda Dam in the northeast of Algeria, in a semiarid area. It is important to fallow the evolution of water resources available for this dam in a variable climatic environment.

We have noted a rupture corresponding to a decrease of precipitations during the mid-1970s over the whole Wadi Mina basin which 58% series present a rupture in 1976. These results confirm as well the occurrence in Morocco of a pluviometric deficit starting with 1970 and its continuation during the 1980s, an extremely severe and long drought episode [106]. At the level of North Africa, the results obtained by Meddi et al. [11] in Algeria and by GIEC [107] in Tunisia indicate the same rupture period and show the spatial extension of the drought, accompanied by a net pluviometric reduction.

In addition, the discharges register a global decrease since 1970. At annual level, the probable rupture dates are 1974, 1975, 1980 and 1994, which corresponds generally to the ruptures detected at month level. In addition, the ruptures on the series of annual daily maximum were for the most part noted in 1970, 1984 and 1995 and the ruptures dates for the annual daily minimum series are detected in 1972, 1981 and 1994. The rupture in the daily maximum flows series is thus never noted the same year as in the average annual flows series, while there are the same ruptures dates for the average annual flows and the minimum daily flows.

As for the majority of watercourses, we found a power relation [127] for Wadi Mina, connecting the concentration of suspended matters in $g \cdot L^{-1}$ to the liquid discharge in $m^3 \cdot s^{-1}$. At seasonal scale, the analysis shows that autumn and spring are distinguished by their strong liquid discharges generating an important flow of suspended particles matters (SPM). The maximum solid flow is of 42,365 $kg \cdot s^{-1}$, achieved in September 1994, resulting from a liquid discharge of more than 351 $m^3 \cdot s^{-1}$.

In contrast, during winter and summer, we note a net regression of liquid discharge that no longer exceed 84 $m^3 \cdot s^{-1}$ in winter and 38 $m^3 \cdot s^{-1}$ in summer.

The average specific sediment discharge of Wadi Mina at the Sidi M'hamed Ben Aouda Dam is 860 $t \cdot km^{-2} \cdot year^{-1}$. We can note that over 90% of sediments of the SMBA Dam appears to come from the contributions of the Wadi Haddad and Wadi Mina upstream of the dam.

In fact, starting in the late 1980s, annual production of sediments became seven times larger than the previous period, with an increase four times greater of contribution of the dry season rates [146].

Author Contributions: F.H. wrote the main paper, and F.H., M.M., G.M., S.T., S.A.R. discussed the results and implications and commented on the manuscript at all stages. F.H., M.M., G.M., S.T., S.A.R. contributed extensively to the work presented in this paper.

Funding: This study was supported by funding from the AUF through the program SIGMED (MeRSI No. 6313PS005) Bureau Western Europe and North Africa.

Acknowledgments: Thanks to ANRH for the hydrological and rainfall data.

Conflicts of Interest: The authors declare no conflict of interest.

References

1. Laborde, J.P. *Carte Pluviométrique de L'ALGÉRIE du Nord à L'échelle du 1/500000*; Projet PNUD/ALG/88/021, Une Carte Avec Notice Explicative; Agence Nationale des Ressources Hydrauliques: Bir Mourad Rais, Algeria, 1993; 44p. (In French)
2. Rossel, F.; Le Goulven, P.; Cadier, E. Répartition spatiale de l'influence de l'ENSO sur les précipitations annuelles de l'Equateur. *Rev. Sci. L'eau J. Water Sci.* **1999**, *12*, 183–200. (In French) [CrossRef]
3. Touazi, M.; Laborde, J.P. Cartographie des pluies annuelles en Algérie du Nord. *Assoc. Int. Climatol.* **2000**, *13*, 192–198. (In French)
4. Meddi, H. *Quantification des Précipitations: Application au Nord-Ouest Algérien Méthodologie Pluvia*; Mémoire de Magister, Centre Universitaire de Mascara: Mascara, Algeria, 2001; 160p. (In French)
5. El Mahi, A. *Déficit Pluviométrique des Dernières Décennies en Algérie du Nord and Son Impact sur les Ressources en Eau*; Mémoire de Magister, Centre Universitaire de Mascara: Mascara, Algeria, 2002; 120p. (In French)
6. Meddi, M.; Hubert, P. Impact de la modification du régime pluviométrique sur les ressources en eau du Nord-ouest de l'Algérie. In Proceedings of the International Symposium on Hydrology of the Mediterranean and Semi Arid Regions, Montpellier, France, 1–4 April 2003; IAHS Publication: Wallingford, UK, 2003; pp. 229–235. (In French)
7. El Mahi, A.; Meddi, M.; Matari, A.; Kattrouci, K. Etat de la pluviométrie en période de sécheresse en Algérie du Nord et sa relation avec le phénomène ENSO. In *Acte du Colloque 'Terre and Eau'*; Algerian Journal of Arid Areas: Annaba, Algeria, 2004; pp. 420–423. (In French)
8. Talia, A.; Meddi, M. La pluvio-variabilité dans le Nord de l'Algérie. In *Actes du Colloque 'Terre and Eau'*; Algerian Journal of Arid Areas: Annaba, Algeria, 2004; pp. 477–480. (In French)
9. Meddi, H.; Meddi, M. Variabilité spatiale and temporelle des précipitations du Nord-Ouest de l'Algérie. *Geogr. Tech.* **2007**, *2*, 49–55. (In French)
10. Bekoussa, B.; Meddi, M.; Jourde, H. Forçage climatique and anthropique sur la ressource en eau souterraine d'une région semi-aride: Le cas de la plaine de Ghriss (Nord-Ouest algérien). *Sécheresse* **2008**, *19*, 173–184. (In French)

11. Meddi, M.; Talia, A.; Martin, C. Evolution récente des conditions climatiques et des écoulements sur le bassin versant de La Macta Nord-Ouest de l'Algérie. *Géogr. Phys. Environ. Physio-Géo* **2009**, *3*, 61–84. (In French) [CrossRef]
12. Meddi, M.; Meddi, H.; Assani, A. Variabilité temporelle des précipitations dans les bassins de la Macta et la Tafna, Nord-Ouest d'Algérie. *Water Ressour. Manag.* **2010**, *24*, 3817. (In French) [CrossRef]
13. Niazi, S.; Snoussi, M. Evaluation des risques d'inondation à une élévation accélérée du niveau de la mer sur le littoral de Tétouan Méditerranée occidentale marocaine. In Proceedings of the 3ème Journée Internationales de Géosciences de l'environnement, El Jadida, Morocco, 8–10 June 2005. (In French)
14. Agence Nationale des Barrages et Transferts (ANBT). *Rapport Annuel D'activité sur L'état des Barrages Algériens*; ANBT: Kouba, Algeria, 2014. (In French)
15. Williams, J.R.; Berndt, H.D. Sediment yield computed with universal equation. *J. Hydraul. Div.* **1972**, *98*, 2087–2098.
16. Lahlou, A. Modèles de prédiction de la sédimentation des retenues de barrages des pays du Grand Maghreb. In Proceedings of the Atelier International UNESCO-AISH-ENIT sur L'Application des Modèles Mathématiques à L'évaluation des Modifications de la Qualité Des Eaux, Tunis, Tunisia, 7–12 May 1990; pp. 312–324. (In French)
17. Wicks, J.M.; Bathurst, J.C. SHE-SED: A physically based, distributed erosion and sediment yield component for the SHE hydrological modeling system. *J. Hydrol.* **1996**, *175*, 213–238. [CrossRef]
18. Kassoul, M.; Abdelgader, A.; Belorgey, M. Caractérisation de la sédimentation des barrages en Algérie. *Rev. Sci. L'eau J. Water Sci.* **1997**, *3*, 339–358. (In French) [CrossRef]
19. Terfous, A.; Meghnounif, A.; Bouanani, A. Etude du transport solide en suspension dans l'oued Mouillah (Nord-Ouest Algérien). *Rev. Sci. L'eau J. Water Sci.* **2001**, *14*, 173–185. (In French) [CrossRef]
20. Abdellaoui, B.; Merzouk, A.; Aberkan, M.; Albergel, J. Bilan hydrologique and envasement du barrage Saboun (Maroc). *Rev. Sci. L'eau J. Water Sci.* **2002**, *15*, 737–748. (In French) [CrossRef]
21. Marzougui, A.; Ben Mammou, A. Le barrage de l'Oued Sejnane: Quantification de l'alluvionnement et évaluation de l'érosion spécifique de son bassin versant. *Geo-Eco-Trop* **2006**, *30*, 57–68. (In French)
22. Remini, B.; Hallouche, W. Prévision de l'envasement dans les barrages Du Maghreb. *Larhyss J.* **2005**, *4*, 69–80. (In French)
23. Lahlou, A. *Envasement des Barrages au Maroc [Siltation of Dams in Morocco]*; Wallada: Casablanca, Morocco, 1994; p. 286. ISBN 9981823074. (In French)
24. Saadaoui, M. *Érosion et Transport Solide en Tunisie. Mesure et Prévision du Transport Solide Dans les Bassins Versants et de L'envasement Dans les Retenues des Barrages*; Direction Générale des Barrages et des Grands Travaux Hydrauliques (DGBGTH), Ministère de L'Agriculture, Direction Générale des Ressources en Eau, Direction des Eaux de Surface: Tunis, Tunisie, 1995; 30p. (In French)
25. Abid, M. *Envasement des Barrages en Tunisie*; Direction Générale des Barrages et des Grands Travaux Hydrauliques (DGBGTH); Ministère de L'Agriculture: Tunis, Tunisie, 1998; 69p. (In French)
26. Bourouba, M. Contribution à l'étude de l'érosion et des transports solides de l'Oued Medjerda supérieur (Algérie orientale). *Bull. Rés. Eros.* **1998**, *18*, 76–97. (In French)
27. Colombani, J.; Olivry, J.C.; Kallel, R. Phénomènes exceptionnels d'érosion et de transport solide en Afrique aride et semi-aride. In *Challenges in African Hydrology and Water, Ressources, Proceedings of the Harare Symposium, Harare, Zimbabwe, 23–27 July 1984*; IAHS Publication: Oxfordshire, UK, 1984; Volume 144, pp. 295–300. (In French)
28. Kettab, A. Les ressources en eau en Algérie: Stratégies, enjeux et vision. *Desalination* **2001**, *136*, 25–33. (In French) [CrossRef]
29. Remini, B. *La Problématique de L'eau en Algérie*; Office des Publications Universitaires: Alger, Algérie, 2005; p. 162. (In French)
30. Boithias, L.; Acuña, V.; Vergoñós, L.; Ziv, G.; Marcé, R.; Sabater, S. Assessment of the water supply: Demand ratios in a Mediterranean basin under different global change scenarios and mitigation alternatives. *Sci. Total Environ.* **2014**, *470*, 567–577. [CrossRef] [PubMed]
31. Surface Water Resources Mobilization. Document of Algerian Ministry of Water Resources. Available online: http://www.mree.gov.dz (accessed on 24 April 2018).
32. Hamiche, A.; Boudghene Stambouli, A.; Flazi, S. A review on the water and energy sectors in Algeria: Current forecasts, scenario and sustainability issues. *Renew. Sustain. Energy Rev.* **2015**, *41*, 261–276. [CrossRef]

33. Bouzid-Lagha, S.; Djelita, B. Study of eutrophication in the Hamman Boughrara Reservoir (Wilaya de Tlemcen, Algeria). *Hydrol. Sci. J.* **2012**, *57*, 186–201. [CrossRef]
34. Tidjani, A.E.; Yebdri, D.; Cherif, E.A. *Ampleur de L'envasement Dans les Barrages Algériens*; Documents Techniques en Hydrologie; UNESCO: Paris, France, 2000; Volume 29, pp. 121–128. (In French)
35. Bourouba, M. Phénomène de transport solide dans les Hauts Plateaux Orientaux. Cas de l'Oued Logmane et oued Leham dans le bassin de la Hodna. *Rev. Sci. Technol.* **1998**, *9*, 5–11.
36. Megnounif, A.; Terfous, A.; Bouanani, A. Production et transport des matières solides en suspension dans le bassin versant de la Haute-Tafna (Nord-Ouest algérien). *Rev. Sci. L'eau J. Water Sci.* **2003**, *16*, 369–380. [CrossRef]
37. Achite, M.; Meddi, M. Variabilité spatio-temporelle des apports liquide et solide en zone semi-aride. Cas du bassin versant de l'oued Mina (nord-ouest algérien). *Rev. Sci. L'eau J. Water Sci.* **2005**, *18*, 37–56. [CrossRef]
38. Achite, M.; Ouillon, S. Suspended sediment transport in semiarid watershed, Wadi abd, Algeria (1973–1995). *J. Hydrol.* **2007**, *343*, 187–202. [CrossRef]
39. Ghenim, A.; Terfous, A.; Seddini, A. Étude du transport solide en suspension dans les régions semiarides méditerranéennes: Cas du bassin versant de l'Oued Sebdou (Nord-Ouest Algérien). *Sécheresse* **2007**, *18*, 39–44. (In French)
40. Ghenim, A. Étude des Écoulements et des Transports Solides Dans les Régions Semi-Arides Méditerranéennes. Ph.D. Thesis, Université de Tlemcen, Chetouane, Algérie, 2008.
41. Zettam, A.; Taleb, A.; Sauvage, S.; Boithias, L.; Belaidi, N.; Sánchez-Pérez, J.M. Modeling Hydrology and Sediment Transport in a Semi-Arid and Anthropized Catchment Using the SWAT Model: The Case of the Tafna River (Northwest Algeria). *Water* **2017**, *9*, 216. [CrossRef]
42. Achite, M.; Ouillon, S. Recent changes in climate, hydrology and sediment load in the Wadi Abd, Algeria (1970–2010). *Hydrol. Earth Syst. Sci.* **2016**, *20*, 1355–1372. [CrossRef]
43. Abdelkader Benkhaled & Boualem Remini. Temporal variability of sediment concentration and hysteresis phenomena in the Wadi Wahrane basin, Algeria. *Hydrol. Sci. J.* **2003**, *48*, 243–255. [CrossRef]
44. Khanchoul, K.; Jansson, M.B.; Lange, Y. Comparison of suspended sediment yield in two catchments, northeast, Algeria. *Z. Geomorphol.* **2007**, *51*, 63–94. [CrossRef]
45. Touaibia, B. Problématique de l'érosion et du transport solide en Algérie septentrionale. *Sécheresse* **2010**, *21*, 333–335. (In French)
46. Houyou, Z.; Bielders, C.L.; Benhorma, H.A.; Dellal, A.; Boutemdjet, A. Evidence of strong land degradation by wind erosion as a result of rainfed cropping in the Algerian steppe: A case study at Laghouat. *Land Degrad. Dev.* **2014**, *27*, 1788–1796. [CrossRef]
47. Hallouz, F.; Meddi, M.; Mahé, G. Modification du régime hydroclimatique dans le basin de l'oued Mina (Nord-Ouest d'Algérie). *Rev. Sci. L'eau J. Water Sci.* **2013**, *26*, 33–38. (In French) [CrossRef]
48. Khomsi, K.; Mahe, G.; Sinan, M.; Snoussi, M.; Cherifi, R.; Nait Said, Z. Evolution des évènements chauds rares et très rares dans les bassins versants du Tensift et du Bouregreg (Maroc) et identification des types de temps synoptiques associés. In *From Prediction to Prevention of Hydrological Risk in Mediterranean Countries*; Ferrari, E., Versace, P., Eds.; University of Calabria: Cosenza, Italy, 2012; pp. 169–182.
49. Mahé, G.; Aksoy, H.; Brou, Y.T.; Meddi, M.; Roose, E. Relationships between man, environment and sediment transport: A spatial approach. *Rev. Sci. L'eau J. Water Sci.* **2013**, *26*, 235–244. [CrossRef]
50. Laouina, A.; Aderghal, M.; Al Karkouri, J.; Chaker, M.; Machmachi, I.; Machouri, N.; Sfa, M. Utilisation des sols, ruissellement et dégradation des terres, le cas du secteur Sehoul, région atlantique, Maroc. *Sécheresse* **2010**, *21*, 309–316. (In French)
51. Mahè, G.; Benabdelfadel, H.; Dieulin, C.; El Baraka, M.; Ezzaouini, M.; Khomsi, K.; Rouche, N.; Sinan, M.; Snoussi, M.; Tra Bi, A.; et al. Evolution des débits liquides et solides du Bouregreg, 2015. In *Gestion Durable des Terres, Proceedings of the Réunion Multi-Acteurs sur le Bassin du Bouregreg, Rabat, Morocco, 28 May 2013*; ARGDT: Rabat, Morocco, 2014; pp. 53–62. ISBN 978-9954-33-482-9. (In French)
52. Touaïbia, B.; Gomer, D.; Ouffar, F.; Geyer, F. Approche Quantitative de l'Erosion Hydrique essais de simulation de pluies sur micro-bassins expérimentaux bassin versant de l'oued mina. W. Relizane. *Bull. Rés. Eros.* **1992**, *12*, 401–407. (In French)
53. Touaïbia, B.; Dautrebande, S.; Gomer, D.; Mostefaoui, M. Quantification de l'érosion à partir d'implantation de quatre retenues collinaires dans la zone des marnes. W. Relizane Algérie. *Bull. Rés. Eros.* **1995**, *15*, 408–416. (In French)

54. Touaïbia, B.; Dautrebande, S.; Gomer, D.; Aidaoui, A. Approche quantitative de l'érosion hydrique à différentes échelles spatiales: Bassin versant de l'Oued Mina. *J. Sci. Hydrol.* **1999**, *44*, 408–416. (In French) [CrossRef]
55. Touaïbia, B.; Gomer, D.; Aidaoui, A. Estimation de l'index d'érosion de wischmeïer dans les micros bassins expérimentaux de l'oued mina en Algérie du Nord. *Bull. Rés. Eros.* **2000**, *20*, 478–484. (In French)
56. Touaïbia, B.; Gomer, D.; Aïdaoui, A.; Achite, M. Quantification et variabilité temporelles de l'écoulement solide en zone semi-aride, de l'Algérie du Nord. *Hydrol. Sci. J.* **2001**, *46*, 41–53. (In French) [CrossRef]
57. Gomer, D. *Programme Hydrométrique: Projet Pilote D'aménagement Intégré*; Rapport D'évaluation: Sidi-M'Hamed Ben Aouda, Algérie, 1990. (In French)
58. Gomer, D.; Vogt, T. Determination of runoff and soil erosion under semiarid conditions using GIS to integrate Landsat TM, DEH und hydrological field data from the Wadi Mina project Algeria. In Proceedings of the International Meeting "Soil Erosion in Semi-Arid Mediterranean Arcas", Taormina, Italy, 28–30 October 1993; European Society for Soil Conservation: Catania, Italy, 1993; pp. 19–93.
59. Kouri, L.; Vogt, H. Détermination de la Sensibilité des Terrains Marneux au Ravinement au Moyen de Système D'information Géographique. B.V de L'oued Mina; Tell Oranais Algérie. *Bull Rés. Eros.* **1997**, *17*, 64–73. (In French)
60. Kouri, L.; Vogt, H.; Gomer, D. Analyse des processus d'érosion hydrique linéaire en terrain marneux. Bassin-versant de l'oued Mina, Tell Oranais, Algérie. *Bull. Rés. Eros.* **1997**, *17*, 64–73. (In French)
61. Gomer, D. *Ecoulement et Érosion Dans les Petits Bassins Versants à sols Marneux Sous Climat Semi-Méditerranéen*. Ph.D. Thesis, Wasserbau, Université de Kajsruhe, Karlsruhe, Germany, 1994.
62. Touaïbia, B.; Achite, M. Contribution à la cartographie de l'érosion spécifique du bassin versant de l'Oued Mina en zone semi-aride de l'Algérie septentrionale. *Hydrol. Sci. J.* **2003**, *48*, 235–242. (In French) [CrossRef]
63. Remini, B.; Bensafia, D. Envasement du Barrage de SMBA Algérie. Comité Scientifique Projet SIGMED Rabat Maroc. 2011. Available online: http://armspark.msem.univ-montp2.fr/SigMED/index.asp?menu=ComiteScientifique (accessed on 12 September 2012). (In French)
64. Bekhti, B.; Errih, M.; Sidi Adda, M. Modélisation de la sédimentation dans les retenues de barrages en Algérie (barrage Es-Saada). *Sécheresse* **2012**, *23*, 38–47. (In French)
65. Remini, B. Evolution spatio-temporelle de l'envasement dans le barrage S.M.B.A. In Proceedings of the Séminaire International, Relizane, Algérie, 2–5 June 2012; pp. 75–89. (In French)
66. Toumi, S.; Meddi, M.; Mahé, G.; Brou, Y.T. Cartographie de l'érosion dans le bassin versant de l'Oued Mina en Algérie par télédétection et SIG. *Hydrol. Sci. J.* **2013**, *58*, 1–17. (In French) [CrossRef]
67. Hallouz, F.; Meddi, M.; Mahe, G. Régimes des matières en suspension dans le bassin versant de l'oued Mina sur l'oued Cheliff (Nord-Ouest algérien). *La Houille Blanche* **2017**, *4*, 61–71. (In French) [CrossRef]
68. Meddi, M. *Hydro-Pluviométrie and Transport Solide Dans le Bassin Versant de L'oued Mina (Algérie)*. Ph.D. Thesis, University Louis Pasteur, Strasbourg, France, 1992. (In French)
69. Dubreuil, P.; Guiscafre, J. La planification du réseau hydrométrique minimal. *Cah. ORSTOM Sér. Hydrol.* **1971**, *8*, 3–38. (In French)
70. Mahieddine, M. *Quantification and Variabilité Parcellaire Sous Simulation de Pluie Dans le Bassin Versant de L'Oued Mina*. Ph.D. Thesis, Ecole Nationale Supérieure des Sciences Agronomiques, Alger, Algeria, 1997.
71. Walling, D.E. Assessing the accuracy of suspended sediment rating curves for a small basin. *Water Resour. Res.* **1977**, *13*, 531–538. [CrossRef]
72. Asselman, N.E.M. Fitting and interpretation of sediment rating curves. *J. Hydrol.* **2000**, *234*, 228–248. [CrossRef]
73. El Mahi, A.; Meddi, M.; Bravard, J.P. Analyse du transport solide en suspension dans le bassin versant de l'Oued El Hammam (Algérie du Nord). *Hydrol. Sci. J.* **2012**, *57*, 1642–1661. (In French) [CrossRef]
74. Tebbi, F.Z.; Dridi, H.; Morris, G.L. Optimization of cumulative trapped sediment curve for an arid zone reservoir: Foum El Kherza (Biskra, Algeria). *Hydrol. Sci. J.* **2012**, *57*, 1368–1377. [CrossRef]
75. Louamri, A.; Mebarki, A.; Laignel, B. Variabilité interannuelle et intra-annuelle des transports solides de l'Oued Bouhamdane, à l'amont du barrage Hammam Debagh (Algérie orientale). *Hydrol. Sci. J.* **2013**, *58*, 1559–1572. (In French) [CrossRef]
76. Hallouz, F. *Transport Solide Dans le Bassin D'El Oued Mina et Sédimentation du Barrage SMBA*. Ph.D. Thesis, Es-Sciences, Ecole Nationale Supérieure d'Hydraulique (ENSH), Soumaâ, Algeria, 2014.

77. Ferguson, R.I. Accuracy and precision of mandhods for estimating river loads. *Earth Surf. Process. Landf.* **1987**, *12*, 95–104. (In French) [CrossRef]
78. Cordova, J.R.; Gonzalez, M. Sediment yield estimation in small watersheds based on stream flow and suspended sediment discharge measurements. *Soil Technol.* **1997**, *11*, 57–69. [CrossRef]
79. Jansson, M.B. Estimating a sediment rating curve of the Reventazon river at Palomo using logged mean loads within discharge classes. *J. Hydrol.* **1996**, *183*, 227–241. [CrossRef]
80. Kendall, S.M.; Stuart, A. *The Advanced Theory of Statistics*, 1969–1976 ed.; Charles Griffin: London, UK, 1943.
81. Lubes-Niel, H.; Masson, J.; Paturel, J.; Servat, E. Variabilité climatique et statistiques. Etude par simulation de la puissance et de la robustesse de quelques tests utilisés pour vérifier l'homogénéité de chroniques. *Rev. Sci. L'eau J. Water Sci.* **1998**, *11*, 383–408. (In French) [CrossRef]
82. Maftei, C.; Bărbulescu, A.; Buta, C.; Șerban, C. Change points detection on variability analysis of some precipitation series, MAMECTIS/NOLASC/CONTROL/WAMUS'11. In Proceedings of the 13th WSEAS International Conference on Mathematical Methods, Computational Techniques and Intelligent Systems, Iasi, Romania, 1–3 July 2011; pp. 232–237.
83. Khomsi, K.; Mahé, G.; Sinan, M.; Snoussi, M. Hydro-climatic variability in two Moroccan basins: Comparative analysis of temperature, rainfall and runoff regimes. In *Climate and Land Surface Changes in Hydrology, Proceedings of the H01, IAHS-IAPSO-IASPEI Assembly, Gothenburg, Sweden, 22–26 July 2013*; IAHS Publication: Oxfordshire, UK, 2013; pp. 193–190, 359.
84. Bonneaud, S. *Méthodes de Détection des Ruptures Dans les Séries Chronologiques*; Rapport de Stage de fin D'études; ISIM: Montpellier, France, 1994. (In French)
85. KHRONOSTAT. *Logiciel D'analyse Statistique de Séries Chronologiques*; IRD Ex: Paris, France, 1998; Available online: http://www.hydrosciences.org/spip.php?article239 (accessed on 14 October 2011). (In French)
86. Lubes, H.; Masson, J.M.; Servat, E.; Paturel, J.-E.; Kouame, B.; Boyer, J.F. *ICCARE: Rapport n 3: Caractérisation de Fluctuations Dans Une Série Chronologique par Applications de Tests Statistiques: Étude Bibliographique*; IRD Ex: Montpellier, France, 1994; 22p. (In French)
87. Meddi, M.; Humbert, J. Etude des potentialités de l'écoulement fluvial dans le Nord de l'Algérie en vue d'une réalimentation des aquifères. In *Eaux Sauvages, Eaux Domestiquées, Hommage à L. DAVY*; University of Provence: Marseille, France, 2000; pp. 177–190.
88. Singla, S.; Mahé, G.; Dieulin, C.; Driouech, F.; Milano, M.; El Guelai, F.Z.; Ardoin-Bardin, S. Evolution des relations pluie-débit sur des bassins versants du Maroc. In *Global Change: Facing Risks and Threats to Water Resources, Proceedings of the Sixth World FRIEND Conference, Fez, Morocco, 25–29 October 2010*; IAHS Publication: Oxfordshire, UK, 2010; Volume 340, pp. 679–687.
89. Hubert, P.; Carbonnel, J.P.; Chaouche, A. Segmentation des séries hydrométéorologiques. Application à des séries de précipitations and de débits de l'Afrique de l'Ouest. *J. Hydrol.* **1989**, *110*, 349–367. (In French) [CrossRef]
90. Chaouche, A. Structure de la Saison des Pluies en Afrique Soudano-Sahélienne. Ph.D. Thesis, J'Ecole des Mines de Paris, Paris, France, 1988. (In French)
91. Moron, V. Variabilité spatio-temporelle des précipitations en Afrique sahélienne and guinéenne de 1933 à 1990. *Météorologie* **1992**, *43–44*, 24–30. (In French)
92. Kebaili Bargaoui, Z. Occurrence des sécheresses dans le bassin de la Medjerda (Tunisie). *Rev. Sci. L'eau J. Water Sci.* **1989**, *2*, 429–447. (In French) [CrossRef]
93. Bobée, B. Extreme flood events valuation using frequency analysis: A critical review. *La Houille Blanche* **1999**, 100–105. [CrossRef]
94. El Adlouni, S.; Ouarda, T.B.M.J. Étude de la loi conjointe débit-niveau par les copules: Cas de la rivière Châteauguay. *Can. J. Civ. Eng.* **2008**, *35*, 1128–1137. (In French) [CrossRef]
95. Fisher, R.A. On the 'Probable Error' of a Coefficient of Correlation deduced from a Small Sample. *Metron* **1921**, *1*, 3–32. (In French)
96. Gumbel, E.J. *Statistics of Extremes*; Columbia University Press: New York, NY, USA, 1958.
97. El Adlouni, S.; Ouarda, T.B. Frequency Analysis of Extreme Rainfall Events. In *Rainfall: State of the Science*; Testik, F.Y., Gebremichael, M., Eds.; American Geophysical Union: Washington, DC, USA, 2010. [CrossRef]

98. El Adlouni, S.; Bobée, B. Système d'Aide à la Décision pour l'estimation du risque hydrologique. In *Numéro Spécial du Journal des Sciences Hydrologiques, Proceedings of the Conference FRIEND, Fès, Morocco, 25–29 October 2010*; IASH Publication: Edinburgh, UK, 2010. (In French)
99. El Adlouni, S.; Bobée, B. Delta diagram based test for the Halphen (A and B) and the Gamma distributions. In Proceedings of the EGU General Assembly Conference, Vienna, Austria, 27 April–2 May 2014; Volume 16.
100. Beirlant, J.; Goegebeur, Y.; Segers, J.; Teugels, J. *Statistics of Extremes: Theory and Applications*; Wiley Series in Probability and Statistics; John Wiley & Sons, Ltd.: Chichester, UK, 2004; 522p, ISBN 13 978-0-471-97647-9.
101. Von Mises, R. *La distribution de la plus grande de n valeurs*; American Mathematical Society: Providence, RI, USA, 1954; Volume II, pp. 271–294.
102. Fisher, R.; Tippett, L. Limiting forms of the frequency distribution of the largest and smallest member of a sample. *Proc. Camb. Philos. Soc.* **1928**, *24*, 180–190. [CrossRef]
103. Jenkinson, A.F. The frequency distribution of the annual maximum (or minimum) values of meteorological elements. *Q. J. R. Meteorol. Soc.* **1955**, *81*, 58–171. [CrossRef]
104. Wischmeier, W.H.; Smith, D.D. *Predicting Rainfall Erosion Losses: A Guide to Conservation Planning*; U.S. Department of Agriculture: Washington, DC, USA, 1978.
105. National Agency for Hydraulic Resources (ANRH). Les changements climatiques et leur impact sur les ressources en eau en Algérie. In Proceedings of the Assises Nationales sur L'Eau, Fez, Morocco, 25–29 October 2010; ANRH: Alger, Algeria, 2010; p. 41. (In French)
106. Bouzaiane, S.; Laforgue, A. *Monographie Hydrologique des oueds Zéroud et Merguellil*; DGRE–ORSTOM: Tunis, Tunisie, 1986. (In French)
107. Le Groupe d'experts intergouvernemental sur l'évolution du climat (GIEC). Impacts Adaptation and Vulnerability, Summary for Policymakers. In *Contribution of Working Group II to the Fourth Assessment Report of the Intergovernmental Panel on Climate Change*; Parry, M.L., Canziani, O.F., Palutikof, J.P., van der Linden, P.J., Hanson, C.E., Eds.; Cambridge University Press: Cambridge, UK, 2007; Available online: www.ipcc.ch (accessed on 2 July 2018).
108. Abdellali, S.A.; Fougrach, H.; Hsain, M.; Saloui, A.; Badri, W. Andude de la variabilité du régime pluviométrique au Maroc septentrional (1935–2004). *Sécheresse* **2011**, *22*, 139–148. (In French)
109. Mahé, G.; Olivry, J.C. Variations des précipitations and des écoulements en Afrique de l'Ouest and centrale de 1951 à 1989. *Sécheresse* **1995**, *6*, 109–117. (In French)
110. Wotling, G.; Mahé, G.; L'Hote, Y.; Lebarbe, L. Analyse par les vecteurs régionaux de la variabilité spatiotemporelle des précipitations annuelles liées à la Mousson africaine. *Veill. Clim. Satell.* **1995**, *52*, 58–73.
111. Kingumbi, A.; Bergaoui, Z.; Bourges, J.; Hubert, P.; Kalled, R. *Étude de L'évolution des Séries Pluviométriques de la Tunisie Centrale*; Documents Techniques en Hydrologie, No. 51 ("Hydrologie des Régions Méditerranéennes" Actes du Séminaire de Montpellier); UNESCO: Paris, France, 2000; pp. 341–345.
112. L'Hôte, Y.; Mahé, G.; Somé, B.; Triboulet, J.P. Analysis of a Sahelian annual rainfall index updated from 1896 to 2000; the drought still goes on. *Hydrol. Sci. J.* **2002**, *47*, 563–572.
113. Sircoulon, J. Variation des débits des cours d'eau et des niveaux des lacs en Afrique de l'ouest depuis le début du vingtième siècle. In *The Influence of Climate Change and Climatic Variability on the Hydrology Regime and Water Resources, Proceedings of the Vancouver Symposium, Vancouver, BC, Canada, 9–22 August 1987*; Solomon, S.I., Beran, M., Hogg, W., Eds.; IAHS Publication: Oxfordshire, UK, 1987; pp. 13–25.
114. Opoku-Ankomah, Y.; Amisigo, B.A. Rainfall and runoff variability in the south western river system of Ghana. In Proceedings of the Abidjan'98 Conference on the Water Resources Variability in Africa during the XXth Century, Abidjan, Cote d'Ivoire, 30 November 1998; IAHS Publication: Oxfordshire, UK, 1998; pp. 307–321.
115. Bello, N.J. Evidence of climate change based on rainfall records in Nigeria. *Weather* **1998**, *52*, 412–418. [CrossRef]
116. Servat, E.; Paturel, J.E.; Lubès-Niel, H.; Kouamé, B.; Masson, J.M.; Travaglio, M.; Marieu, B. De différents aspects de la variabilité de la pluviométrie en Afrique de l'ouest et centrale non sahélienne. *Rev. Sci. L'eau J. Water Sci.* **1999**, *12*, 363–387. (In French) [CrossRef]
117. Alpert, P.; Ben-Gai, T.; Bahard, A.; Benjamini, Y.; Yekutieli, D.; Colacino, M.; Diodato, L.; Ramis, C.; Homar, V.; Romero, R.; et al. The paradoxical increase of Mediterranean extreme daily rainfall in spite of decrease in total values. *Geophys. Res. Lett.* **2002**, *29*, 31-1–31-4. [CrossRef]

118. Norrant, C.; Douguédroit, A. Monthly and daily precipitation trends in the Mediterranean (1950–2000). *Theor. Appl. Climatol.* **2006**, *83*, 89–106. [CrossRef]
119. Hulme, M.; Osborn, T.J.; Johns, T.C. Precipitation sensitivity to global warming: Comparisons of observations with HadCM2 simulations. *Geophys. Res. Lett.* **1998**, *25*, 3379–3382. [CrossRef]
120. Laborde, J.P.; Goubesville, P.; Assaba, M.; Demmak, A.; Belhouli, L. Climate evolution and possible effects on surface water resources of North Algeria. *Curr. Sci.* **2010**, *98*, 1056–1062.
121. Mahé, G.; Singla, S.; Driouech, F.; Khomsi, K. Analyse de la persistance de ruptures dans des séries pluviométriques au Maroc en fonction de l'échelle spatiale and de la reconstitution des données. In Proceedings of the Conférence CIREDD4, Blida, Algeria, 22–23 February 2011; pp. 22–23.
122. Ghenim, A.N.; Megnounif, A. Ampleur de la sécheresse dans le bassin d'alimentation du barrage Meffrouche (Nord-Ouest de l'Algérie). *Géogr. Phys. Environ. Physio-Géo* **2013**, *7*, 35–49. (In French) [CrossRef]
123. Ghenim, A.N.; Megnounif, A. Analyse des précipitations dans le Nord-Ouest Algérien. *Sécheresse* **2013**, *24*, 107–114. (In French) [CrossRef]
124. Jerolmack, D.J.; Paola, C. Shredding of environmental signals by sediment transport. *Geophys. Res. Lett.* **2010**, *37*, L19401. [CrossRef]
125. Coulthard, T.J.; Ramirez, J.; Fowler, H.J.; Glenis, V. Using the UKCP09 probabilistic scenarios to model the amplified impact of climate change on drainage basin sediment yield. *Hydrol. Earth Syst. Sci.* **2012**, *16*, 4401–4416. [CrossRef]
126. Knight, J.; Harrison, S. The impacts of climate change on terrestrial Earth surface systems. *Nat. Clim. Chang.* **2013**, *3*, 24–29. [CrossRef]
127. Probst, J.L.; Bazerbachi, A. Transports en solution et en suspension par la Garonne Supérieure. *Sci. Géol. Bull.* **1986**, *39*, 79–98. (In French)
128. Asnouni, F. Etude du Transport Solide en Suspension Dans le Bassin Versant D'oued AL ABD. Master's Thesis, Hydraulique Université de Tlemcen, Tlemcen, Algeria, 2014.
129. Roose, E. *Introduction à la Gestion Conservatoire de l'eau, de la Biomasse et de la Fertilité des Sols (GCES)*; Bulletin Pédologique de la FAO No. 70; United Nations Food and Agriculture Organization: Rome, Italy, 1999; Available online: http://www.fao.org/docrep/T1765F/t1765f00.htm (accessed on 22 July 2016).
130. Roose, É.; Chebbani, R.; Bourougaa, L. Ravinement en Algérie: Typologie, facteurs de contrôle, quantification et réhabilitation. *Sécheresse* **2000**, *11*, 317–326.
131. Silva, A.M. Rainfall Erosivity Map for Brazil. *Catena* **2004**, *57*, 251–259. [CrossRef]
132. Yu, B. Rainfall Erosivity and its Estimation for Australia's Tropics. *Aust. J. Soil Res.* **1998**, *36*, 143–166. [CrossRef]
133. Isikwue, M.O.; Ocheme, J.O.; Aho, M.I. Evaluation of Rainfall Erosivity Index for Index for Abuja Lombardi Method. *Niger. J. Technol.* **2015**, *34*, 56–63. [CrossRef]
134. Bourouba, M. Les variations de la turbidité and leurs relations avec les précipitations et les débits des oueds semi arides de l'Algérie orientale. *Bull. ORSTOM* **1997**, *17*, 345–360.
135. Megnounif, A. Étude du Transport des Sédiments en Suspension Dans les Écoulements de Surface. Ph.D. Thesis, Universit de Tlemcen, Tlemcen, Algérie, 2007; p. 164.
136. Demmak, A. Contribution à L'étude de L'érosion et des Transports Solides en Algérie Septentrionale. Ph.D. Thesis, Université de Pierre et Marie Curie, Paris, France, 1982. (In French)
137. La CROIX. Inondations Meurtrières en Algérie, Parue le 12/11/2001 à 12h00, L'explication Météorologique. Available online: http://www.la-croix.com/Archives/2001-11-12/INondations-meurtrieres-en-Algerie-_NP_-2001-11-12-144826 (accessed on 3 March 2018).
138. Maleval, V. Premiers résultats des mesures d'érosion ravinaire sur les versants lacustres du barrage Sidi Mohammed Ben Abdellah (Maroc) et perspectives de recherche. In *Gestion Durable des Terres, Proceedings of the Réunion Multi-Acteurs sur le Bassin du Bouregreg CERGéo, Rabat, Morocco, 28 May 2013*; Laouina, A., Mahe, G., Eds.; ARGDT: Rabat, Morocco, 2014; pp. 53–62. ISBN 978-9954-33-482-9. (In French)
139. Toumi, S. Application des Techniques Nucléaires et de la Télédétection à L'étude de L'érosion Hydrique Dans le Bassin Versant de L'oued Mina. Ph.D. Thesis, Es-Sciences, Ecole Nationale Supérieure d'Hydraulique (ENSH), Soumaâ, Algeria, 2014; p. 189. (In French)
140. Larfi, B.; Remini, B. Le transport solide dans le bassin versant de l'oued Isser. Impact sur l'envasement fu barrage de Beni Amrane (Algérie). *LARHYSS J.* **2006**, *5*, 63–73. (In French)

141. Deutsche Gesellschaft fur Technische Zusammenabeit (GTZ). *Carte du Danger D'érosion. Projet D'aménagement Intégré du Bassin Versant de L'oued Mina. Schéma Directeur D'aménagement*; Office Allemand de la Coopération Technique, S.A.R.L (GTZ), IFG, Institut des Géosciences Appliquées: Offenbach, Germany, 1987. (In French)
142. Kouidri, R. *Premiers Résultats de Quantification de Ruissellement et de L'érosion en Nappe sur Jachère en Algérie (Région de Médéa)*; Annales de la Recherche Forestière en Algérie; INRF: Alger, Algérie, 1993. (In French)
143. Diallo, D. Erosion des sols en Zone Soudanienne du Mali Transfert des Matériaux Érodés Dans le Bassin Versant de Djitiko (Haut Niger). Ph.D. Thesis, L'université Joseph Fourier, Grenoble, France, 2000; p. 212. (In French)
144. Naimi, M.; Tayaa, M.; Ouzizi, S.; Choukra-Ilha, R.; Kerby, M. Dynamique de l'érosion par ravinement dans un bassin versant du Rif occidental au Maroc. *Sécheresse* **2003**, *14*, 95–100. (In French)
145. Lienou, G.; Mahé, G.; Paturel, J.E.; Servat, E.; Sighomnou, D.; Ecodeck, G.E.; Dezeter, A.; Dieulin, C. Changements des régimes hydrologiques en région équatoriale camerounaise: Un impact du changement climatique en Afrique équatoriale? *Hydrol. Sci. J.* **2008**, *53*, 789–801. (In French) [CrossRef]
146. Megnounif, A.; Nekkache Ghenim, A. Influence des fluctuations hydro-pluviométriques sur la production des sédiments: Cas du bassin de la Haute Tafna. *Rev. Sci. L'eau J. Water Sci.* **2013**, *26*, 53–62. [CrossRef]

 © 2018 by the authors. Licensee MDPI, Basel, Switzerland. This article is an open access article distributed under the terms and conditions of the Creative Commons Attribution (CC BY) license (http://creativecommons.org/licenses/by/4.0/).

Article

Anthropogenic Reservoirs of Various Sizes Trap Most of the Sediment in the Mediterranean Maghreb Basine

Mahrez Sadaoui [1,2,*], **Wolfgang Ludwig** [1,2], **François Bourrin** [1,2], **Yves Le Bissonnais** [3] **and Estela Romero** [2,4]

1. Centre de Formation et de Recherche sur les Environnements Méditerranéens, Université de Perpignan Via Domitia, UMR 5110, 52 Avenue Paul Alduy, F-66860 Perpignan CEDEX, France; ludwig@univ-perp.fr (W.L.), francois.bourrin@univ-perp.fr (F.B.)
2. Centre de Formation et de Recherche sur les Environnements Méditerranéens, CNRS, UMR 5110, 52 Avenue Paul Alduy, F-66860 Perpignan CEDEX, France; estela.romero@univ-perp.fr
3. Laboratoire d'Étude des Interactions Sol-Agrosystème-Hydrosystème, INRA, UMR LISAH INRA-IRD-SupAgro-Univ. Montpellier, 2 Place Viala, F-34060 Montpellier, France; y.le-bissonnais@orange.fr
4. Global Ecology Unit, Center for Ecological Research and Forestry Applications, CREAF-Universitat Autònoma de Barcelona, Edifici C, Campus UAB, 08193 Bellaterra, Spain
* Correspondence: mahrez.sadaoui@univ-perp.fr; Tel.: +33-68-66-20-93

Received: 16 June 2018; Accepted: 10 July 2018; Published: 12 July 2018

Abstract: The purpose of this study is to obtain a spatially explicit assessment of the impact of reservoirs on natural river sediment fluxes to the sea in the Mediterranean Maghreb Basin (MMB), a region where both mechanical erosion rates and the anthropogenic pressure on surface water resources are high. We combined modeling of riverine sediment yields (sediment fluxes divided by the drainage basin area) and water drainage intensities in a 5′ × 5′ grid point resolution (~10 km × 10 km) with a compilation of existing reservoirs in the area, and calculated sediment trapping based on average water residence time in these reservoirs. A total number of 670 reservoirs could be assembled from various sources (including digitization from Google maps), comprising large-scale, small-scale and hillside reservoirs. 450 of them could be implemented in our modeling approach. Our results confirm that natural sediment yields are clearly above the world average, with the greatest values for Morocco (506 t km^{-2} year^{-1}), followed by Algeria (328 t km^{-2} year^{-1}) and by Tunisia (250 t km^{-2} year^{-1}). Including dams in the downstream routing of suspended sediments to the sea reduces the natural sediment flux in the entire MMB to the sea from 96 to 36 Mt km^{-1} year^{-1}, which corresponds to an average sediment retention of 62%. Trapping rates are highest in the Tunisian basin part, with about 72%, followed by the Algerian (63%) and the Moroccan basin parts (55%). Small reservoirs and hillside reservoirs are quantitatively important in the interception of these sediments compared to large reservoirs. If we only considered the dams included in the widely used Global Reservoir and Dam (GRanD) database which comprises mainly large reservoirs sediment retention behind dams would account for 36% of the natural suspended particulate matter (SPM) flux to the Mediterranean Sea. Our data reveal negative correlation between sediment retention and natural erosion rates among the three Maghreb countries, which can be explained by the greater difficulties to build dams in steep terrains where natural sediment yields are high. Although the lowest sediment retention rates are found in the Moroccan part of the MMB, it is probably here where riverine sediment starvation has the greatest impacts on coastline dynamics. Understanding the impact of dams and related water infrastructures on riverine sediment dynamics is key in arid zones such as the MMB, where global warming is predicted to cause important changes in the climatic conditions and the water availability.

Keywords: Mediterranean Maghreb Basin; water fluxes; sediment fluxes; reservoirs; hillside reservoirs; sediment retention

1. Introduction

Because of the strong seasonal and inter-annual variability of flow regimes in inland waters, water stress is endemic in most parts of the Mediterranean drainage basin. Water is relatively scarce throughout most of the year, whereas high flows threaten lives and property on a small number of days per year or decade. Under such conditions, the rising demand for water is met by steadily increasing reservoir constructions to store seasonal or annual water surpluses [1]. These reservoirs impact the natural transfer and cycling of water, sediments and associated elements in the corresponding drainage basins. They retain considerable amounts of riverine nutrients and suspended sediments [2,3], preventing them to be supplied to the coastal waters in order to sustain biological productivity and to allow geomorphological edification of beaches and deltaic systems.

Sediment starvation in rivers through damming has been recognized as a major environmental problem in the Mediterranean area, as it exacerbates coastal erosion and sea level rise in relation to global warming [4,5]. In Mediterranean rim countries most of the population is concentrated in the coastal areas, which also represent the dominant economic resources via tourism and fishery. Erosion in these areas is hence a major threat. Rapid sediment accumulation in the reservoirs has also negative economic impacts as it reduces their performance and lifetime. Detailed knowledge on the riverine sediment dynamics in relation to the natural and anthropogenic forcings is therefore mandatory for sustainable development strategies. This is particularly true in the Mediterranean area because natural riverine sediment fluxes are high in its drainage basins [6] and environmental conditions respond rapidly to anthropogenic alterations.

The impact of reservoir construction on riverine sediment fluxes to the Mediterranean Sea was examined in a recent study [7]. By combining spatially explicit modeling of natural sediment and water fluxes with the precise locations and properties of dams as they are documented in the GRanD database (Global Reservoir and Dam, [8]), they calculated that about 35% of the natural sediment discharges should nowadays be retained in reservoirs. This underlines their significant impact on the basin-wide budgets, although the percentage is comparable to what has been estimated at the global scale [9,10], which questions the idea of the Mediterranean being a hot spot area for anthropogenic perturbations. Nevertheless, Mediterranean rivers are often small in size [11] and GRanD is far from being complete in this area.

We therefore attempted to go a step further and to examine the role of the "missing dams". We selected the Maghreb region of Northern Africa as a regional focus and intended to assemble a much more complete database from various sources, also including small-scale and hillside reservoirs. This allowed us to produce for the first time a detailed regional assessment on dams in the area, which is not only useful for general budget considerations, but which may also serve as a management tool for the evaluation of potential environmental risks in terms of coastal erosion and/or reservoir silting. Our method may further be applied to other regions of the Mediterranean drainage basin in order to put together a more complete picture of human perturbations in this key environment for global change studies.

2. Data and Methods

2.1. Study Area

The Maghreb in northwestern Africa, also known as the small Maghreb, is a mountainous region including the countries of Morocco, Algeria and Tunisia. In particular, the Mediterranean Maghreb basin (MMB) covers the northern part of Morocco excluding the terrains that are drained to the Atlantic, Algeria, and the northern and eastern parts of Tunisia (Figure 1). The Maghreb is among the regions that are most vulnerable to erosion in the world. Several researchers [12–14] have attempted to explain the mechanisms of solid transport in this region, and Probst and Suchet proposed an overall estimate for the riverine sediment fluxes in the Maghreb basin of 100 Mt year^{-1} [15], corresponding to a specific sediment flux (sediment yield) of about 400 t km^{-2} year^{-1}. These are probably among the highest values in Africa.

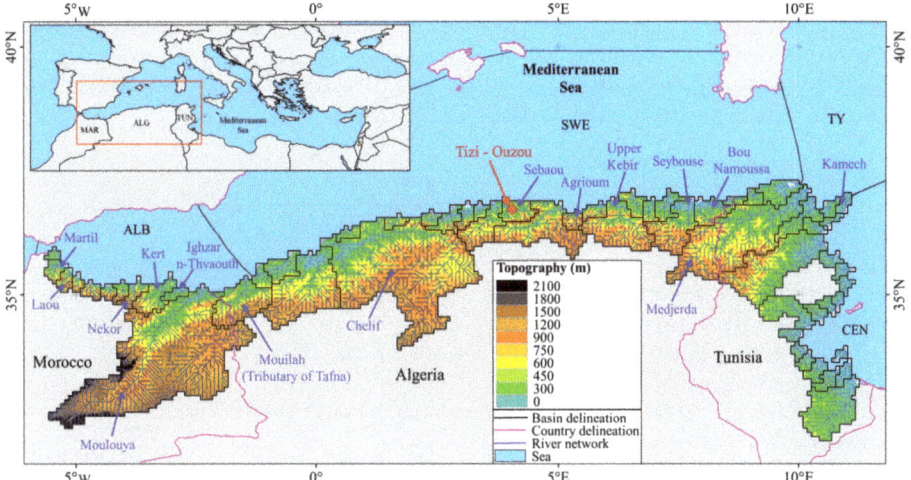

Figure 1. Location map of the Mediterranean Maghreb Basin (MMB) and delimitation of its drainage basins at a spatial resolution of 5 arc-minutes. Topography was extracted from the digital elevation model of [16]. All rivers and locations which are mentioned in the text have been included in this figure. Abbreviations: ALB: Alboran sub-basin, CEN: Central sub-basin, SWE: Southwestern sub-basin, TY: Tyearrhenian sub-basin.

These average values, however, cannot be transposed locally because their spatial variability can be significant. For example, rather moderate sediment yields of 165 t km^{-2} year^{-1} have been reported for the Mouilah basin [17] in Algeria; intermediate values of about 600 t km^{-2} year^{-1} for the Medjerda basin [15] in Tunisia and 884 t km^{-2} year^{-1} for the Upper Kebir basin in Algeria [18]; high values of 1500 and 2000 t km^{-2} year^{-1} for the Kamech basin in Tunisia [19] and the Martil basin in Morocco [20], respectively; and peak values of 5000 to 5900 t km^{-2} year^{-1} for the Agrioum basin [21] in Algeria and the Nekor basin [22] in Morocco. Several factors combine in order to explain these high erosion rates. The Maghreb is a fragile mountainous area. The mountains extend over large areas and are characterized by high altitudes as well as steep slopes. Lithology is often made of soft sedimentary rocks susceptible to erosion and the Maghreb is subject to contrasted climatic conditions. In summer, climate is dominated by Saharan aridity with high temperature and low precipitation. In winter, climate is colder and wetter due to the oceanic influence [23]. Hence, climate is irregular and characterized by a strong contrast between dry and wet years, with often intense and devastating rains [24].

Due to the aridity of the climate, water is an important factor for economic development and the construction of dams represents a major issue for the Maghreb countries. North Africa (here considered as including the totality of these three countries: Morocco, Algeria and Tunisia) has currently over 230 large dams with a total storage capacity of 23 km^3 [25]. Large dams serve in general to store drinking water and for industrial and electrical energy production. In addition, many small dams and hillside reservoirs exist and serve to mitigate floods, contribute to water table recharge and improve water storage for irrigation and livestock watering and domestic uses. A detailed census of the different reservoir categories is nevertheless difficult. The only existing database is the GRanD database [8], which includes 53 dams in the Maghreb basin with a total storage capacity of about 6.5 km^3 [7].

Anthropogenic degradation in MMB is very high, increasing erosion and enhancing early silting of reservoirs in Morocco [20,26], in Algeria [24,27] and in Tunisia [28]. According to [25], who monitored and evaluated silting in the 230 dams of North Africa, the average capacity loss in these dams can be estimated to about 0.54% year^{-1}. Silting of these dams poses many problems at the level of the reservoir itself, but also in the basin parts upstream and downstream the reservoir. The authors of [25,29] cite several dams in North Africa where major degradations occurred, such as the reduction of the dam capacity, the destabilization of the structure (e.g., Oued El Fodda, Algeria), the filling of drainage ditches (e.g., Zardézas dams, Algeria), the silting of irrigation channels (e.g., bypass channel, Morocco) and the degradation of water quality (e.g., El Khattabi Dam, Morocco).

2.2. Modeling Framework

The modeling framework we used to calculate river sediment fluxes in the MMB and their retention in dams was set up in the study by [7]. It is based on the numerical definition of the drainage basin delineations and hydrological networks around the Mediterranean Sea in a 5' × 5' grid point resolution (~10 km × 10 km), and was produced by aggregation of the Hydrosheds digital elevation model of [30] with manual corrections of the resulting hydrological networks (for further details, see [7]). A total of 549 river basins could be distinguished for the entire Mediterranean drainage basin. In each basin all internal grid points are linked to one single mouth point that represents the river estuary according to a realistic river routing scheme (multi-branch estuaries that often exist for large river systems cannot be reproduced this way).

Quantification of specific riverine suspended matter (SPM) and specific water (Q) fluxes was obtained through empirical modeling relating these parameters at the grid point scale to a series of environmental factors that are considered to represent their major controls. The corresponding multi-regression models could be established on the basis of the average long-term basin characteristics from a large data set of 126 (Q) and 80 (SPM) Mediterranean rivers [7]. In short, specific water discharge (Q) was produced as a combination of annual precipitation, potential evapotranspiration, seasonal variability of precipitation (all climatic parameters were taken from [31]) and grid point slope [30], whereas specific SPM fluxes could be linked to specific water discharge, grid point slope, the percentage of sedimentary rocks in altitudes above 600 m [32] and the percentage of erodible land use classes (agriculture, grasslands and bare soils—[33]).

Both the produced Q and SPM fluxes represent average long-term values that do not account for temporal fluctuations in relation to climate variability and/or recent climate change, although we are aware that seasonal and inter-annual variability of SPM fluxes can be very important in the Mediterranean area and short and violent floods may represent more than 90% of the annual sediment load [34]. Moreover, the SPM fluxes are considered to represent natural (i.e., "pre-damming") values, since most of the literature estimates that were used to fit the regressions correspond to the pre-1950–1980 period, when human impoundments on rivers were not as frequent as today. This allowed simulating the impact of the dams constructed along the river courses through conservative routing of the grid-point SPM fluxes into the reservoirs and retaining them as a function of the average water residence time according to the equation initially proposed by [35]. The whole method has been described with more detail in [7].

2.3. Inventory of Dams

We started our inventory with the geo-referenced dams included in the GRanD database of [8]. It contains 53 dams in the Maghreb basin (6 for Morocco, 28 for Algeria and 19 for Tunisia). To this, 48 dams could be added from repertoires from various national agencies in the Maghreb, including 36 dams from the ANBT (National Agency for Dams and Transfers, [36]) in Algeria, 7 dams from the ABHL (Loukkos Hydraulic Basin Agency, [37]) in Morocco and 5 dams from the DGBGTH (General Directorate for Dams and Major Hydraulic Works, [38]) in Tunisia. Generally, these agencies provide information on storage capacity, water discharge and the year of reservoir construction, but no digitized files in a GIS format for the location of the dams. Manual location of these dams from Google satellite maps was necessary (consulted in June 2016). A total of 101 dams were consequently assembled

(Table 1). The majority of these dams were constructed in recent decades. More than 30% were built after 2000, and 60% between 1950 and 2000. Less than 10% were erected over the period from 1860 to 1950.

Table 1. Characteristics of the 101 dams of the Mediterranean Maghreb Basin (MMB) considered in this study.

NB	Dams	Country	Year of Construction	CAP (Mm^3)	S (km^2)	H (m)	Source
1	Ain Dalia	Algeria	1987	82	3.4	24	GRanD
2	Ain zada	Algeria	1986	125	6.9	18	GRanD
3	Ajras	Morocco	1969	3	0.4	8	GRanD
4	Al Thelat-Laou	Morocco	1934	30	0.9	33	GRanD
5	Asmir	Morocco	1991	43	2.6	17	ABHL
6	Bakhada	Algeria	1935	56	3.5	16	GRanD
7	Barbara-Zoutina	Tunisia	1999	58	3.9	15	GRanD
8	Beni Amrane	Algeria	1988	12	1.0	12	ANBT
9	Beni Bahdel	Algeria	1946	63	1.2	53	GRanD
10	Beni Boussaid	Algeria	2007	1	0.2	7	ANBT
11	Beni Haroune	Algeria	2003	998	44.4	22	ANBT
12	Beni metir	Tunisia	1954	73	1.8	41	GRanD
13	Beni zid	Algeria	2000	40	3.5	12	ANBT
14	Bezrik	Tunisia	1959	7	0.7	9	GRanD
15	Bir M'Cherga	Tunisia	1971	160	4.3	37	GRanD
16	Bou Hanifa	Algeria	1948	73	2.7	27	GRanD
17	bou heurtna	Tunisia	1976	118	7.1	17	GRanD
18	Bou Roumi	Algeria	1985	188	1.7	111	GRanD
19	Boughzoul	Algeria	1934	55	16.3	3	GRanD
20	Bougous	Algeria	2010	65	2.8	24	ANBT
21	Boukourdene	Algeria	1992	110	3.1	36	ANBT
22	Boussiaba	Algeria	2000	120	10.3	12	ANBT
23	Cap Djinet	Algeria	1990	13	0.4	34	ANBT
24	Cheffia	Algeria	1965	171	7.0	24	GRanD
25	Chelif	Algeria	2009	130	7.2	18	ANBT
26	Cheufra	Algeria	1935	14	1.3	11	GRanD
27	Chiba	Tunisia	1963	8	1.1	7	GRanD
28	Colonel bougara	Algeria	1988	13	2.3	6	GRanD
29	Dahmouni	Algeria	1987	41	3.4	12	GRanD
30	Deurdeur	Algeria	1984	115	4.8	24	ANBT
31	Djoumine	Tunisia	1983	130	5.4	24	GRanD
32	Douera	Algeria	2014	87	1.3	65	ANBT
33	Draa Diss	Algeria	2016	151	0.5	80	ANBT
34	El Abid	Tunisia	2002	10	1.4	7	DGBGTH
35	El Agrem	Algeria	2002	34	2.3	15	ANBT
36	El Djomoaa	Morocco	1992	7	0.5	12	ABHL
37	El habib-ouareb	Tunisia	1988	110	5.0	22	GRanD
38	El Hamiz	Algeria	1990	16	0.9	17	ANBT
39	El hamma	Tunisia	2002	12	0.6	21	DGBGTH
40	El Masri	Tunisia	1968	6	0.7	8	ANBT
41	Ennakhela	Morocco	1961	13	0.6	22	GRanD
42	Erraguen	Algeria	1961	200	11.0	18	ANBT
43	Fergoug	Algeria	1970	18	0.7	26	GRanD
44	Gargar	Algeria	1988	450	19.0	24	GRanD
45	GHrib	Algeria	1938	280	7.1	39	GRanD
46	GuenitraOumToub	Algeria	1984	125	1.3	96	GRanD
47	H.M.Bouhemdene	Algeria	1987	220	5.1	43	GRanD
48	Hammam Grouz	Algeria	1987	45	2.3	20	GRanD
49	Harreza	Algeria	1984	70	1.0	70	GRanD
50	Hassan II	Morocco	2000	400	11.0	36	ABHL
51	Hassan II Midelt	Morocco	2006	400	11.8	34	ABHL
52	Ighil Emda	Algeria	1953	156	2.9	53	ANBT
53	kasseb	Tunisia	1969	81	4.0	20	GRanD
54	Kebir	Tunisia	1925	26	1.6	16	GRanD
55	Keddara	Algeria	1990	142	5.6	26	ANBT
56	Kef Eddir	Algeria	2015	150	0.5	77	ANBT
57	Kerrada	Algeria	2010	70	3.6	19	ANBT
58	Kissir	Algeria	2010	68	3.6	19	ANBT
59	Koudiat Acerdoun	Algeria	2009	400	16.5	24	ANBT
60	Koudiat resfa	Algeria	2007	75	3.7	20	ANBT
61	Kramis	Algeria	2005	45	2.1	21	ANBT
62	Ladrat	Algeria	1988	10	0.5	20	GRanD
63	Lakhmess	Tunisia	1966	8	0.7	11	GRanD

Table 1. Cont.

NB	Dams	Country	Year of Construction	CAP (Mm^3)	S (km^2)	H (m)	Source
64	Lekhal	Algeria	1985	30	0.9	33	GRanD
65	lepna	Tunisia	1986	30	6.6	5	GRanD
66	M.B.A.L.khattabi	Morocco	1981	43	2.1	20	GRanD
67	Mahouane	Algeria	2014	148	0.3	76	ANBT
68	Martil	Morocco	2014	100	0.8	118	ABHL
69	Mechra Hammadi	Morocco	1955	42	1.2	35	GRanD
70	Mefrouch	Algeria	1960	15	1.8	8	ANBT
71	Merdja sidi abed	Algeria	1984	55	7.0	8	GRanD
72	Meurad	Algeria	1860	0.2	0.1	26	ANBT
73	Mexa	Algeria	1998	47	5.4	9	ANBT
74	Mlaabi	Tunisia	1967	4	0.8	4	DGBGTH
75	Mohammed V	Morocco	1967	730	16.3	45	GRanD
76	M. Bouchta	Morocco	2014	13	0.7	18	ABHL
77	Nabeur-Mellegue	Tunisia	1954	300	8.5	24	GRanD
78	Nebhana	Tunisia	1965	86	2.6	33	GRanD
79	Oued Cherf	Algeria	1995	157	10.2	15	ANBT
80	Oued el hajar	Tunisia	1999	6	2.0	3	DGBGTH
81	Oueder mel tunis	Tunisia	1999	22	7.8	3	DGBGTH
82	Oued Fouda	Algeria	1932	228	2.9	79	GRanD
83	Oued Rmel	Morocco	2008	25	1.1	22	ABHL
84	Ouizert	Algeria	1985	100	2.0	50	GRanD
85	S.M.B.Aouda	Algeria	1978	235	8.0	29	GRanD
86	Sarno	Algeria	1953	22	0.2	110	GRanD
87	Sejnane	Tunisia	1990	130	5.8	22	GRanD
88	Sekkak	Algeria	2004	30	2.0	15	ANBT
89	Sidi abdelli	Algeria	1988	110	0.8	138	GRanD
90	Sidi el barek	Tunisia	2002	275	9.2	30	GRanD
91	Sidi saad	Tunisia	1981	209	9.8	21	GRanD
92	Sidi salem	Tunisia	1981	555	40.1	14	GRanD
93	Sidi Yakoub	Algeria	1983	286	6.4	45	GRanD
94	Siliana	Tunisia	1987	70	4.7	15	GRanD
95	SMB taiba	Algeria	2005	75	2.9	25	ANBT
96	Souani	Algeria	1988	14	2.0	7	GRanD
97	Teksebt	Algeria	2001	180	4.6	39	ANBT
98	Tichy haff	Algeria	2009	80	4.6	18	ANBT
99	Tilsedit	Algeria	2009	83	9.7	9	ANBT
100	Zardezaz	Algeria	1973	31	1.3	24	ANBT
101	Zit Emba	Algeria	2002	117	7.7	15	ANBT

Note: NB: number of dams; CAP: reservoir storage capacity; S: reservoir surface; H: height of dike; ABHL: Loukkos Hydraulic Basin Agency, Morocco; ANBT: National Agency for Dams and Transfers, Algeria; DGBGTH: General Directorate for Dams and Major Hydraulic Works, Tunisia; GRanD: Global Reservoir and Dam database.

In a second step, we enlarged our database through manual digitization of additional reservoirs from Google satellite maps. In this approach, we mainly focused on the inclusion of small and hillside reservoirs, but also a few large reservoirs previously not identified could be added that way. To facilitate our work, we first imported our river network map and the previously collected dams (GRanD and Maghreb agencies) in order to avoid the re-digitization of existing reservoirs. The maps were imported in kml-format at a resolution of 5 arc-minutes and consisted of polygons, vectors and points.

We then created our own layer on which we digitized the rest of the reservoirs by zooming in on the Google maps to their maximum resolution and following the river network from upstream to downstream in each river basin. Reservoirs were identified initially by the blue color of the waters and the presence of a dike downstream. Two layers were created, the first one to digitize the surface of the reservoirs with polygons and the second one to locate the dams by fixing a geographical reference point downstream the reservoir (near the dike), which represents the outlet of the dam. It was sometimes difficult to distinguish between small dams, hillside reservoirs, and natural lakes. Besides checking the legend of reservoir names provided by Google maps, we used the shape of the water surfaces as a distinguishing criterion. Reservoirs are generally characterized by their longitudinal shapes and the existence of a dike downstream the flow direction, whereas natural lakes are usually characterized by a round shape and the absence of the dike. Hillside reservoirs are not implemented on major river courses and typically found in agricultural fields.

For the manually-digitized reservoirs we had no information on the storage capacity. In order to produce an estimate for this parameter too, we calculated a linear regression between the storage capacity (CAP) and the surface (S) of the reservoir. In fact, two regression models were established, one for the GRanD database reservoirs (Figure 2A)

$$CAP = 34.79\, S \tag{1}$$

$$n = 323,\ r^2 = 0.82,\ p < 0.001$$

And one for the hillside reservoirs (Figure 2B)

$$CAP = 9.29\, S \tag{2}$$

$$n = 25,\ r^2 = 0.89,\ p < 0.001$$

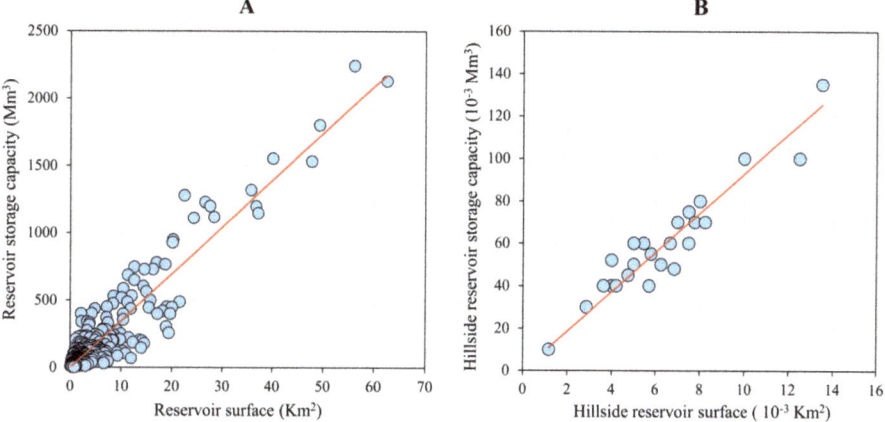

Figure 2. Linear relationships between storage capacity and reservoir surface (**A**) using the Mediterranean reservoirs of the GRanD database [8] and (**B**) using the hillside reservoir database of [39].

Both regressions were forced to pass through the origin (i.e.; no intercept). Equation (1) is based on information for all Mediterranean reservoirs in the GRanD database, and Equation (2) on the compilation of hillside reservoirs in the Tizi-Ouzou region (Algeria) produced by [39], which is the only available study in the area (Table 2).

Table 2. Characteristics of the hillside reservoirs of the Tizi-Ouzou region (Algeria) described by the authors of [39] and considered in this study.

NB	Name of Hillside Reservoir	S (10^3 m^2)	H (m)	CAP (10^{-3} Mm^3)	Year of Construction
1	Bouzeguene N° 2	4.0	13	52	1999
2	Bouzeguene N° 3	7.5	10	75	1999
3	Iloula oumalou N° 1	3.6	11	40	2000
4	Iloula oumalou N° 399	8.0	10	80	2000
5	Timizart N° 201	6.7	9	60	1989
6	Timizart N° 360	5.0	10	50	1988
7	Timizart N° 363	7.8	9	70	1986
8	Timizart N° 366	13.5	10	135	1985
9	Timizart N° 388	7.5	8	60	1985
10	Freha N° 389	6.3	8	50	2000

Table 2. Cont.

NB	Name of Hillside Reservoir	S (10^3 m^2)	H (m)	CAP (10^{-3} Mm^3)	Year of Construction
11	Freha N° 395	5.8	9.5	55	1985
12	Freha N° 396	12.5	8	100	1987
13	Freha N° 397	4.7	9.5	45	1986
14	Freha N° 398	10.0	10	100	1989
15	Mekla N° 367	1.2	8.5	10	1985
16	Mekla N° 368	8.2	8.5	70	1986
17	Mekla N° 391	4.0	10	40	1989
18	Mekla N° 393	6.9	7	48	1985
19	Mekla N° 394	5.7	7	40	1986
20	Souomaâ N° 1	5.5	11	60	1999
21	Azzefoun N° 10	4.2	9.5	40	1986
22	Azzefoun N° 11	5.0	12	60	1997
23	Azzefoun N° 12	7.0	10	70	1988
24	Azzefoun N° 14	3.6	11	40	1994
25	Azzefoun N° 15	2.9	10.5	30	1995

Note: S: hillside reservoir surface; H: Dike height; CAP: reservoir storage capacity.

It should be noted that the database produced by [39] was only used to determine Equation (2). Since the database does not include the geographic coordinates of the reservoirs, it is difficult to differentiate them from other newly digitalized hillside reservoirs. To avoid confusion and duplications, we kept this database only to determine Equation (2).

The storage capacity of the reservoirs can be strongly influenced by their morphology, and notably by the slope of the reservoir surface. The slope of hillside reservoirs is generally very low compared to that of the dams, in particular large dams. Two reservoirs that have the same surface but a different slope can have a very different storage capacity; this is the reason why we kept two different regression models (Figure 2A,B).

Overall, we were able to digitize the surface of 569 additional reservoirs, mostly consisting of small dams and hillside reservoirs. This expanded our database to 670 reservoirs with areas larger than one hectare. On the basis of Equations (1) and (2), we then calculated the storage capacity of these additional reservoirs (i.e., when the newly digitized reservoirs were identified as hillside reservoirs, we applied Equation (2); elsewhere, we used Equation (1). All the reservoirs were then classified according to their storage capacities into seven classes: C_1 (CAP > 200 Mm^3), C_2 (100 < CAP < 200 Mm^3), C_3 (50 < CAP < 100 Mm^3), C_4 (20 < CAP < 50 Mm^3), C_5 (5 < CAP < 20 Mm^3), C_6 (1 < CAP < 5 Mm^3), and C_7 (CAP < 1 Mm^3). It should be noted that hillside reservoirs never exceeded a storage capacity of 1 Mm^3, and almost 90% of them had a storage capacity lower than 0.8 Mm^3. For small dams, we considered that the maximum capacity limit was 20 Mm^3 (see Section 3.2) and found that only 21% of these dams had a storage capacity lower than 1 Mm^3. Although there is some overlap with the maximum size of hillside reservoirs, we considered this value to be a reasonable lower limit for small dams.

The integration of all these reservoirs in our modeling framework suffered however from the limited spatial resolution of this framework. In many cases several reservoirs fell within one single 5 arc-minutes grid cell. For the calculation of sediment retention, in addition to the storage capacity we need the amount of water entering and leaving the reservoirs annually. This parameter can only be calculated at the grid point level from our empirical modeling of mean annual water discharge (see Section 2.2). We therefore decided to retain for our modeling of sediment retention only the largest reservoir in a given grid point cell, to which we attributed the corresponding average annual water discharge following the assumption that this reservoir has been constructed along the principal grid point drainage direction. This reduced the total number of reservoirs which could be properly used for modeling to 450.

3. Results and Discussion

3.1. Natural Water and Sediment Fluxes

Average water fluxes in the MMB are rather low (Figure 3A). Drainage intensity (water discharge divided by the basin area) for the entire region is only 59 mm per year (Table 3), with little differences between the three countries (52, 62 and 62 mm year^{-1} for Morocco, Algeria and Tunisia, respectively). More humid conditions are restricted to mountainous areas in the vicinity of the coast and exist in the NW part of Morocco (e.g., the Martil and Laou rivers with Q values of 332 mm year^{-1} and 252 mm year^{-1}, respectively) and along the central and NE coast of Algeria (e.g., the Sebaou and Bou-Namoussa rivers with Q values of 387 mm year^{-1} and 290 mm year^{-1}, respectively). The driest conditions are found in the Saharan part of Tunisia and the drainage basins and basin parts which stretch far to the South. All MMB rivers together discharge about 16.2×10^3 Mm^3 year^{-1} of water to the Mediterranean Sea, which corresponds to less than 4% of the total freshwater flux exported by the Mediterranean drainage basin [7]. About 48% of the MMB water flux stems from the rivers in Algeria. The mean annual water discharge for individual rivers ranges from 1 Mm^3 for the Tunisian Saharan rivers to 1720 Mm^3 for the Moulouya in Morocco, the largest river in the MMB (55,713 km^2). Only about 7% of all rivers have an average water discharge greater than 1000 Mm^3 year^{-1}, but they provide more than 32% of the total water flux. On the other hand, 50% of the rivers have a water flux which does not exceed 100 Mm^3 year^{-1} and represent only 7% of the total water flux.

Table 3. Summary of average water and sediment fluxes in the Mediterranean Maghreb Basin (MMB) before and after dam construction, and total number of reservoirs considered in this study.

Parameters	Morocco	Algeria	Tunisia	Maghreb
A (10^3 km^2)	69	126	78	274
A-drained by dams (10^3 km^2)	59	110	40	209
Q (mm year^{-1})	52	62	62	59
FSPMs pre-damming (t km^{-2} year^{-1})	506	328	250	351
FSPMs post-damming (t km^{-2} year^{-1})	229	120	69	133
NB LD tot	12	62	25	99
NB SD tot	18	126	80	224
NB Hil tot	26	190	131	347
NB Dams tot	56	378	236	670
NB LD mod	12	62	25	99
NB SD mod	18	126	80	224
NB Hil mod	8	87	32	127
NB Dams mod	38	275	137	450

Note: A: Basin area; Q: Drainage intensity; FSPMs pre-damming: specific natural suspended particulate matter flux; FSPMs post-damming: specific anthropogenic suspended particulate matter flux; NB LD tot: Total number of large dams; NB SD tot: Total number of small dams; NB Hil tot: Total number of hillside dams; NB Dams tot: Total number of dams; NB LD mod: number of large dams used for modeling; NB SD mod: number of small dams used for modeling; NB Hil mod: number of hillside dams used for modeling; NB Dams mod : number of dams used for modeling.

Mountainous regions have also a dominant role for the average sediment fluxes (Figure 3B). High sediment yields (specific suspended particulate matter fluxes, FSPMs) are associated with high Q values, although peak values can also stretch further to the South to lower discharge regions in combination with higher elevations (e.g., in the larger basins such as the Moulouya and Medjerda basins). On average, sediment yields (Table 3) are greatest in Morocco (506 t km^{-2} year^{-1}), followed by Algeria (328 t km^{-2} year^{-1}) and by Tunisia (250 t km^{-2} year^{-1}). Peak values are obtained for the Agrioum River in Algeria (4001 t km^{-2} year^{-1}) and the Nekor River in Morocco (2956 t km^{-2} year^{-1}), whereas the lowest values are found for the Bou-Namoussa River in Algeria (15 t km^{-2} year^{-1}), which is located east of the Seybouse River, and for the Ighzar n-Thyaouth River in Morocco (21 t km^{-2} year^{-1}), which lies between the Moulouya and Kert rivers. Notice that for the former two rivers, which are probably the most erosive rivers in Africa, our empirical modeling resulted in somewhat lower values than those found in the literature (see Section 2.1). This is due to the fact that we excluded from our regressions a few rivers (notably the Agrioum and Nekor rivers) that had average SPM concentrations greater than 15 g L^{-1}. We considered that such extreme values might indicate possible errors in the literature citations or refer to highly peculiar erosion processes that cannot be reproduced by general regression models (for a more detailed discussion, see [7]).

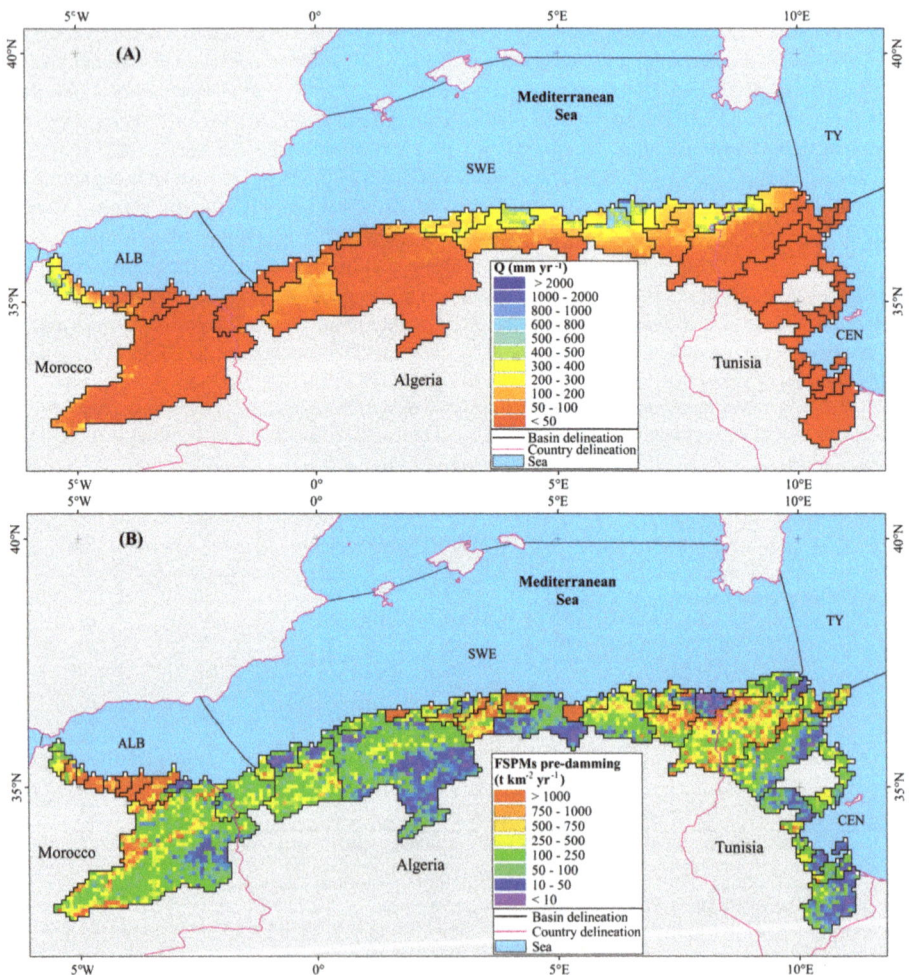

Figure 3. Spatial distribution of (**A**) specific riverine water discharge (Q-mod) and (**B**) specific suspended particulate matter fluxes (FSPMs pre-damming) in the Mediterranean Maghreb Basin according to the study of [7].

We estimate that the specific natural suspended particulate matter flux (FSPMs pre-damming) for the entire MMB is 351 t km^{-2} year^{-1} and corresponds to an overall sediment delivery to the Mediterranean Sea of 96 Mt year^{-1}. This value is very close to the estimate of 100 Mt year^{-1} presented by [15] for the same study area. About 60% of this value is exported by only seven rivers (Moulouya, Medjerda, Agrioun, Kert, Nekor, Chelif and Seybousse), and Algerian rivers alone account for about 43% of it. Compared to the world average FSPMs of 150 t km^{-2} year^{-1} [40], it becomes evident that the

Mediterranean Maghreb basin average erosion rates are higher than those in other regions of the world. This is also apparent when compared to other key environments in the Mediterranean Sea. Taking the Gulf of Lions (GoL) drainage basin as an example (which includes the Rhone River, nowadays the largest Mediterranean river in terms of its average water discharge and also characterized by remarkable sediment yields—[34]), we calculate that, despite the fact that the GoL rivers discharge into the Mediterranean a water flux that is four times greater than the water flux of the MMB rivers, their SPM flux represents only 40% of the SPM flux from the MMB rivers.

3.2. Sediment Retention by Dams

The values cited above represent average long-term means that do not take into account sediment retention by recent damming of the MMB rivers. In order to quantify the latter at the scale of our study region, we combined our inventory of dams (Figure 4A and Table 3) with our modeling of water and SPM fluxes according to the method described by [7]. Dams of various sizes stretch over the entire study region (Figure 4A), except in the Saharan part of Tunisia, where conditions are too dry for the sustainable exploitation of surface water resources. Consequently, about 76% of the MMB area is drained into dams (87% for Algeria, 86% for Morocco and 51% for Tunisia). The average density of all types of dams is about 2.5 dams per 10^3 km^2, with increasing densities as the size of the dams decreases (0.4, 0.8 and 1.3 dams per 10^3 km^2 for large dams, small dams and hillside reservoirs, respectively). When relating the densities to the areas which are drained into the dams (hence omitting to some extent the Saharan part of Tunisia), it is clear that the average dam densities increase from Morocco to Tunisia (1.0 dams per 10^3 km^2 for Morocco, 3.4 for Algeria and 5.9 for Tunisia), which is opposite to the average sediment yields in these countries (see Section 3.1). Construction of dams is obviously constrained by the availability of existing water resources and the natural erodibilities of the corresponding land surfaces. Notice also from Table 3 that while all large and small dams in our inventory could be included in the modeling approach, we could only include 37% of the hillside reservoirs (46%, 31% and 24% for Algeria, Morocco and Tunisia, respectively) due to the model limited resolution (see Section 2.3). Sediment retention by hillside reservoirs may therefore be somewhat underestimated in our study.

Including dams in the routing of suspended sediments to the sea reduces the average SPM flux from the MMB to the sea to 36 Mt year^{-1}, which corresponds to a post-damming FSPMs value of 133 t km^{-2} year^{-1}. In other words, 62% of the natural SPM flux (Section 3.1) is retained behind dams. Retention rates are highest in the Tunisian basin part (72%), followed by the Algerian (63%) and the Moroccan basin parts (55%). This is in agreement with the density of dams, which is highest in Tunisia and decreases westward toward Morocco. In terms of the absolute SPM fluxes, the greatest retention is observed for Algeria (26 Mt year^{-1}), followed by Morocco (19 Mt year^{-1}) and Tunisia (14 Mt year^{-1}). Interestingly, when only considering the dams included in the GRanD database (as was the case in the study by [7]), sediment retention behind the dams accounts for only 36% of the natural SPM flux to the Mediterranean Sea. The present results therefore demonstrate that small reservoirs are quantitatively crucial for the interception of sediments in river basins compared to large reservoirs. The majority of the MMB reservoirs (~85%) in the GRanD database have a storage capacity greater than or equal to 20 Mm^3. If we consider this value as an appropriate threshold between small and large reservoirs, we can make the following observations: small reservoirs have very limited capacities (cumulative capacity of 1.1 versus 12.8 km^3 for large dams), intercept annually less water discharge (cumulative interception of 8.4 versus 20.1 km^3 year^{-1} for large dams), drain smaller areas (mean area of 151 versus 1525 km^2 for large dams), and have shorter water residence time in the reservoir (mean of 0.13 versus 0.63 years for large dams). They nevertheless assign a significant anthropogenic signature to the global sediment flux in the MMB (Figure 5) and enhance mean retention by 13% (from 49 to 62%) as compared to large reservoirs. It can also be seen from Figure 5 that the different size classes we created increase about linearly the average sediment retention in the MMB from large to small basins (except for very large reservoirs which are underrepresented in the MMB), whereas the total number of these reservoirs increase more or less exponentially.

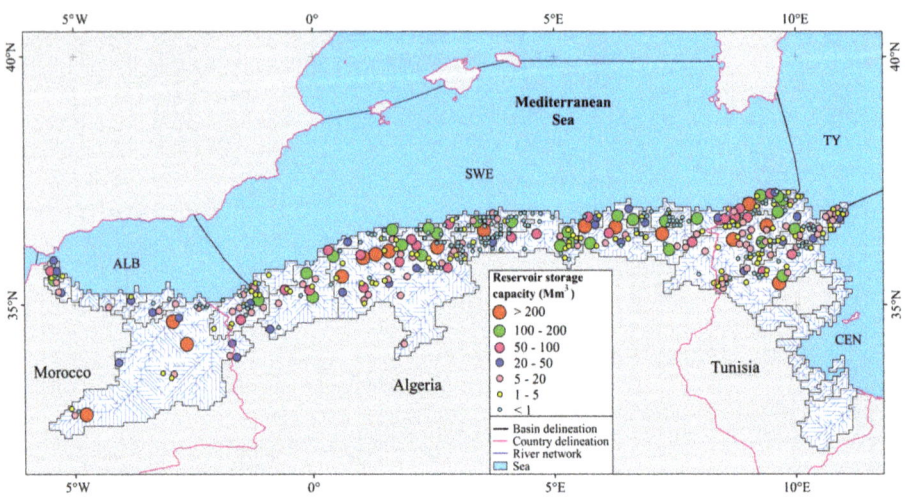

(**A**) Average reservoir storage capacity (Mm³)

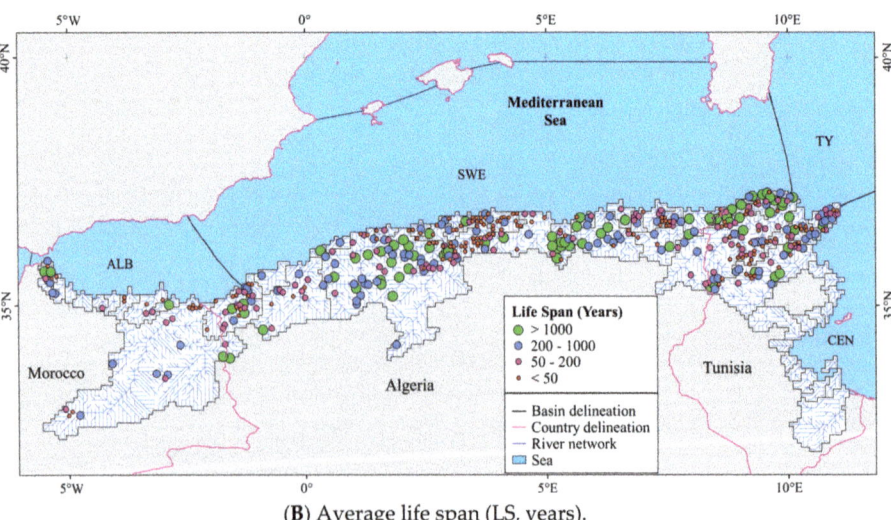

(**B**) Average life span (LS, years).

Figure 4. Geographical distribution of the 450 dams and hillsides implemented in the modeling part of this study.

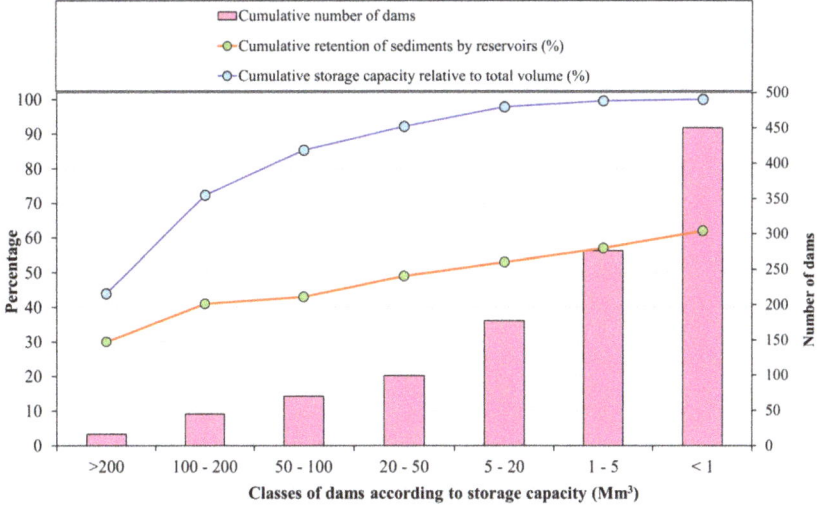

Figure 5. Cumulative reservoir storage capacity (blue line) and sediment retention (orange line) according to various classes of reservoir storage capacities. The cumulative number of dams in each class is shown as well (purple bars). For further explanations, see text.

3.3. Economic and Environmental Implications

Our combination of a spatially explicit mapping of natural sediment yields with an inventory of reservoirs in the MMB also allows evaluating the loss of storage capacity and the life spans of the reservoirs. On the one hand, this can be compared with observed accumulation rates behind dams for validation purposes. On the other hand, it may also help to assess the loss of performance and economic value of existing dams and to evaluate the feasibility of future damming projects.

For this purpose, we followed the approach of [7] and converted the calculated mass accumulation of sediments into volumetric values by assuming an average density of 0.96 t m^{-3}, as proposed by [41]. The reservoirs of the MMB consequently retain a volume of 54 Mm^3 of sediment each year, which corresponds to an annual loss of storage capacity of about 0.39%. This volume can be further split into 22 Mm^3 for the Algerian part (annual loss of 0.27%), 14 Mm^3 for the Tunisian part (annual loss of 0.39%) and 18 Mm^3 for the Moroccan part (annual loss of 0.89%). These values match well with other estimates from the literature. Remini estimated the sediment retention in reservoirs behind Algerian dams to be 20 Mm^3 [42], which accounts for a reduction of their capacity (6.2 Mm^3 according to [43]) of 0.3% each year. This value is identical to our estimate. Mammou and Louati proposed a capacity loss for Tunisian dams between 0.5 and 1% each year [28], compared to the 0.4% we found. Due to its greater relief and higher natural sediment yields, the values proposed for the Moroccan part of the MMB generally exceed 0.7% [20,22]. We estimate an annual capacity loss of about 0.9%. At larger scales, Remini, W. and Remini, B. evaluated the annual capacity loss of the 230 dams in North Africa to be 0.54% per year [25], which is only slightly higher than our results for the MMB. But many of the reservoirs in their study are located outside our study region, especially for Morocco. Notice that only the three Moroccan dams Eddahabi, El Massira and Ben El Ouidane in the Atlantic drainage basin of the Maghreb retain annually 18.3 Mm^3 of sediment, which is equivalent to one third of the overall sediment retention in the entire MMB.

The life span of reservoirs (LS, year) can consequently be estimated via the ratio between the reservoir capacity and the annual volume of sediments that is retained behind the reservoir dams. Figure 4B shows the average LS for the 450 reservoirs considered in our modeling approach. It varies from less than 10 years to 143 × 10^3 years, with a median of 150 years and lower and upper quartiles of 50 and 550 years. About 35% of these reservoirs have a LS of less than 50 years in relation to reduced storage capacities (90% have a storage capacity of less than 2.5 Mm^3). The majority is in the Algerian basin part (62%), followed by the Tunisian (28%) and the Moroccan basin parts (10%). On the other side, 25% of all reservoirs have LS between 50 and 200 years, and 23% between 200 and 1000 years. Only for 17% of the reservoirs LS exceeds 1000 years.

For the 101 dams in our database which were collected from the GRanD database and the national agencies of the Maghreb countries (see Section 2.3), the precise construction dates were known (Table 1) and therefore we could predict their filling. In 2016, the total sediment volume accumulated in these dams was 1.5 km^3, equivalent to a fill of 12% of the total storage capacity (11.7 km^3). The average silting is thus 37 Mm^3 year^{-1}, and if nothing changes, about 30% of their initial capacity would be lost by 2060. Interestingly, with our modeling tools we can also calculate that, if these were the only dams in the MMB and no smaller and hillside dams existed to catch part of the sediment fluxes further upstream, the average silting would increase to 42 Mm^3 year^{-1} and the 30% capacity loss would arrive by the year 2050. These results confirm the findings of [24], who suggested that the delay of siltation of large dams is possible by installing smaller structures further upstream.

Table 4 lists the siltation rates for the 11 dams in the MMB that are the most affected by rapid filling. Two examples are worth being cited more in detail. One is the Mohamed Ben Abd Lekrim El Khattabi dam in the Nekor basin of Morocco, with a storage capacity of 43 Mm^3. Our study shows that average annual silting is here 3.1 Mm^3, which corresponds to a storage capacity loss of 7.2%. These figures are well confirmed by the study of [26], who estimated an average silting rate of 2.7 Mm^3 year^{-1} and an annual storage capacity loss of 6.3%. The accelerated siltation of this dam is closely linked to the extremely high erosion rates in the Nekor basin, one of the highest in the world (see Section 2.1). Without measures to prevent siltation, this dam would have been completely filled only 15 years after its construction in 1981.

Table 4. Calculated volume of sediments retained in some Mediterranean reservoirs that are threatened by rapid silting in comparison with other literature estimates.

Dams	Count	Source	Year-Cons	CAP (Mm^3)	Vsed (Mm^3 year^{-1})	Vsed (% year^{-1})	Vsed 2016 (Mm^3)	Vsed 2016 (%)	LS (Years)	Vsed-lit (%)	Source-lit
Bakhada	ALG	GRanD	1935	56	0.1	0.1	4	8	1033	20	2
Beni Amrane	ALG	ANBT	1988	12	0.3	2.3	8	64	44	80	2
Beni Bahdel	ALG	GRanD	1946	63	0.1	0.2	9	15	469	17	2
Bou Hanifa	ALG	GRanD	1948	73	0.5	0.7	36	50	137	57	2
Boughzoul	ALG	GRanD	1934	55	0.5	1.0	43	78	105	56	2
Fergoug	ALG	GRanD	1970	18	0.3	1.7	14	77	60	100	2
Ighil Emda	ALG	ANBT	1953	156	1.0	0.7	66	42	149	54	2
M.B.A.L.khattabi	MAR	GRanD	1981	43	3.1	7.2	108	251	15	59	3
MohammedV	MAR	GRanD	1967	730	8.2	1.1	403	55	89	56	1
Mellegue	TUN	GRanD	1954	300	3.3	1.1	203	68	92	85	1
Sidi Salem	TUN	GRanD	1981	555	2.9	0.5	100	18	194	21	1

Note: Count: Country; Year-Cons: Year of Construction; CAP: reservoir storage capacity; Vsed: volume of sediment trapped by dams; Vsed 2016: volume of sediment trapped by dams until 2016; LS: life span of the dams; Vsed-lit: volume of sediment trapped by dams from the literature; Source-lit: source according to the literature. Source-lit: 1: [25] (silting until 2002); 2: [24] (silting until 2006); 3: [26] (silting until 1990). For the Source abbreviations, see Table 1.

The second example is the Ighil Emda dam in Algeria, located in a drainage basin with the same name. We calculated here an average sediment filling of 1.1 Mm^3 year^{-1} and an average capacity loss of 0.67% year^{-1}. Since its commissioning in 1953, the Ighil Emda dam has accumulated nearly 66 Mm^3 of sediments and lost 42% of its capacity. These results are very close to those presented in the study by [44], who estimated the cumulative silting during the period 1953–1993 from bathymetric surveys to be 52 Mm^3. During this 40-year period, the dam has lost nearly 35% of its initial capacity. This corresponds to an average capacity loss of 0.87% year^{-1}. The main reason for this rapid silting is the nature of its drainage basin, formed by soft rocks, in combination with a thin plant cover and a predominance of the agricultural domain [44].

It is finally also interesting to look at the theoretical sediment starvation in the coastal areas, which can be estimated from the difference between the pre-damming river fluxes of SPM and the present-day SPM fluxes normalized to the coastline length. The latter is about 422 km, 1210 km, and 1225 km for Morocco, Algeria and Tunisia, respectively. Although this information does not take into account the integration of sediment starvation over time (which depends on the precise construction date of dams), neither the existence of littoral drifts which redistribute the sediments into the coastal domain, it is nevertheless indicative for the potential risk of coastal erosion in relation to decreasing river sediment supply. Maximum values of 45×10^3 t km^{-1} year^{-1} are obtained for the coast of Morocco, which should consequently be the most threatened by beach erosion and sea level rise. Note that this value per kilometer of coastline corresponds to about the average SPM flux produced annually within the whole basin of a typical coastal river in the Gulf of Lions—like the Tet River, with a basin area of 1400 km^2 [34,45]. In other words, the sediments produced elsewhere in a hinterland area of 1400 km^2 are equivalent to the amount of sediments missing every km along the Moroccan coast. The corresponding values for Algeria and Tunisia are also rather high: 21×10^3 t km^{-1} year^{-1} and 12×10^3 t km^{-1} year^{-1}, respectively.

4. Conclusions

For the first time a spatially explicit assessment of the impact of river damming on the natural sediment fluxes to the sea could be produced for the Mediterranean Maghreb basin (MMB), a region which is characterized by a critical state of the available surface water resources and by strong mechanical erosion rates. Our data are based on modeling, which often represents simplifications with respect to reality and which is always constraint by the validity of the basic assumptions leading to the model formulations. Modeling on a grid point scale is limited by the quality of the input data and the adequacy of the regression models used to predict the riverine sediment fluxes. It is furthermore difficult to assess the quality of the literature estimates we used to adjust our regressions and which define the natural (pre-damming) state of the river systems, because anthropogenic perturbations can have a long history in the Mediterranean area. Nevertheless, the values we produced fit very well with other literature estimates, both in terms of the general sediment budgets and with regard to the measured reservoir silting rates, which makes us confident on their reliability.

Our results clearly demonstrate that, because of damming, nowadays only slightly more than one third of the natural river sediment fluxes reach the coastal Mediterranean waters of the Maghreb. Contrary to other hot spot areas of river damming, such as the East, Southern and Southeast Asian region where most of the sediment retention is related to the existence of mega dams [46], a considerable part of the sediment retention is here related to the existence of small dams and hillside reservoirs. Consequently, when only considering the large dams as those documented in the GRanD database, sediment retention is strongly underestimated. Using the latter database alone results in an overall sediment retention of about 36%, while the supplementary dams and hillside reservoirs we collected increase this value to 62%.

Large and small reservoirs in regulated basins trap 30% and 23% of the global riverine sediment fluxes according to [9]. Our retention estimates are therefore substantially higher than those given at the global scale, but the differences are not as striking as one may expect. In the [9] estimate, large reservoirs correspond to CAP > 500 Mm^3, which are rare in the MMB, and the 62% retention we calculated should rather be compared to the 23% retention that [9] found for small dams. The major difference is that small dams have shorter life spans than large dams, and the economic exploitation of these structures is limited in time. Notice that according to our assessment, 35% of the MMB reservoirs have life spans of less than 50 years in relation to reduced storage capacities, which requires frequent renewal and replacement of these structures for further development. It should be also mentioned that sediment retention rates alone are not necessarily an appropriate measure of the potential impact of damming on natural systems. Our data reveal some negative correlation between sediment retention and natural erosion rates in the three Maghreb countries (see Section 3.2), which could be explained by the greater difficulties to build dams in terrains with steep slopes and high natural sediment yields. Although sediment retention rates are "only" 55% in the Moroccan part of MMB, it is probably here where riverine sediment starvation may have the greatest impacts on the coastline dynamics.

Author Contributions: M.S. conducted the data analysis and wrote the draft of the manuscript. M.S. and W.L. worked on subsequent drafts of the manuscript. M.S., W.L., F.B., Y.L. and E.R. contributed to the final version of the manuscript.

Funding: This work was supported by the National Centre of Scientific Research—France (Centre National de la Recherche Scientifique—France, CNRS) through the attribution of a PhD grant to M.S., and by the DEMEAUX project funded by the regional council Occitania and the FEDER.

Acknowledgments: We are grateful to three anonymous reviewers for their valuable and useful comments during the review process.

Conflicts of Interest: The authors have no conflict of interest to declare.

References and Note

1. García-Ruiz, J.M.; Ignacio López-Moreno, J.I.; Vicente-Serrano, S.M.; Lasanta-Martínez, T.; Beguería, S. Mediterranean water resources in a global change scenario. *Earth Sci. Rev.* **2011**, *105*, 121–139.
2. Van Cappellen, P.; Maavara, T. Rivers in the Anthropocene: Global scale modifications of riverine nutrient fluxes by damming. *Ecohydrol. Hydrobiol.* **2016**, *16*, 106–111. [CrossRef]
3. Walling, D.E.; Fang, D. Recent trends in the suspended sediment loads of the World's rivers. *Glob. Planet. Chang.* **2003**, *39*, 111–126. [CrossRef]
4. Brunel, C.; Certain, R.; Sabatier, F.; Robin, N.; Barusseau, J.P.; Alemant, N.; Raynal, O. 20th century sediment budget trends on the Western Gulf of Lions shoreface (France): An application of an integrated method for the study of sediment coastal reservoirs. *Geomorphology* **2014**, *204*, 625–637. [CrossRef]
5. Besset, M.; Anthony, E.J.; Sabatier, F. River delta shoreline reworking and erosion in the Mediterranean and Black Seas: the potential roles of fluvial sediment starvation and other factors. *Elem. Sci. Anth.* **2017**, *5*, 54. [CrossRef]
6. Vanmaercke, M.; Poesen, J.; Verstraeten, G.; de Vente, J.; Ocakoglu, F. Sediment yield in Europe: Spatial patterns and scale dependency. *Geomorphology* **2011**, *130*, 142–161. [CrossRef]
7. Sadaoui, M.; Ludwig, W.; Bourrin, F.; Romero, E. The impact of reservoir construction on riverine sediment and carbon fluxes to the Mediterranean Sea. *Prog. Oceanogr.* **2018**, *163*, 94–111. [CrossRef]
8. Lehner, B.; Reidy Liermann, C.; Revenga, C.; Vorosmarty, C.; Fekete, B.; Crouzet, P.; Doll, P.; Endejan, M.; Frenken, K.; et al. *Global Reservoir and Dam Database, Version 1 (GRanDv1): Dams, Revision 01*; NASA Socioeconomic Data and Applications Center (SEDAC): Palisades, NY, USA, 2011. [CrossRef]
9. Vörösmarty, C.J.; Meybeck, M.; Fekete, B.; Sharma, K.; Green, P.; Syvitski, J.P.M. Anthropogenic sediment retention: major global impact from registered river impoundments. *Glob. Planet. Chang.* **2003**, *39*, 169–190. [CrossRef]
10. Syvitski, J.P.M.; Vörösmarty, C.; Kettner, A.J.; Green, P. Impact of humans on the flux of terrestrial sediment to the global coastal ocean. *Science* **2005**, *308*, 376–380. [CrossRef] [PubMed]
11. Ludwig, W.; Meybeck, M.; Abousamra, F. *Riverine Transport of Water, Sediments, and Pollutants to the Mediterranean Sea*; UNEP/MAP: Athens, Greece, 2003; Available online: http://web.unep.org/unepmap/mts-141-riverine-transport-water-sediments-and-pollutants-mediterranean-sea (accessed on 12 July 2018).
12. Fournier, F. *Climat et Erosion*; Presses Universitaires de France: Paris, France, 1960; p. 201. (In French)

13. Dedkhov, A.P.; Mozzherin, D.I. Erosiya I stok nanosov na zemle, Izdatelstvo Kasagskogo Universsiteta. 1984.
14. Walling, D.E.; Webb, B.W. Material Transport by the World's Rivers: Evolving Perspectives. Available online: http://hydrologie.org/redbooks/a164/iahs_164_0313.pdf (accessed on 1 May 2016).
15. Probst, J.L.; Suchet, P.A. Fluvial suspended sediment transport and mechanical erosion in the Maghreb (North Africa). *Hydrol. Sci. J.* **1992**, *37*, 621–636. [CrossRef]
16. Hijmans, R.J.; Cameron, S.E.; Parra, J.L.; Jones, P.G.; Jarvis, A. Very high resolution interpolated climate surfaces for global land areas. *Int. J. Climatol.* **2005**, *25*, 1965–1978. [CrossRef]
17. Ghenim, A.; Seddini, A.; Terfous, A. Variation temporelle de la dégradation spécifique du bassin versant de l'Oued Mouilah (nord-ouest algérien). *Hydrol. Sci. J.* **2008**, *53*, 448–456. (In French) [CrossRef]
18. Tourki, M.; Khanchoul, K.; Le Bissonnais, Y.; Belala, F. Sediment yield assessment in the Upper Wadi Kebir catchment, Kébir Rhumel River, Northeast of Algeria (1973–2006). *Rev. Sci. Technol.* **2017**, *34*, 122–133.
19. Inoubli, N.; Raclot, D.; Mekki, I.; Moussa, R.; Le Bissonnais, Y. A spatiotemporal multiscale analysis of runoff and erosion in a Mediterranean marly catchment. *Vadose Zone J.* **2017**, *16*. [CrossRef]
20. Badraoui, A.; Hajji, A. Envasement des retenues de barrages. *La Houille Blanche* **2001**, *6–7*, 72–75. [CrossRef]
21. Demmak, A. Contribution à l'Étude de l'Érosion et du Transport Solide en Algérie Septentrionale. Thèse de Docteur-Ingénieur, Université Paris-VI, Paris, France, 1982. (In French)
22. Lahlou, A. Etude actualisée de l'envasement des barrages au Maroc. *Rev. Sci. Eau.* **1988**, *6*, 337–356. (In French)
23. Estienne, P.; Godard, A. *Climatologie*; Librairie Armand Colin: Paris, France, 1970. (In French)
24. Remini, B.; Leduc, C.; Hallouche, W. Évolution des grands barrages en régions arides: Quelques exemples algériens. *Sécheresse* **2009**, *20*, 96–103. (In French)
25. Remini, W.; Remini, B. La sédimentation dans les barrages de l'Afrique du Nord. *Larhyss* **2003**, *2*, 45–54. (In French)
26. Lahlou, A. Envasement du barrage Mohamed Ben Abdelkrim Al Khattabi et lutte anti-érosive du bassin versant montagneux situe à l'amont. *IAHS Publ.* **1990**, *194*, 243–252. (In French)
27. Kassoul, M.; Abdelkader, A.; Belorgey, M. Caractérisation de la sédimentation des barrages en Algérie. *Revue des Sciences de l'Eau* **1997**, *10*, 339–358. (In French) [CrossRef]
28. Mammou, B.A.; Louati, M.H. Evolution temporelle de l'envasement des retenues de barrages de Tunisie. *Rev. Sci. Eau* **2007**, *20*, 201–210. (In French) [CrossRef]
29. Seklaoui-Oukid, O. Valorisation des Sédiments du Barrage d'El Merdja Sidi Abed: Etude Technoéconomique. Available online: http://www.secheresse.info/spip.php?article60431 (accessed on 1 April 2016). (In French)
30. Lehner, B.; Verdin, K.; Jarvis, A. New global hydrography derived from spaceborne elevation data. *Eos Trans. AGU* **2008**, *89*, 93–94. [CrossRef]
31. Harris, I.; Jones, P.D.; Osborn, T.J.; Lister, D.H. Updated high resolution grids of monthly climatic observations—The CRU TS3.10 Dataset. *Int. J. Climatol.* **2014**, *34*, 623–642. [CrossRef]
32. Dürr, H.; Meybeck, M.; Dürr, S.H. Lithologie composition of the earth's continental surfaces derived from a new digital map emphasing riverine material transfer. *Glob. Biochem. Cycles* **2005**, *19*. [CrossRef]
33. Mayaux, P.; Bartholomé, E.; Cabral, A.; Cherlet, M.; Defourny, P.; Di Gregorio, A.; Diallo, O.; Massart, M.; Nonguierma, A.; Pekel, J.F.; et al. A new land-cover map of Africa for the year 2000. *J. Biogeogr.* **2004**, *31*, 861–877. [CrossRef]
34. Sadaoui, M.; Ludwig, W.; Bourrin, F.; Raimbault, P. Controls, budgets and variability of riverine sediment fluxes to the Gulf of Lions (NW Mediterranean Sea). *J. Hydrol.* **2016**, *540*, 1002–1005. [CrossRef]
35. Vörösmarty, C.J.; Sharma, K.; Fekete, B.; Copeland, A.H.; Holden, J.; Marble, J.; Lough, J.A. The storage and aging of continental runoff in large reservoir systems of the world. *Ambio* **1997**, *26*, 210–219.
36. ANBT. National Agency for Dams and Transfers. 2016. Available online: http://www.soudoud-dzair.com (accessed on 1 June 2016).
37. ABHL. Loukkos Hydraulic Basin Agency/Tétouan. 2016. Available online: http://www.abhloukkos.ma/abhl/index.php/fr/ (accessed on 1 June 2016).
38. DGBGTH. General Directorate for Dams and Major Hydraulic Works. 2016. http://www.semide.tn/ (accessed on 1 June 2016).
39. Saradouni, F. Contribution à l'Etude de la Vulnérabilité des Retenues Collinaires vis-à-vis des Aléas Naturels, dans un Système d'Information Géographique (SIG). Available online: https://dl.ummto.dz/handle/ummto/910 (accessed on 11 July 2018). (In French)
40. Ludwig, W.; Probst, J.L. River sediment discharge to the oceans: Present-day controls and global budgets. *Am. J. Sci.* **1998**, *296*, 265–295. [CrossRef]

41. Minear, J.T.; Kondolf, G.M. Estimating reservoir sedimentation rates at large spatial and temporal scales: A case study of California. *Water Resour. Res.* **2009**, *45*. [CrossRef]
42. Remini, B. L'envasement des Barrages: Quelques Exemples Algériens. Revue Techniques Sciences Méthodes. 1999. Available online: http://www.beep.ird.fr/collect/bre/index/assoc/HASH6487.dir/20-165-171.pdf (accessed on 1 April 2016).
43. Riad, S.; Salih, A. Options for future water security in the Arab Countries. In Proceeding of the First International Conférence on the Geology of Africa, Assiut, Egypt, 23–25 November 1999; pp. 459–466.
44. Remini, B.; Kettab, A.; Hmat, H. Envasement du Barrage d'IGHIL EMDA (Algérie). 1995. Available online: https://www.shf-lhb.org/articles/lhb/pdf/1995/02/lhb1995008.pdf (accessed on 1 March 2016).
45. Serrat, P.; Ludwig, W.; Navarro, B.; Blazi, J.L. Variabilité spatio-temporelle des flux de matières en suspension d'un fleuve côtier méditerranéen: la têt (France). *Earth Planet. Sci. Lett.* **2001**, *333*, 389–397. [CrossRef]
46. Gupta, H.; Kao, S.-J.; Dai, M. The rôle of mega-dams in reducing sediment fluxes: A case study of large Asian Rivers. *J. Hydrol.* **2012**, *464–465*, 447–458. [CrossRef]

© 2018 by the authors. Licensee MDPI, Basel, Switzerland. This article is an open access article distributed under the terms and conditions of the Creative Commons Attribution (CC BY) license (http://creativecommons.org/licenses/by/4.0/).

Article

Evaluating the Erosion Process from a Single-Stripe Laser-Scanned Topography: A Laboratory Case Study

Yung-Chieh Wang * and **Chun-Chen Lai**

Department of Soil and Water Conservation, National Chung Hsing University, Taichung 402, Taiwan; final898y@gmail.com
* Correspondence: wangyc@nchu.edu.tw; Tel.: +886-4-2284-0831 (ext. 105)

Received: 30 May 2018; Accepted: 17 July 2018; Published: 19 July 2018

Abstract: Topographies during the erosion process obtained from the single-stripe laser-scanning method may provide an accurate, but affordable, soil loss estimation based on high-precision digital elevation model (DEM) data. In this study, we used laboratory erosion experiments with a sloping flume, a rainfall simulator, and a stripe laser apparatus to evaluate topographic changes of soil surface and the erosion process. In the experiments, six slope gradients of the flume (5° to 30° with an increment of 5°) were used and the rainfall simulator generated a 30-min rainfall with the kinetic energy equivalent to 80 mm/h on average. The laser-scanned topography and sediment yield were collected every 5 min in each test. The difference between the DEMs from laser scans of different time steps was used to obtain the eroded soil volumes and the corresponding estimates of soil loss in mass. The results suggest that the collected sediment yield and eroded soil volume increased with rainfall duration and slope, and quantified equations are proposed for soil loss prediction using rainfall duration and slope. This study shows the applicability of the stripe laser-scanning method in soil loss prediction and erosion evaluation in a laboratory case study.

Keywords: soil erosion; rill development; erosion topography; sloping flume experiments

---

1. Introduction

Rill and interrill erosion are the major intermediate processes between sheet and gully erosion on hillslopes. As the main source that causes soil loss on sloping croplands and rangelands, rill erosion provides 76% of the total sediment load eroded from hillslopes [1]. In an eroding rill, the flow concentrates and results in hydrodynamic characteristics with an increasing scouring power, and thus leads to a significant increase in soil erosion [2]. Meanwhile, the development of a rill network results in a threefold soil loss increment [3], and the eroding rills become the primary channels for sediment transport. It is reported that more than 80% of the eroded soil is transported through rills [4,5]. Therefore, rill development not only has its significance in erosion process evaluation, but also shows an implication for drainage network evolution on hillslopes.

The evolution of rill networks and rill morphology progresses in time and space [6]. Rill networks develop with varying complexity, leading to flow concentration and increasing runoff connectivity along the networks [7]. Variations in rill morphology interact with soil erosion processes, including detachment, transport, and deposition, which are dependent on the hydrodynamic features and transport capacity of the flow in rills [8]. Complex feedback exists in rill erosion mechanics, since the flow erodes soil and changes the rill bed surface, which in turn alters the hydrodynamic features and transport capacity of the flow [9]. In general, a rill flow with a higher velocity and deeper depth provides a larger transport capacity for carrying sediments, leading to more variations in rill morphology.

In erosion process evaluation, the evolution of a rill network and rill morphology has been a focus to develop and improve process-based erosion models and prediction [5,10]. While the topography caused by rill erosion is often complicated and irregular, a rill-by-rill survey is difficult and even impractical in the field [1], and field-scale experiments are limited to qualitative or semi-quantitative descriptions [11]. Therefore, quantification of the rill erosion process has become an essential but challenging issue. At the laboratory scale, rainfall simulators are widely applied in research on infiltration and soil erosion processes for mimicking the process and characteristics of natural rainfall [12]. In order to replicate rill erosion processes, rainfall simulation has its advantages in producing rainfall events at an arbitrary intensity and duration on demand; thereby, the erratic and unpredictable variability of natural rains can be eliminated [12]. Most of the previous studies on erosion process evaluation have developed quantification methods for rill erosion and rill network development using rainfall simulators to produce rainfall events on laboratory flumes or field plots.

Among the rill erosion research, many types of rill erosion quantification methods have been applied, such as collecting sediment outflow at the flume end [12], a volumetric or volume replacement method [2,13,14], manual measurements of rills using photo images [1,15], photogrammetry methods [16–19], a terrestrial laser scanner (TLS) [5,20–23], an airborne laser scanner (ALS) [24,25], and other remote-sensing technologies. In [12], laboratory erosion experiments were carried out in a two-dimensional tilting flume with a pre-forming rill before the rainfall application on the soil surface; then, rill and interrill erosion were assessed by flow measurements taken from the two outlets, corresponding to the rill and interrill area, at the end of the flume.

The volumetric method measures and calculates the cross-sectional areas along the eroded rills and estimates the eroded soil volume [14]. The volume replacement method uses other materials, such as soil, tiny foam particles, rice grains, and water, to refill rills and measure the rill volume by the filling materials to estimate the erosion volume [2,13]. Specifically, water is used to refill the eroded rills in order to minimize the measurement errors due to the difficulty of rill identification in the cases of not-well-confined rills and various rill sizes [2]. However, sealing of the soil surface, which prevents infiltration and further erosion, is the key factor for precise measurements of the rill volume and may become the major source of error in practice. Manual measurements of rills using photo images are applied in laboratory studies [1,15] for obtaining each rill's width, depth, and location coordinates along with rainfall duration from photographs taken of the soil surface in 1- or 2-min intervals throughout each rainfall event. The manual measurement method is usually time- and force-consuming and can be dependent on subjective judgement of rill boundaries.

Methods of photogrammetry are applied to assess soil erosion in laboratory-scale experiments [16–19]. Estimation of soil erosion using small-scale laboratory erosion experiments was carried out by the method of structure-from-motion (SfM) photogrammetry [16]. Digital elevation models (DEMs) of a 0.5 m^2 soil box were built by the photogrammetric SfM technique, and the computed sediment yield was obtained by the DEM of difference (DoD) technique [16]. They proposed that the choice of DoD threshold was a key point that affects the computed sediment yield; the results show that the computed sediment yield was 13% greater than the measured sediment yield [16]. Rill development and soil erosion was studied using a gravity-type rainfall simulator, a tilting box filled with sand and soil material, and photogrammetric equipment [17]. In this research, stereo imagery was taken from two positions to generate DEMs of the soil surface, which were used to analyze soil surface and rill network changes between the time steps of rainfall events. The results suggest that rill density and energy expenditure decreased with time [17]. The application of a close-range photogrammetry method for soil erosion assessment and a comparison between photogrammetry and laser-scanner technology in producing DEMs and soil surface elevation differences are shown in previous studies [18,19]. While both of the studies were carried out using laboratory flumes, they suggest the potential of the photogrammetry method in field applications provided the limitations of the technique are considered [18,19].

The application of the TLS technique in soil erosion and rill development evaluation can be found in laboratory, plot, and field studies [5,20–23]. The study of the morphological characteristics of rill evolution was performed using laboratory soil pan rainfall simulations [5] and the TLS technique following [20]. They proposed quantitative descriptions of a rill network using the fractal dimension, which was analyzed from the TLS data and the resulting DEMs. To evaluate soil erosion at the plot scale, uses three types of TLS are used to measure the topography before and after the manual removal of some soil volume (to imitate soil erosion) of a field plot about 30 m × 30 m in size with a 20° slope [21]. The results show that soil erosion measured with TLS varies considerably when different data processing software is used, and the laser-scanning technique is applicable in measuring soil erosion at the plot scale when adequate calibration and spatial resolution are performed [21]. In field investigations of soil erosion using TLS [20,22,23], rill morphology and soil loss evaluation at the Masse experimental station in Italy were studied [20]. The triangulated irregular network model was used to quantify the eroded volume and rill morphological characteristics; corresponding manual measurements were also carried out using a profilometer [20]. A basin-scale assessment of riverbank erosion in Northern Italy was monitored, and the eroded and deposited volume in the surveyed area were measured using TLS [22]. The effects of TLS to analyze the intensity of soil erosion in a mountainous forest area with sufficient measurement stations are shown [23]. As the results indicate, choices of the coordinate system for the object or the scanner significantly influence the analysis of erosion phenomena when applying TLS in forested areas [23].

The technique of ALS has also been applied in field studies to assess riverbank erosion [24] and characterize land degradation processes [25]. ALS was used to obtain detailed topographic data for characterizing sediment and phosphorous contributions from riverbank erosion of the Blue Earth River in southern Minnesota, USA [24]. Since the source of error and the uniformity within and between the scans were difficult to determine, no elevation or planimetric corrections of the laser data were made in the study. Nevertheless, the soil loss estimates obtained by ALS locate in the range of the reported literature values for the same study site. The land degradation processes, including soil erosion, channel incision, and collapse sinkholes development in the Dead Sea region, are quantified using ALS [25]. The results demonstrate the ability of ALS to detect sub-metric geomorphic features, such as gullies, headcuts, and embryonic sinkholes, which can occur in land degradation processes [25]. Other remote-sensing technologies, such as an Unmanned Aerial Vehicle (UAV), Time-of-Flight (ToF) cameras (or range cameras), and airborne and terrestrial Light Detection and Ranging (LiDAR) [20], have been applied to investigate topographic evolution in the field, and have their potential and applicability in soil erosion and land degradation process evaluation, but are usually expensive in cost and/or require delicate operating and data-processing skills.

In this study, we took advantage of the high resolution provided by the laser-scanning technique but used a more affordable stripe laser apparatus, compared to a three-dimensional (3D) laser scanner, to construct the erosion topography for erosion process evaluation. At the laboratory scale, erosion processes on slopes were reproduced in an adjustable slope flume with a rainfall simulator, and a series of laser-scanned topographies during the erosion process was obtained by the stripe laser apparatus for evaluating the rill development, erosion progress, and the eroded soil volume. The effects of rainfall duration and slope on the eroded soil volume and sediment yield were quantified using a multiple regression analysis. By comparing the eroded volume with the outflow sediments collected from the flume outlet, a quantified relation was proposed between the eroded soil volume and the sediment yield at the outlet of the sloping flume. The results of this study present a quantified evaluation of the influences of rainfall duration and slope on the erosion process and a prediction of the sediment yield using single-stripe laser-scanned topography in a laboratory flume case study.

2. Materials and Methods

2.1. Soil Specimens and Geotechnical Tests

The soils used for the erosion experiments were collected from the watershed of Agongdian Reservoir in Kaohsiung city, which is located in southwestern Taiwan (Figure 1). The parent rock in the watershed belongs to green-gray mudstone, which melts easily when absorbing water; most of the top soils are weathered loam or sandy loam based on the USDA (U.S. Department of Agriculture) soil classification system. Due to the high percentage of fine sediments and the melting characteristic of mudstone, the soils are hardened when dry but easily eroded when subjected to continuous rainfall during storm events. In addition, the high content of salts of mudstone soils is not suitable for vegetation, and thus soil loss from hillslopes has led to environmental issues in the watershed, such as reservoir siltation.

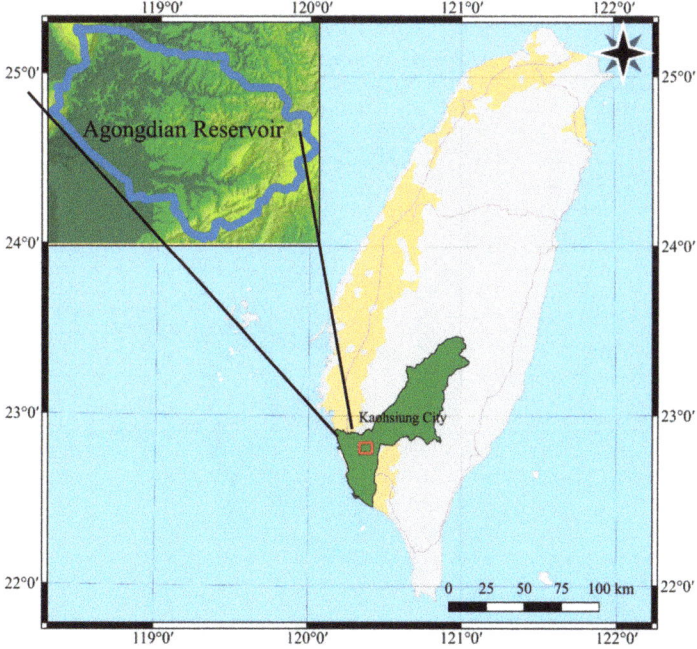

Figure 1. Map of the Agongdian Reservoir watershed in Kaohsiung city, Taiwan.

After the soils were collected from the field, preliminary treatments, including air drying, gravels and plants picking, peds crumbling, and sieving through a 2-mm sieve, were carried out before the conventional geotechnical tests. As background information, the specific gravity and grain size distribution of the mudstone soils were determined before the erosion experiments by following procedures outlined in American Society for Testing and Materials (ASTM) standards, including ASTM D854-14 [26], ASTM C136/C136M-14 [27], ASTM D1140-14 [28], and ASTM D422-63-07 [29], referring to the pycnometer test, dry sieve analysis, wet sieve analysis, and hydrometer test, respectively. As the results indicate, the mudstone soils have an average specific gravity of 2.69 g/cm^3 and consist of 52.15% of sand, 45.75% of silt, and 5.91% of clay on average, and belong to sandy loam based on the USDA soil classification system. The moisture content shows the ratio of pore water mass to the mass of dry solids, and was conducted following the ASTM D2216-10 [30] procedure. Before each trial of the erosion experiments, 24 soil specimens were sampled to estimate the spatial variation of moisture contents of the soils filled in the flume to determine the antecedent soil moisture condition. The soil specimens for soil moisture content measurements were sampled following the method for group B in the ASTM D4220/D4220M-14 [31] procedure. Based on [1,15], the soils were filled to the flume as three 0.05-m-thick layers, and eight soil specimens were sampled at an equal distance for each soil layer.

2.2. Sloping Flume Erosion Experiment

The experimental setup of the erosion experiment, including the rainfall simulator, the sloping flume, and the stripe laser apparatus, is shown in Figure 2. The flume is 0.75 m in length, 0.25 m in width, and 0.25 m in depth, and with an adjustable slope gradient range from 0° to 30°. Below the flume bottom, a plastic container was used to collect the infiltrated outflow and sediments during the erosion process. At the outlet of the flume, a funnel-shape opening was designed for collecting the outflow water and sediment yields. To minimize experimental uncertainty and obtain good controls on experimental factors, we carried out the erosion experiments and recorded the erosion topography using a small-scale rainfall simulator and a flume with adjustable slopes. The rainfall simulator, stripe laser apparatus, soil specimen preparation, and test procedure and measurements of the laboratory erosion experiment are described in the following subsections.

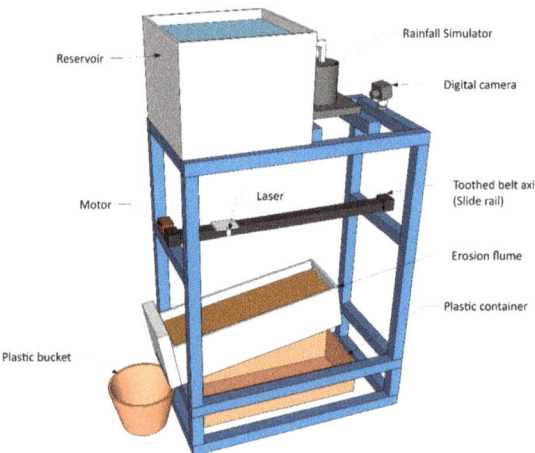

Figure 2. Setup of the sloping flume erosion experiment.

2.2.1. Rainfall Simulator

The mini rainfall simulator used for the erosion test is manufactured by Eijkelkamp Soil & Water (EM Giesbeek, the Netherlands) [32], with a rainfall area of 0.25 m × 0.25 m produced by 49 probes with a constant water head. Compared to natural precipitation, simulated rainfalls usually have a shorter falling distance, which results in a lower terminal velocity of rain drops and thus a smaller kinetic energy. This concept was described in the previous study, in which the same rainfall simulator of this study was applied [33]. Therefore, to compensate for the short falling distance of the simulated rainfall (ranging from 0.61 m to 0.69 m with the slope gradient ranging from 5° to 30°, which covered most of the slopes in the study area), the high intensity of the simulated rainfall was applied. From the equation of time-specific kinetic energy [34], the time-specific kinetic energy produced by the simulated rainfalls equals that of natural rainfall events with the intensity ranging from 76 mm/h to 84 mm/h when the slope varies from 5° to 30°. The rainfall simulator was located at the upstream side of the flume and generated simulated rainfalls covering the range of 0.25 m × 0.25 m. The remaining downstream region of the soil surface was used to observe rill development during the erosion process.

2.2.2. Stripe Laser Apparatus

The stripe laser apparatus used for obtaining erosion topographies consists of a 100-mW diode laser (FLEXPOINT® Machine Vision Lasers Mvnano Series, Olching, Germany), an electric slide rail (Igus drylin® ZLW, Köln, Germany), and a digital camera (Nikon D7200, Tokyo, Japan). By applying pattern projection methods [35], the projected single stripe of light onto the soil surface, which was produced by the diode laser moving with a constant speed along the flume on the electric slide rail, was recorded by the digital camera to reconstruct a whole surface from digital camera images. Detailed discussions of the stripe laser apparatus, the basic principles of the scan processing method, and scan resolution and error estimation are provided in previous studies [35,36]. The reconstruction of a scanned surface was achieved using the images of the stripe laser rays captured by the digital camera for each cross-section of the soil surface apart by 1 mm. The images were processed in the MATLAB (2017a, MathWorks Inc., Natick, MA, USA) program developed the previous study [36] and the erosion topographies were obtained.

2.2.3. Soil Specimen Preparation

For the soil specimen preparation of each erosion test, the air-dried and pretreated soils were mixed with tap water thoroughly by human force; then, the moisturized soil was placed in the storage container and the soil surface was covered and allowed to stand for at least 16 h as suggested by the wet preparation method in the ASTM D4318-10 procedure [37]. To target an antecedent soil moisture content around 20% and a bulk density of 1400 kg/m^3, 39.375 kg of dried soil and 1.875 kg of tap water were used. The moisturized soils were filled by three layers, with each layer 0.05-m thick [1,15], and a total of 0.15 m in soil depth. After a layer was filled, soils sampled at the eight sites distributed in equal spaces were used to determine the actual antecedent soil moisture conditions. Therefore, a total of 24 soil specimens for the 3 filled soil layers were sampled to determine the average antecedent soil moisture content for each test of the erosion experiments. After the soils were filled in the flume, an initial laser scan was carried out through the soil surface to obtain the topography before the simulated rainfall [36]. Then, the flume was tilted by a screwing jack to a designed slope and the simulated rainfall began.

2.2.4. Test Procedure and Measurements

After a rainfall simulation began, the outflows of water and sediments were collected by the bottom container and the outlet bucket every 5 min until the end of the test. To collect infiltrations during experiments [38,39], the bottom container collected the infiltrated outflow in this study, while the outlet bucket collected the surface outflow. Then, the total sediment yield can be obtained as the summation of the sediments transported by the infiltrated and surface outflows. The collected outflows were dried in an oven at 105 °C for 24 h [1,15,40], and the remnant soils were weighed as the measurements of sediment yields. In this study, the collected sediment yields are considered as the "measured" soil loss. Meanwhile, to record the soil surface topography for each 5-min increment of the rainfall, the simulated rainfall was temporarily stopped for 20 min [17], the flume was lowered to the horizontal position, and the laser scan was carried out. The duration of each rainfall event was 30 min [15]. Therefore, seven laser-scanned topographies were obtained for a single rainfall event. To account for the effect of steepness on hillslope erosion, six slope gradients (5°, 10°, 15°, 20°, 25°, and 30°) were used in the experiments and two replicates [15] were applied for each gradient.

2.3. Erosion Volume Generation and Data Treatment

By applying the stripe laser-scanning technique, we obtained a DEM of the soil surface through each test of the scanning procedure at each time step. Then, the DEMs were subtracted on a cell-by-cell basis to obtain the difference of depths throughout the whole soil surface. The soil volume differences among the erosion topographies were obtained; then, the mass of soil loss can be estimated by the soil volume differences, the moisture content, and the bulk density of the soil. The resultant soil loss in mass is considered to be the "estimated" soil loss. Since the computation of the difference of depths from two DEMs can propagate the errors of the initial DEMs [16,35,36], we applied an error analysis by comparing the elevation precision using check points on the soil surface of the laser-scanned DEMs to determine the minimum level of detection for the depth difference. The error in depth difference of two DEMs was estimated as the quadratic composition of errors [35,36] in each of the original DEMs being compared [16], and it can be calculated as:

$$E_{dod\_i} = \sqrt{E_{DEM\_i}^2 + E_{DEM\_i+1}^2} \tag{1}$$

where E_{dod_i} means the error of depth difference between two DEMs of the erosion experiments; E_{DEM_i} means the error of the DEM at time step i calculated from the check points calibration; and E_{DEM_i+1} means the error of the DEM at time step $i+1$ calculated from the check points calibration. For the erosion experiments, the detection threshold for the depth differences between two DEMs was set as the E_{dod} value. From the calibrations and error analyses of the check points, the minimum value of E_{dod} [16] was 1.28 mm. Therefore, a difference of depths was accounted for in the calculation of a soil volume difference between two DEMs when its absolute value is ≥ 1.28 mm.

3. Results

3.1. Soil Loss Measured by the Sediment Yields

During a test of a 30-min simulated rainfall event, the sediment yield was collected every 5 min until the end of the test, and the cumulative sediment yields are shown in Table 1. The cumulative sediment yields increase with the rainfall duration, and the total sediment yields at the end of the test range from 1.228 kg to 8.300 kg of oven-dried soil mass. Based on the infiltrated outflow collected by the plastic container below the flume bottom during the erosion experiments, outflows of infiltrated sediments occurred in the tests with 5°, 10°, and 15° of slopes but not in the tests with steeper slopes. On average, the cumulative infiltrated sediment outflows account for 18.4% of the total cumulative sediment yields after the 30-min rainfall events.

Table 1. Sediment yield measurements collected at the outlet of the flume.

Index	Cumulative Sediment Yield [b] (kg)					
	Rainfall Duration (min)					
Slope	5	10	15	20	25	30
30° (1) [a]	1.802	3.335	4.404	5.546	6.923	8.3
30° (2) [a]	1.471	2.815	4.06	5.149	5.948	6.766
25° (1)	1.829	2.98	4.052	4.946	5.714	6.439
25° (2)	2.287	3.422	4.687	5.975	7.382	8.919
20° (1)	1.018	1.937	2.816	3.799	4.484	5.376
20° (2)	0.802	1.604	2.315	3.396	4.289	5.086
15° (1)	0.173	0.695	1.195	1.68	2.099	2.46
15° (2)	0.285	0.596	0.836	1.062	1.334	1.646
10° (1)	0.167	0.675	1.108	1.395	1.75	2.071
10° (2)	0.467	1.116	1.678	2.15	2.708	3.1
5° (1)	0.088	0.303	0.504	0.903	1.213	1.507
5° (2)	0.118	0.374	0.646	0.807	1.065	1.228

Note: [a] (1) and (2) refers to the first and the second erosion test of each slope gradient, respectively. [b] Sediment yield = sediment outflow collected by the outlet bucket + infiltrated sediment outflow collected by the bottom container.

3.2. Soil Loss Estimated by the Laser-Scanned Soil Volume Differences

From the differences of depths generated by the subtraction of any two DEMs (with a spatial resolution of 1 mm × 1 mm), the soil volume differences were calculated by summing up all the depth difference values exceeding the detection threshold. Then, the estimated soil loss in mass, the volume-difference soil loss, was the product of the soil volume difference and the soil bulk density resulting from the average soil moisture contents. During a test of a 30-min simulated rainfall event, the volume-difference soil loss was obtained every 5 min until the end of the test, and the cumulative volume-difference soil loss is shown in Table 2. The cumulative volume-difference soil loss increases with the rainfall duration, and the total volume-difference soil loss at the end of the test ranges from 1.833 kg to 10.775 kg. From observations during the experiments, the depression of the soil surface under rain drop compaction and collapse along the side-wall of rills were the main factors for overestimates in the soil loss, especially during the first 5 min of the rainfalls.

Table 2. Estimates of soil loss using laser-scanned topographies.

Index	Cumulative Volume-Difference Soil Loss (kg)					
Slope	Rainfall Duration (min)					
	5	10	15	20	25	30
30° (1) [a]	3.100	5.058	6.087	7.367	8.540	9.518
30° (2)	1.981	4.706	4.990	5.443	6.880	6.992
25° (1)	3.836	5.012	5.820	6.813	7.081	7.725
25° (2)	3.982	6.120	7.407	8.123	9.594	10.775
20° (1)	3.080	4.414	5.074	5.652	6.822	6.798
20° (2)	3.113	3.854	4.397	5.176	6.057	6.593
15° (1)	2.630	3.206	3.553	4.025	4.114	4.552
15° (2)	1.831	1.947	2.418	2.630	2.661	2.917
10° (1)	2.937	3.432	4.097	4.106	4.568	4.740
10° (2)	2.327	3.143	3.471	3.877	4.216	4.403
5° (1)	0.979	1.394	1.374	1.861	2.030	2.082
5° (2)	0.858	1.382	1.520	1.698	1.862	1.833

Note: [a] (1) and (2) refers to the first and the second erosion test of each slope gradient, respectively.

3.3. Laser-Scanned Topographies during Erosion

For each test of the simulated rainfall erosion experiment, seven laser-scanned topographies, including the initial soil surface topography, were carried out. As an example, the laser scanner topographies of the erosion test with a 30° slope are shown in Figure 3. From the soil surface topographies, the propagation of erosion processes can be described qualitatively. In Figure 3, an increase in the soil surface depression areas suggested soil erosion propagation caused by rain drop compaction/rain splash erosion, sheet flow erosion, and sediment entrainment of concentrated flows. Specifically, rain drop compaction occurred in the rainfall region at 45 cm to 70 cm from the outlet; erosion of the overland sheet flow occurred in the rainfall region and the middle stream area around $x = 30$ cm to $x = 45$ cm during the first 10 min of the rainfall, and then the formation of rills was observed as the effect of rill erosion. In the area with a distance from the outlet less than 30 cm, the main erosion process was sediment entrainment of the concentrated flows. The depths of the eroded channel for the concentrated flow in $x = 0$ to $x = 30$ increased with the rainfall duration, which are shown in Figure 3b–f. However, comparison between Figure 3f,g suggests that deposition occurred in the region of $x = 0$ to $x = 15$ during the last 5 min of the rainfall event, which may be caused by collapse of the rill sides during sediment entrainments or the outflow being too small and plugged.

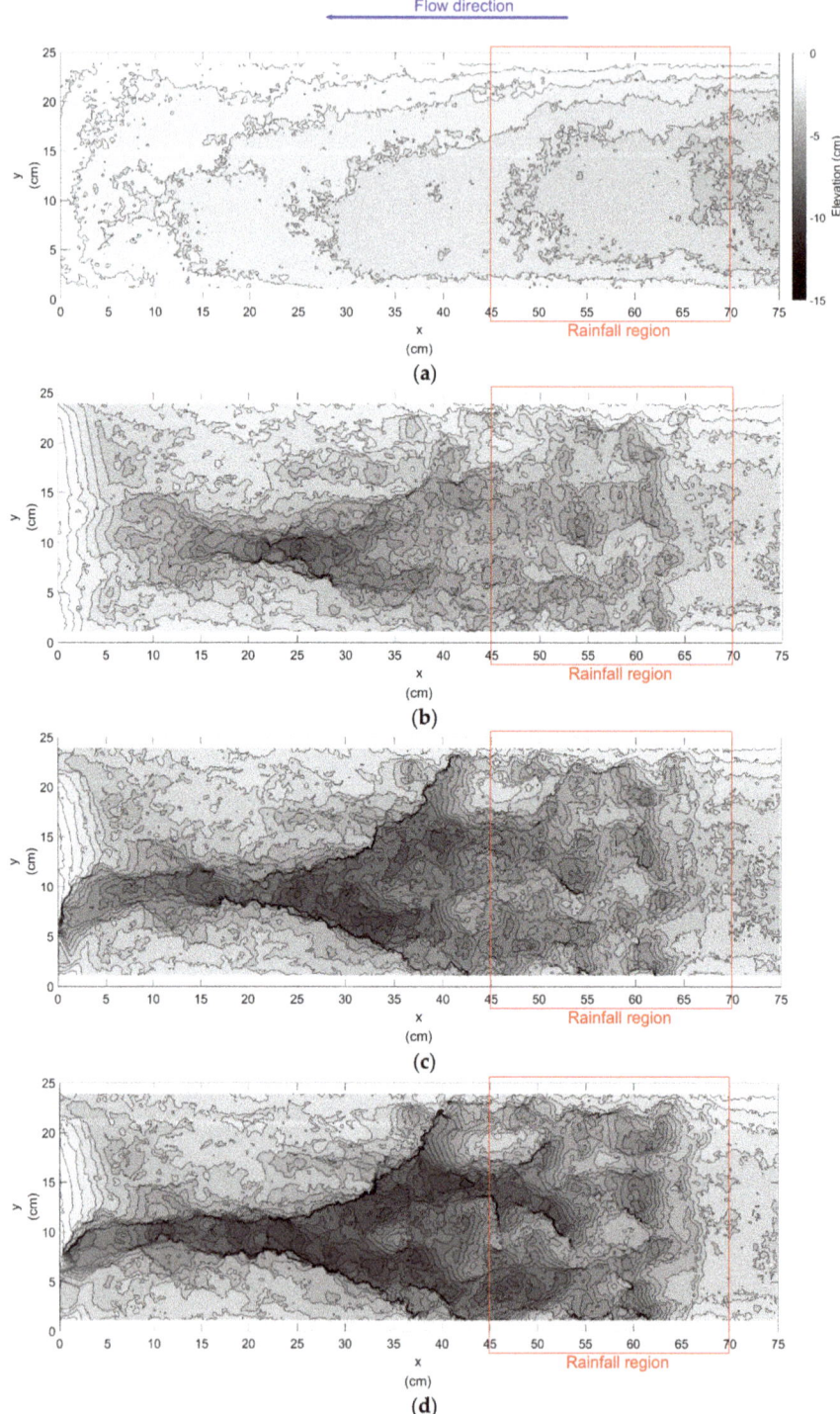

Figure 3. *Cont.*

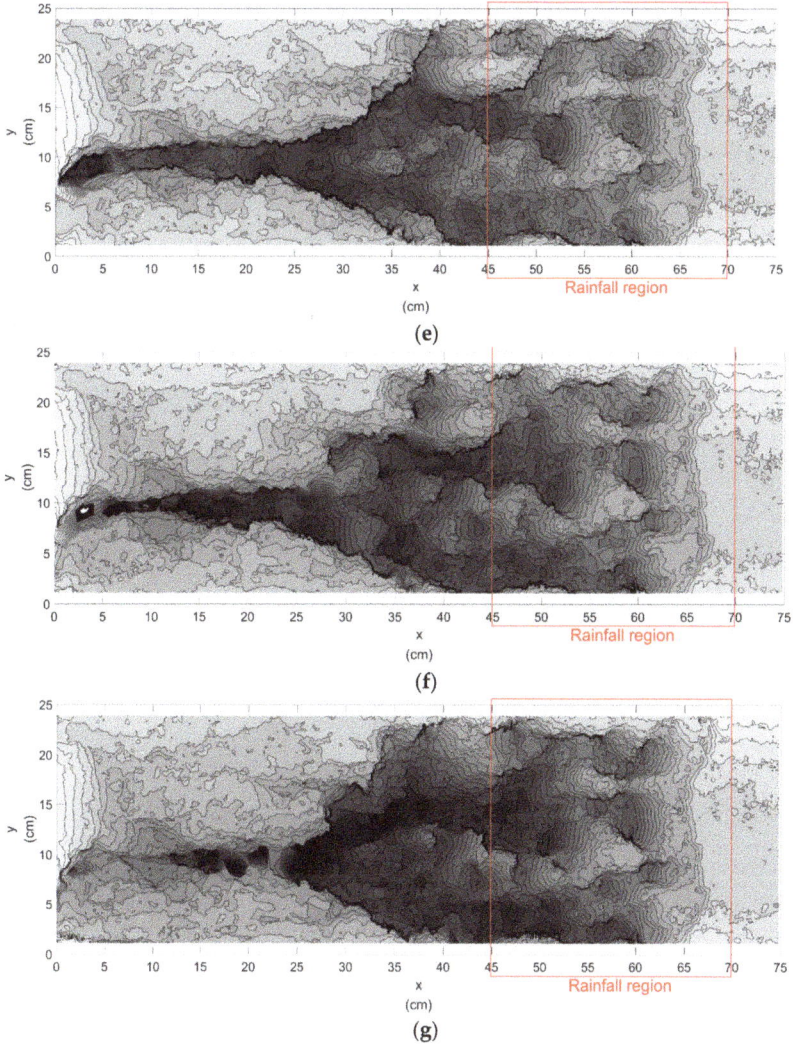

Figure 3. The laser-scanned topographies during the simulated rainfall erosion test of a 30° slope: (**a**) the initial topography before the rainfall started; the erosion topographies after (**b**) 5 min; (**c**) 10 min; (**d**) 15 min; (**e**) 20 min; (**f**) 25 min; and (**g**) 30 min of rainfall.

4. Discussion

4.1. Effects of Rainfall Duration and Slope on Soil Loss

Effects of rainfall duration and slope on the soil loss mass are illustrated in Figure 4. From the figure, the mass of the soil loss, both measured from the sediment yields and estimated from the soil volume differences, increased with the rainfall duration and the slope. From previous studies [41,42], similar results have been proposed regarding the positive correlation between rainfall or constant flow duration and the eroded soil volume quantified by rill dimensions. Under a constant rainfall intensity, the increase in rainfall duration suggests the increase of cumulative kinetic energy provided by rain drop compaction and the surface flow, which thus results in a larger eroded soil volume and mass of soil loss [41,42]. The increase of soil loss with slope has also been presented in many previous studies [2,15,43], which may be attributed to the decrease of soil particle stability as the slope increases [44].

In Figure 4, the slopes of the trend lines decreased from 0.22 to 0.04 as the slope gradient of the flume decreased from 30° to 5°, indicating that the effect of the topography on soil loss became less significant in cases of mild slopes. In other words, the effect of rainfall duration on the soil loss mass can be influenced by the slope, and thus a collective effect of the rainfall duration and the slope on the soil loss should be considered. We applied a multiple regression analysis to quantify the effects of rainfall duration, slope, and their product term (to account for the collective effect of rainfall duration and slope) on the measured/estimated soil loss in mass, and the equations are as follows:

$$SL_m = -0.023t + 0.018S + 0.009(S \times t) \qquad R^2 = 0.96 \qquad (2)$$

$$SL_e = 0.048t + 0.112S + 0.005(S \times t) \qquad R^2 = 0.96 \qquad (3)$$

where SL_m = measured soil loss in mass (kg); SL_e = estimated soil loss in mass (kg); S = slope of the flume (°); t = rainfall duration (min). In Equations (2) and (3), positive coefficients before the product terms ($S \times t$) suggest that the positive effects of rainfall duration on soil loss manifested as the slope increased.

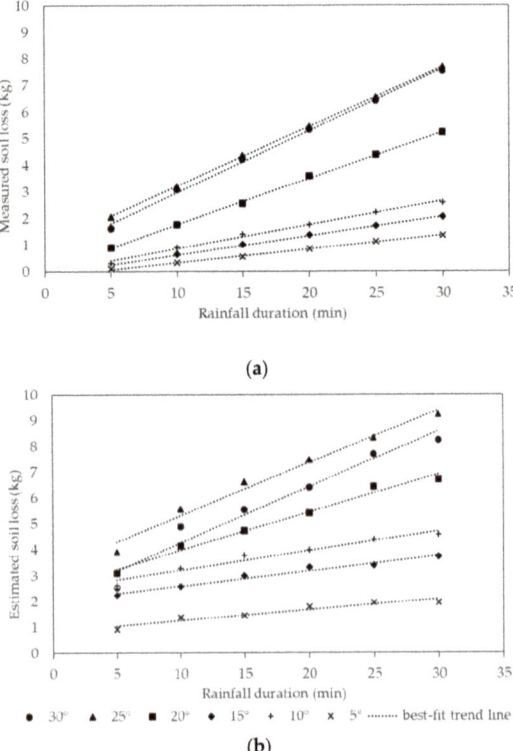

Figure 4. Cumulative mass of soil erosion obtained by: (a) the cumulative sediment yields as the measured soil loss; (b) the cumulative volume-differences soil loss as the estimated soil loss.

4.2. Relation between the Estimated and Measured Soil Loss

In Figure 5, data of the estimated soil loss, calculated based on the soil volume difference between two DEMs, are plotted versus the measured soil loss, which comprised the sediment yields collected during the experiments. The data points located above the 1:1 line suggest a systematic overestimation, which might occur within the first 5 min of rainfalls for all of the cases. Soil compaction and surface depression caused by the impact of rain drops might be the main reason for the overestimated soil loss. After the compaction of the surface soil stopped when the large soil pores collapsed, the soil volume difference increased as soil erosion propagated and more soil particles were eroded and transported to the flume outlet. The linear relation between the estimated soil loss (SL_e) and the measured soil loss (SL_m) shown in Equation (4) suggests a similar increasing rate in SL_e with SL_m by the trend line slope close to unity. The positive intercept in Equation (4) indicates an overestimation in soil loss resulting from soil compaction and effects of heterogeneity in soil structure and soil density or moisture content.

$$SL_e = 0.98 SL_m + 1.73 \qquad R^2 = 0.91 \qquad (4)$$

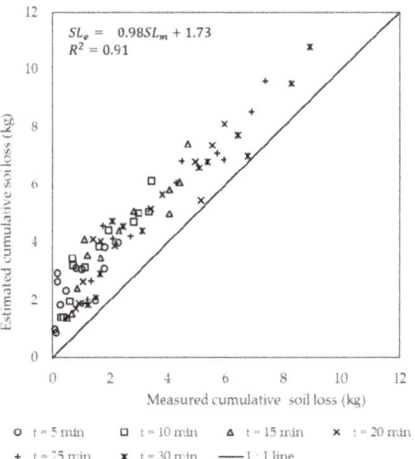

Figure 5. Estimated soil loss versus measured soil loss.

For application, the best-fit least square trend line is proposed by plotting the measured cumulative soil loss versus the soil volume differences subtracted from two DEMs, which were obtained by the stripe laser-scanned topographies, and the equation is:

$$SL_m = 1281.9V - 1.36 \qquad R^2 = 0.91 \qquad (5)$$

where V means the eroded soil volume (m^3) calculated as the soil volume differences subtracted from two DEMs. From Equation (5), the overestimation of soil volume differences due to soil compaction by rain drops was accounted for and offset by the negative intercept. Therefore, once an eroded soil volume can be obtained, the soil loss in mass may be predicted using Equation (5).

4.3. Erosion Processes and Rill Develoopllemt

Different types of soil erosion were reviewed, including splash erosion, sheet flow (interrill) erosion and rill erosion, and sediment transport of flow [45]. Based on the descriptions and characteristics of the soil erosion types [45], we classified the erosion processes for the case of the 30° slope erosion test shown in Figure 3. From the soil surface topographies in Figure 3, we quantified the estimated soil loss from soil volume differences for three cross sections in the flume following the descriptions of the laser-scanned topographies in Section 3.3, which are 45–70 cm (the rainfall region), 30–45 cm, and 0–30 cm from the outlet, to show the percentages of soil loss attributed to different erosion processes. From Table 3, the rain drop compaction and rain splash erosion that occurred in the rainfall area (45 cm to 70 cm from the outlet) accounted for 23% to 33% of the estimated total soil loss as the rainfall duration increased. However, the soil surface depression observed during the experiments suggested that the larger volumetric soil differences resulted from soil surface depression and compaction by the rain drops in addition to the actual soil loss by rain splash erosion; this is considered to be the main source of error for overestimations of the measured soil loss. The region of 30 cm to 45 cm from the outlet provided 28% to 33% of the estimated total soil loss. In this region, sheet erosion and interill erosion of the overland flow were observed as the dominant erosion processes during the first 10 min. Afterwards, rill erosion became more significant as rills developed. In the region close to the outlet ($x = 0$ cm to $x = 30$ cm), sediment entrainment by the concentrated flow in the rills was the dominant erosion process, and it provided 34% to 49% of the estimated total soil loss. While the transitions among different erosion processes were observed during the experiments and judged based on the laser-scanned topographies, more experiments designed for observing each erosion type and relative further studies are desired to validate the quantitative descriptions of the erosion process propagation.

Table 3. Soil loss percentage of regions with different distances from the outlet: results of the erosion test with a 30° slope.

Index	Soil Loss Percentage to the Total Soil Loss		
	Distance from the Outlet		
Rainfall Duration (min)	45–70 cm	30–45 cm	0–30 cm
5	23%	28%	49%
10	25%	32%	43%
15	29%	32%	39%
20	32%	31%	37%
25	33%	31%	36%
30	33%	33%	34%

From Figure 3, the development of a rill can be observed with the propagation of erosion processes and be described qualitatively. Specifically, the areas with dense contours suggest an incision in the soil surface and the development of rills. For the case of the 30° slope erosion test (Figure 3b), the initiation of the rill occurred at $x = 20$ cm to $x = 30$ cm where the concentrated flow was formed on the soil surface, which was around 20 cm apart from the lower boundary of the rainfall region (i.e., $x = 45$ cm), after 5 min of rainfall. Afterwards, the development of the dendritic structure indicated by the dense contours in Figure 3c–g suggested that the rill continuously developed upstream by headcutting towards the rainfall region as the rainfall duration increased. Meanwhile, the concentrated flow eroded the soil surface continuously towards the outlet and resulted in a deep and wide main stream in the region of $x = 0$ cm to $x = 30$ cm. Critical conditions for rill initiation were studied [46]. They proposed that headcutting occurred following rill formation and thus the upper ends of the rills were not the exact positions of rill initiation. While the rill development process observed in this study was similar to that reported in the literature [46], further research is required to obtain quantified evaluations of rill network development and the effects of topographic factors, such as slope.

5. Conclusions

In this study, erosion processes on a small-scale soil surface were observed and evaluated by erosion experiments using an adjustable slope flume, a rainfall simulator, and a stripe laser-scanning apparatus. From the results, the effects of rainfall duration, slope, and the combined influences of the two factors on the measured/estimated soil loss mass were quantified by Equations (2) and (3), suggesting that the positive effects on soil loss of rainfall duration were manifested as the slope increased. By comparing the measured and the estimated soil loss, Equation (4), with the slope near to unity and the R^2 value of 0.91, indicates good agreements between the soil loss measurements and the estimations. The stripe laser-scanning method was applied successfully to evaluate soil loss by subtracting the DEMs generated from scanned topographies. In practice, Equation (5) suggests that soil loss can be predicted by the soil volume differences between two DEMs given information of soil bulk density/moisture content and the soil compaction by rain drops.

From the laser-scanned soil surface topographies, the propagation of erosion processes and development of rills can be described qualitatively. From the case of the 30° slope erosion experiment, erosion processes, including rain drop compaction and rain splash erosion, sheet erosion, interrill erosion, and rill erosion, were observed. Based on the observation, the initiation of a rill occurred around 20 cm apart from the lower boundary of the rainfall region after 5 min of rainfall and progressed both upstream and downstream as the rainfall continued. This phenomenon coincides with the headcutting reported in the literature [46]. While more experiments and further studies are desired to obtain a whole picture of erosion process propagation and rill development during rainfall events, the results in this study provide: (1) quantified equations to predict soil loss by rainfall duration and slope and by eroded soil volume from DEM data; and (2) a potential method to quantitatively evaluate an erosion process in laboratory-scale experiments.

Author Contributions: Y.-C.W. conceived and designed the experiments; C.-C.L. performed the experiments; Y.-C.W. and C.-C.L. analyzed the data; Y.-C.W. wrote the paper.

Funding: The authors would like to thank the Ministry of Science and Technology, and the Soil and Water Conservation Bureau of the Republic of China, Taiwan, for financially supporting this research under the Contract No. 106-2625-M-005-013 and the Report No. SWCB-107-042, respectively.

Acknowledgments: The authors would like to thank Chi-Yao Hung, who is currently a faculty member in the Department of Soil and Water Conservation, National Chung Hsing University, Taiwan, and provided professional advice and technical support for the stripe laser-scanning method and DEM data transformation.

Conflicts of Interest: The authors declare no conflict of interest.

References

1. Shen, H.; Zheng, F.; Wen, L.; Lu, J.; Jiang, Y. An experimental study of rill erosion and morphology. *Geomorphology* **2015**, *231*, 193–201. [CrossRef]
2. Chen, X.-Y.; Zhao, Y.; Mi, H.-X.; Mo, B. Estimating rill erosion process from eroded morphology in flume experiments by volume replacement method. *Catena* **2016**, *136*, 135–140. [CrossRef]
3. Meyer, L.D.; Foster, G.R.; Romkens, M.J.M. Source of soil eroded by water from upland slopes. In *Present and Prospective Technology for Predicting Sediment Yields and Sources, Proceedings of the Sediment-Yield Workshop, Oxford, MS, USA, 28–30 November 1972*; pp. 177–189.
4. Mutchler, C.K.; Young, R.A. Soil detachment by raindrops. In *Present and Prospective Technology for Predicting Sediment Yields and Sources, Proceedings of the Sediment-Yield Workshop, Oxford, MS, USA, 28–30 November 1972*; pp. 113–117.
5. Zhang, P.; Tang, H.; Yao, W.; Zhang, N.; Xizhi, L.V. Experimental investigation of morphological characteristics of rill evolution on loess slope. *Catena* **2016**, *137*, 536–544. [CrossRef]
6. Boardman, J. Soil erosion science: Reflections on the limitations of current approaches. *Catena* **2006**, *68*, 73–86. [CrossRef]
7. Moreno-de Las Heras, M.; Espigares, T.; Merino-Martín, L.; Nicolau, J.M. Water-related ecological impacts of rill erosion processes in Mediterranean-dry reclaimed slopes. *Catena* **2011**, *84*, 114–124. [CrossRef]
8. Lei, T.W.; Zhang, Q.; Zhao, J.; Tang, Z. A laboratory study of sediment transport capacity in the dynamic process of rill erosion. *Am. Soc. Agric. Eng.* **2001**, *44*, 1537–1542.
9. Lawrence, W.; Gatto, L.W. Soil freeze-thaw-induced changes to a simulated rill: Potential impacts on soil erosion. *Geomorphology* **2000**, *32*, 147–160.

10. Dong, Y.Q.; Zhuang, X.H.; Lei, T.W.; Yin, Z.; Ma, Y.Y. A method for measuring erosive flow velocity with simulated rill. *Geoderma* **2014**, *232–234*, 556–562. [CrossRef]
11. Bewket, W.; Sterk, G. Assessment of soil erosion in cultivated fields using a survey methodology for rills in the Chemoga watershed, Ethiopia. *Agric. Ecosyst. Environ.* **2003**, *97*, 81–93. [CrossRef]
12. Aksoy, H.; Unal, N.E.; Cokgor, S.; Gedikli, A.; Yoon, J.; Koca, K.; Inci, S.B.; Eris, E. A rainfall simulator for laboratory-scale assessment of rainfall-runoff-sediment transport processes over a two-dimensional flume. *Catena* **2012**, *98*, 63–72. [CrossRef]
13. Zheng, F.L. A research on method of measuring rill erosion amount. *Bull. Soil Water Conserv.* **1989**, *9*, 41–49.
14. Casalí, J.; Loizu, J.; Campo, M.A.; De Santisteban, L.M.; Álvarez-Mozos, J. Accuracy of methods for field assessment of rill and ephemeral gully erosion. *Catena* **2006**, *67*, 128–138. [CrossRef]
15. Shen, H.; Zheng, F.; Wen, L.; Han, Y.; Hu, W. Impacts of rainfall intensity and slope gradient on rill erosion processes at loessial hillslope. *Soil Tillage Res.* **2016**, *155*, 429–436. [CrossRef]
16. Balaguer-Puig, M.; Marqués-Mateu, Á.; Lerma, J.L.; Ibáñez-Asensio, S. Estimation of small-scale soil erosion in laboratory experiments with Structure from Motion photogrammetry. *Geomorphology* **2017**, *295*, 285–296. [CrossRef]
17. Berger, C.; Schulze, M.; Rieke-Zapp, D.; Schlunegger, F. Rill development and soil erosion: A laboratory study of slope and rainfall intensity. *Earth Surf. Proc. Land.* **2010**, *35*, 1456–1467. [CrossRef]
18. Heng, B.C.P.; Chandler, J.H.; Armstrong, A. Applying close range digital photogrammetry in soil erosion studies. *Photogramm. Rec.* **2010**, *25*, 240–265. [CrossRef]
19. Nouwakpo, S.; Huang, C.-H.; Frankenberger, J.; Bethel, J. A simplified close range photogrammetry method for soil erosion assessment. In Proceedings of the 2nd Joint Federal Interagency Conference, Las Vegas, NV, USA, 27 June–1 July 2010.
20. Vinci, A.; Brigante, R.; Todiscoa, F.; Mannocchi, F.; Radicioni, F. Measuring rill erosion by laser scanning. *Catena* **2015**, *124*, 97–108. [CrossRef]
21. Afana, A.; Solé-Benet, A.; Pérez, J.L. Determination of soil erosion using laser scanners. In Proceedings of the 19th World Congress of Soil Science, Soil Solutions for a Changing World, Brisbane, Australia, 1–6 August 2010.
22. Longoni, L.; Papini, M.; Brambilla, D.; Barazzetti, L.; Roncoroni, F.; Scaioni, M.; Ivanov, V.I. Monitoring riverbank erosion in mountain catchments using terrestrial laser scanning. *Remote Sens.* **2016**, *8*, 241. [CrossRef]
23. Dąbek, P.B.; Patrzałek, C.; Ćmielewski, B.; Żmuda, R. The use of terrestrial laser scanning in monitoring and analyses of erosion phenomena in natural and anthropogenically transformed areas. *Cogent Geosci.* **2018**, *4*, 1437684. [CrossRef]
24. Thoma, D.P.; Gupta, S.C.; Bauer, M.E.; Kirchoff, C.E. Airborne laser scanning for riverbank erosion assessment. *Remote Sens. Environ.* **2005**, *95*, 493–501. [CrossRef]
25. Filin, S.; Baruch, A.; Morik, S.; Avni, Y.; Marco, S. Characterization of land degradation processes using airborne laser scanning. In Proceedings of the International Archives of the Photogrammetry, Remote Sensing and Spatial Information Science, Kyoto, Japan, 9–12 August 2010.
26. ASTM. *Standard Test Methods for Specific Gravity of Soil Solids by Water Pycnometer*; D854-14; ASTM: West Conshohocken, PA, USA, 2014.
27. ASTM. *Standard Test Method for Sieve Analysis of Fine and Coarse Aggregate*; ASTM C136/C136M-14; ASTM: West Conshohocken, PA, USA, 2015.
28. ASTM. *Standard Test Methods for Determining the Amount of Material Finer than 75-µm (No. 200) Sieve in Soils by Washing*; D1140-14; ASTM: West Conshohocken, PA, USA, 2014.
29. ASTM. *Standard Test Method for Particle-Size Analysis of Soils*; ASTM D422-63; ASTM: West Conshohocken, PA, USA, 2007.
30. ASTM. *Standard Test Methods for Laboratory Determination of Water (Moisture) Content of Soil and Rock by Mass*; D2216-10; ASTM: West Conshohocken, PA, USA, 2010.
31. ASTM. *Standard Practices for Preserving and Transporting Soil Samples*; D4220/D4220M-14; ASTM: West Conshohocken, PA, USA, 2014.
32. Eijkelkamp Soil & Water. Available online: https://en.eijkelkamp.com/products/field-measurement-equipment/rainfall-simulator.html (accessed on 3 July 2018).
33. Nciizah, A.D.; Wakindiki, I.I.C. Rainfall pattern effects on crusting, infiltration and erodibility in some South African soils with various texture and mineralogy. *Water SA* **2014**, *40*, 57–64. [CrossRef]

34. Salles, C.; Poesen, J.; Sempere-Torres, D. Kinetic energy of rain and its functional relationship with intensity. *J. Hydrol.* **2002**, *257*, 256–270. [CrossRef]
35. Hung, C.-Y.; Capart, H. Rotating laser scan method to measure the transient free-surface topography of small-scale debris flows. *Exp. Fluids* **2013**, *54*, 1544. [CrossRef]
36. Hung, C.-Y. Relation between Debris Flow Rheology and Fan Deposit Morphology Investigated Using Small-Scale Experiments. Master's Thesis, National Taiwan University, Taipei, Taiwan, June 2011.
37. ASTM. *Standard Test Methods for Liquid Limit, Plastic Limit, and Plasticity Index of Soils*; D4318-10; ASTM: West Conshohocken, PA, USA, 2010.
38. Bu, C.-F.; Wu, S.-F.; Yang, K.-B. Effects of physical soil crusts on infiltration and splash erosion in three typical Chinese soils. *Int. J. Sediment Res.* **2014**, *29*, 491–501. [CrossRef]
39. Di Prima, S.; Concialdi, P.; Lassabatere, L.; Angulo-Jaramillo, R.; Pirastru, M.; Cerdà, A.; Keesstra, S. Laboratory testing of Beerkan infiltration experiments for assessing the role of soil sealing on water infiltration. *Catena* **2018**, *167*, 373–384. [CrossRef]
40. Lin, L.-L.; Tung, H.-P.; Wan, H.-S. *Handbook of Soil Physics Experiments*; Department of Soil and Water Conservation, National Chung Hsing University: Taichung City, Taiwan, 2002.
41. Shen, H.O.; Zheng, F.L.; Wen, L.L.; Lu, J.; Han, Y. An experimental study on rill morphology at loess hillslope. *Acta Ecol. Sin.* **2014**, *34*, 5514–5521. [CrossRef]
42. Niu, Y.B.; Gao, Z.L.; Li, Y.H.; Luo, K.; Yuan, X.H.; Du, J.; Zhang, X. Rill morphology development of engineering accumulation and its relationship with runoff and sediment. *Trans. Chin. Soc. Agric. Eng.* **2016**, *32*, 154–161.
43. He, J.-J.; Lu, Y.; Gong, H.-L.; Cai, Q.-G. Experimental study on rill erosion characteristics and its runoff and sediment yield process. *Shuili Xuebao* **2013**, *44*, 398–405.
44. Chen, J.J.; Sun, L.Y.; Cai, C.F.; Liu, J.T.; Cai, Q.G. Rill erosion on different soil slopes and their affecting factors. *Acta Pedol. Sin.* **2013**, *50*, 281–288.
45. Liu, Q.Q.; Li, J.C.; Chen, L.; Xiang, H. Dynamics of overland flow and soil erosion (II): Soil erosion. *Adv. Mech.* **2004**, *34*, 493–506.
46. Yao, C.; Lei, T.; Elliot, W.J.; McCool, D.K.; Zhao, J.; Chen, S. Critical conditions for rill initiation. *Trans. ASABE* **2008**, *51*, 107–114. [CrossRef]

© 2018 by the authors. Licensee MDPI, Basel, Switzerland. This article is an open access article distributed under the terms and conditions of the Creative Commons Attribution (CC BY) license (http://creativecommons.org/licenses/by/4.0/).

Article

Effects of Climate Change and Human Activities on Soil Erosion in the Xihe River Basin, China

Shanshan Guo, Zhengru Zhu and Leting Lyu *

College of Urban and Environment, Liaoning Normal University, Dalian 116029, China; shanshanguo126@126.com (S.G.); Zhengruzhu@gmail.com (Z.Z.)
* Correspondence: lvleting@lnnu.edu.cn; Tel.: +86-411-8425-8364

Received: 27 June 2018; Accepted: 13 August 2018; Published: 15 August 2018

Abstract: Climate change and human activities are the major factors affecting runoff and sediment load. We analyzed the inter-annual variation trends of the annual rainfall, air temperature, runoff and sediment load in the Xihe River Basin from 1969–2015. Pettitt's test and the Soil and Water Assessment Tool (SWAT) model were used to detect sudden changes in hydro-meteorological variables and simulate the basin hydrological cycle, respectively. According to the simulation results, we explored spatial distribution of soil erosion in the watershed by utilizing ArcGIS10.0, analyzed the average soil erosion modulus by different types of land use, and quantified the contributions of climate change and human activities to runoff and sediment load in changes. The results showed that: (1) From 1969–2015, both rainfall and air temperature increased, and air temperature increased significantly ($p < 0.01$) at 0.326 °C/10 a (annual). Runoff and sediment load decreased, and sediment load decreased significantly ($p < 0.01$) at 1.63×10^5 t/10 a. In 1988, air temperature experienced a sudden increased and sediment load decreased. (2) For runoff, R^2 and Nash and Sutcliffe efficiency coefficient (*Ens*) were 0.92 and 0.91 during the calibration period and 0.90 and 0.87 during the validation period, for sediment load, R^2 and *Ens* were 0.60 and 0.55 during the calibration period and 0.70 and 0.69 during the validation period, meeting the model's applicability requirements. (3) Soil erosion was worse in the upper basin than other regions, and highest in cultivated land. Climate change exacerbates runoff and sediment load with overall contribution to the total change of −26.54% and −8.8%, respectively. Human activities decreased runoff and sediment load with overall contribution to the total change of 126.54% and 108.8% respectively. The variation of runoff and sediment load in the Xihe River Basin is largely caused by human activities.

Keywords: climate change; human activities; soil erosion; SWAT model; Xihe River Basin

1. Introduction

Soil erosion can be differentiated into natural erosion, which is a geographical phenomenon affecting the surface of the Earth, and human activities-accelerated erosion. The speed of soil erosion arising from human activities significantly exceeds the rate of soil formation. Soil erosion leads to the loss of soil nutrients, decline in soil fertility [1], deterioration of the ecological environment, and changes to the ecological landscape [2]. Soil erosion can also result in the accumulation of sediment in river channels, rising of river beds, sediment accumulation in reservoirs, and increased risk of flooding [3]. The dangers of soil erosion are compelling, and it is among the most significant global environmental concerns [4,5]. Scholars have conducted extensive research in various regions around the world with the aim of distinguishing the factors that influence soil erosion and developing new control methods [6–14]. The Soil and Water Assessment Tool (SWAT) [15,16] is a large-scale watershed hydrological model developed by the Agricultural Research Center of the US Department of Agriculture. The model contains three sub-models, namely the hydrological process, soil erosion, and pollution load and can simulate runoff and sediment load, nonpoint source pollution, and pesticide diffusion [17]. The tool is widely applied today in studying the water and sediment simulation and soil erosion of the basin [18–23]. Using the SWAT model, Wu et al. [24] quantified the impacts of climate change and human activities on runoff and sediment in the Yanhe River Basin and found that climate change and human activities are the main factors affecting runoff and sediment, respectively. Alighalehbabakhani et al. [25] used the SWAT model to evaluate sediment accumulation rates within the Lake Rockwell and Ballville reservoirs and concluded that sediment accumulation efficiency is closely related to the characteristics of the inflow, reservoir inflow time, and reservoir size. Yen et al. [26] built the Arroyo Colorado watershed SWAT model and studied the effects of the four different sediment transport functions to predict sediment. They demonstrated that without complete observation data, four sediment transport equations might produce similar sediment yield and the default model might not necessarily simulate the best results. At present quantitatively distinguish the effects of climate change and human activities on runoff and soil erosion has attracted much attention from scholars and identifying the impacts of each factor has important applications value in controlling soil erosion [27].

China is one of the countries with the most serious soil erosion in the world. The soil erosion types are various, and the erosion process is complex. It mainly occurs in the Loess Plateau of the middle and upper reaches of the Yellow River, the hilly areas in the middle and upper reaches of the Yangtze River, and the Northeast Plain area. According to the second national remote sensing survey of soil erosion in 2001, the area of water and wind erosion in China is 3.569 million km^2, accounting for 37% of the land area. These phenomena seriously affect China's ecological environment construction, hindering social and economic development.

The continuously increasing population of the basin and the rapid development of companies such as Benxi Steel Group and Bei Steel Group along the Xihe River have increased domestic as well as industrial and agricultural water use, resulting in a reduction of runoff in the basin. In addition, the accelerated urbanization rate has increased the urban land area, the ground is hardened and the soil is not easily eroded. From 2006–2008, the Nanfen District of the Xihe River Basin implemented numerous control measures including automation of dams, flood dikes, and dredging, therefore sediment discharge from the upper stream into the channel was intercepted and cleared, resulting in a significant reduction in sediment load [28]. The Nanfen tailings pool has also achieved ecological benefits. The forest surface has increased, effectively decreased runoff and soil erosion in the tailings reservoir area. In 2012, to restore the natural environment of the Xihe River Basin, the Nanfen District implemented the "Blue Water Project" for Xihe River and its tributaries, including river closure and protection. A land area of 1299 ha of land was preserved, and 264 ha of forest were planted, which effectively reduced soil erosion and sediment transported downstream.

The overall objective of this study is to investigate changes in runoff and sediment load from climatic change and human activities in Xihe River Basin of Northeast China. The primary objectives are: (1) to identify abrupt change points in annual hydro-meteorological series from 1969–2015; (2) to apply a SWAT model in the Xihe River Basin; (3) to evaluate the impacts of climate change and human activities on runoff and soil erosion. The results can provide valuable reference information for controlling soil erosion in this basin.

2. Materials and Methods

2.1. Research Area Overview

The Xihe River Basin is in the eastern region of Liaoning Province (China), centered in Benxi City and is a tributary of the Taizihe River, which is located between $40°47'$–$41°16'$ N and $123°32'$–$123°59'$ E. The diagram of the research area is shown in Figure 1. Its source lies in Fengcheng County, Dandong City, Liaoning Province and the basin has a total area of 1047 km^2. It is the main source of water for industrial, agricultural, and residential consumption in Benxi City. The area is dominated by mountains ranging from 75 to 1157 m a.s.l (above sea level). The characteristic of the area is high lands in the east and south, and low lands in the center and northwest. The east belongs to the Changbai Mountains. The region is in the semi-humid and semi-arid monsoon climate zones. The months with the highest and lowest temperature are July and January respectively, with an annual average temperature and annual rainfall of 8.4 °C and 772.18 mm, respectively. There are five types of land uses in the basin: woodland, cultivated land, towns, water areas, and unused land. The soil types are Calcaric Regosols, Haplic Phaeozem, Haplic Luvisols, Gleyic Luvisol, and Eutric Cambisols.

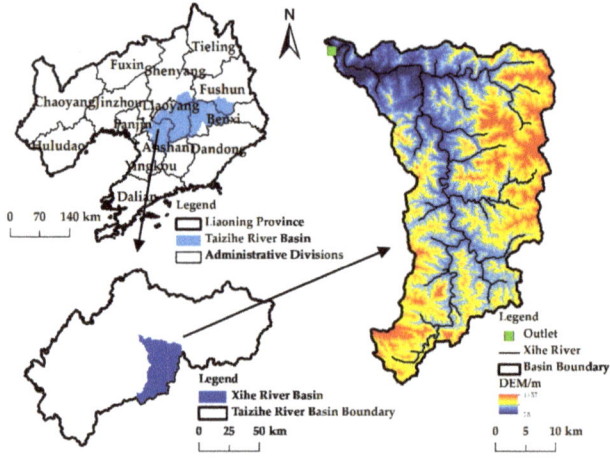

Figure 1. Research Area Overview.

2.2. Data Sources and Research Methodology

2.2.1. Data Sources

SWAT requires topographical, land use, soil, and meteorological inputs. Thematic layers and climatic data were developed using the sources specified in Table 1. The land use map and soil map are shown in Figure 2.

Table 1. Input data of Soil and Water Assessment Tool (SWAT) model in the Xihe River Basin.

Data Type	Data Input	Accuracy	Data Source
Spatial data	Digital Elevation Model (DEM)	30 m	Geospatial Data Cloud. http://www.gscloud.cn/ 16 August 2017
	Land use map	30 m	Geospatial Data Cloud Remote Sensing Map
	Soil type map	1000 m	Harmonized World Soil Database, HWSD
Meteorological data	Precipitation, Temperature, Wind, Solar radiation, Humidity	Daily 1969–2015	China Meteorological Data Center. http://data.cma.cn/ 2 October 2017
Hydrological data	Runoff, Sediment	Daily	Hydrological Yearbook
Soil data	Soil moisture density, Effective water content	1969–2015	Harmonized World Soil Database, HWSD

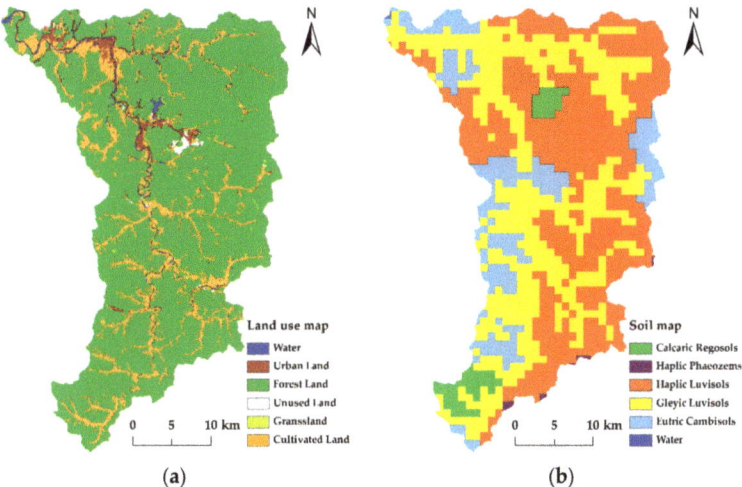

Figure 2. Land use map (**a**) and Soil map (**b**).

2.2.2. Analysis of the Changed Characteristics of Hydro-Meteorological Elements

In this study, linear regression analysis and the Pettitt sudden change point detection method [29] were used to analyze the inter-annual variation trends and sudden changes in rainfall, air temperature, runoff, and sediment load in the Xihe River Basin from 1969–2015. The Pettitt sudden change detection method is a non-parametric test that can quantify the statistically significant level of change points while varying the hydrological and meteorological factors, which has been widely used by scholars [30–32]. This method uses the Mann–Whitney statistical function $U_{t,N}$, and considers the samples $x_1, x_2, \ldots, x_t$ and $x_{t+1}, \ldots, x_N$ to be independently and identically distributed. N represents the sample size. $U_{t,N}$ is calculated as:

$$U_{t,N} = U_{t-1,N} + \sum_{i=1}^{N} \text{sgn}(x_t - x_i)(t = 2, \ldots, N) \qquad (1)$$

If time t is satisfied by $k_t = \max_{1 \leq t < N} |U_{t,N}|$ then t represents a sudden change point and the formula for the significant level of p of the change point is as follows:

$$p = 2\exp(\frac{-6k_t^2}{N^3 + N^2}) \qquad (2)$$

If $p \leq 0.05$ then the detected variations are considered statistically significant.

According to the sudden change analysis, the time series of runoff and sediment load from 1969–2015 were divided into two periods. The period prior to the sudden change was considered as 'the baseline period' and used as the calibration period. The period after the sudden change was considered as 'the impacted period' and used as the validation period [33–35].

2.2.3. SWAT Model Set Up

The SWAT model is a physically based distributed model that can evaluate the impact of land management practices on water resources, sediments and agricultural chemical yields in large and complex basins. The major components of SWAT include climate, hydrology, sediment movement, crop growth, nutrient cycling and agricultural management. SWAT model divide the basin into numerous sub-basins based on topography, and then further subdivided into a series of Hydrologic Response Units (HRU) with unique soil, land cover and slope characteristics [16].

In this study, we chose the Soil Conservation Service curve number (SCS-CN) procedure that SWAT model provides [36] to estimate surface runoff. SWAT calculates the peak runoff rate using a modified rational method. The potential evapotranspiration is estimated by using the Penman–Monteith method [37]. It is calculated by the following formula:

$$\lambda E = \frac{\Delta \cdot (H_{net} - G) + \rho_{air} \cdot C_p \cdot (e_z^o - e_z)/r_a}{\Delta + \gamma \cdot (1 + \frac{r_c}{r_a})} \qquad (3)$$

where λE is latent heat flux density (MJ/(m^2·d)); E is evaporation rate (mm/d); Δ is saturation vapor pressure-the slope of the temperature curve; H_{net} is net radiation (MJ/(m^2·d)); G is ground heat flux density (MJ/(m^2·d)); ρ_{air} is air density (kg/m^3); C_p is specific heat at fixed air pressure (MJ/kg ·°C); e_z^o is saturated vapor pressure at z height (kPa); e_z is the vapor pressure at z height (kPa); γ is psychrometric constant (kPa/°C); r_c is the impedance of the vegetation canopy (s/m); r_a is the diffusion impedance of the air layer (s/m).

The SWAT model uses the Modified Universal Soil Loss Equation (MUSLE) to simulate sediment yield [38]. Sediment transport is a function of deposition and degradation, which are determined through comparing the sediment concentration and maximum sediment concentration [39]. Further details of hydrological and sediment transport processes can be found in the SWAT Theoretical Documentation [39].

We used the ArcSWAT 2009 interface to establish and parameterize the model. First, we chose a threshold drainage area of 22.13 km^2, which is recommended by SWAT model, to delineate the basin. And the basin was divided into 27 sub-basins. Furthermore 960 HRUs were generated according to the thresholds of land use (0%), soil type (0%), and slope class (0%), respectively. The HWSD database provides partial physical attribute data for the soil, the effective field water content (SOL_AWC) and other attributes were calculated using the soil water characteristic software "SPAW". The minimum infiltration rate was calculated according to the empirical formula introduced in literature [40], and the soil erosion factor USLE_K was calculated according to the basic principle of SWAT [39]. Finally, we set up the SWAT model using daily meteorological data from 1969–2015, and calibrated the parameters using runoff and sediment load data of the baseline period, therefore the natural rainfall-runoff process was simulated.

2.2.4. Model Sensitivity Analysis, Calibration and Validation

There are many parameters in SWAT model that affect the results of runoff and sediment load simulations. In order to select the main parameters to improve the usability of the model, the sensitivity analysis of the parameters should be carried out first. We use the sequential uncertainty fitting algorithm (SUFI-2) algorithm [41] in SWAT-CUP 2012 to carry out sensitivity analysis. The 'baseline period' was used as the calibration period, and the 'impacted period' was used as the validation period. The model performance was assessed by the coefficient of determination (R^2) and Nash and Sutcliffe efficiency coefficient (Ens) [42]. R^2 is the square of the correlation coefficient, and the closer it is to 1, the higher the degree of agreement between the simulated value and the observed value is. Ens is a statistical measure that determines the relative magnitude of the residual variance compared to the measured data variance, and the closer it is to 1, the closer the simulated value is to the observed value. When $R^2 \geq 0.6$ and $Ens \geq 0.5$, the simulation results may be deemed satisfactory [43].

2.2.5. Quantitative Analysis of the Effects of Climate Change and Human Activities on Runoff and Sediment load

Runoff and sediment load are simultaneously affected by climate change and human activities, the total changes in runoff and sediment load is the difference between the baseline period and the impacted period of the observed value, and was calculated as follows:

$$\Delta Q_t = Q_v - Q_b \qquad (4)$$

where ΔQ_t represents total change, while Q_v and Q_b represent the observed annual average values of the baseline period and the impacted period, respectively. Change in total runoff and sediment load can be separated into climate change and human activities, as follows:

$$\Delta Q_t = \Delta Q_c + \Delta Q_h \tag{5}$$

where ΔQ_c represents hydrological changes attributable to climate change, and ΔQ_h represents hydrological changes attributable to human activities, calculated as follows:

$$\Delta Q_c = Q_{vs} - Q_{bs} \tag{6}$$

where Q_{vs} represents the SWAT model simulated baseline period average annual values, and Q_{bs} represents the SWAT model simulated impacted period annual average values. The difference between Q_{vs} and Q_{bs} represents the amount of runoff and sediment load attributable to climate change. According to Equation (5), the difference between ΔQ_t and ΔQ_c is the amount of change attributable to human activities. The percent contributions of change in runoff and sediment load attributable to climate change and human activities can be calculated as follows:

$$P_c = \frac{\Delta Q_c}{\Delta Q_t} \times 100\% \tag{7}$$

$$P_h = \frac{\Delta Q_h}{\Delta Q_t} \times 100\% \tag{8}$$

3. Results

3.1. Temporal Trends and Sudden Changes Analysis of Hydrological and Meteorological Variables

3.1.1. Inter-Annual Variation Trends in Hydrological and Meteorological Variables

The variation trends of annual precipitation, temperature, runoff, and sediment load in the Xihe River Basin from 1969–2015 can be found in Figure 3a,d. In the basin, the annual precipitation showed an insignificant increasing trend and the annual runoff showed an insignificant decreasing trend. The annual average precipitation and runoff was 772.18 mm and 3.16×10^8 m^3, respectively. The average annual temperature was 8.4 °C and annual temperature showed a significantly increasing trend ($p < 0.01$) at a rate of 0.326 °C/10 a. The annual average sediment load was 5.02×10^5 t and annual sediment load showed a significantly decreasing trend ($p < 0.01$) at a rate of 1.63×10^5 t/10 a.

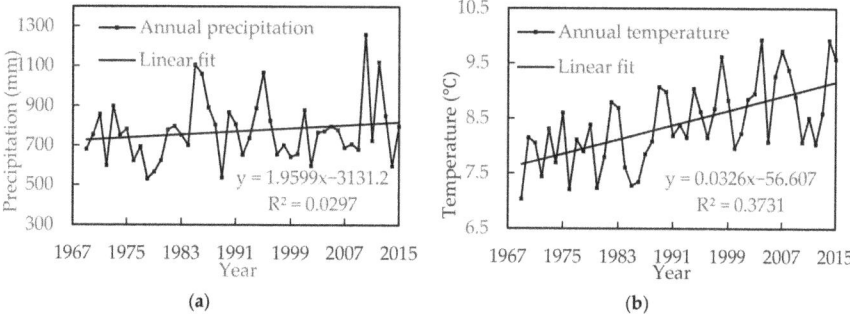

Figure 3. *Cont.*

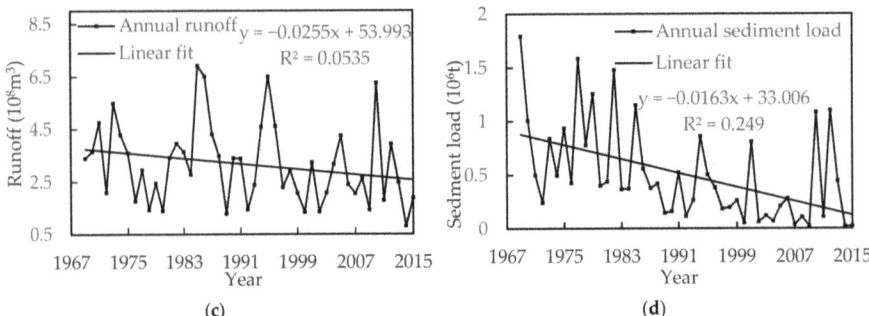

Figure 3. Inter-annual variation trend of the annual (**a**) precipitation, (**b**) temperature, (**c**) runoff and (**d**) sediment load from 1969–2015 in the Xihe River Basin.

3.1.2. Hydrological and Meteorological Variables Sudden Changes Analysis

The results of the Pettitt sudden changes analysis of annual precipitation, temperature, runoff, and sediment load in the Xihe River Basin from 1969–2015 can be found in Figure 4a–b. The results showed no sudden changes in precipitation and runoff. The results of the temperature tests $|U_{t,N}|$ showed that the largest year was 1988 ($p < 0.01$). Prior to the change point, the annual average temperature was 7.88 °C, and afterwards it was 8.78 °C. Following the change point, the annual average temperature increased by 11.56%. Sediment load in 1988 also experienced a sudden change ($p < 0.05$). Prior to the change point, the annual average sediment load was 7.89×10^5 t and after the change point was 3.75×10^5 t. After the change point, the annual average sediment load decreased by 52.47%. To study the effects of climate change and human activities on runoff and sediment load, the study period was divided into two periods, the baseline period 1971–1987 and the impacted period 1988–2015.

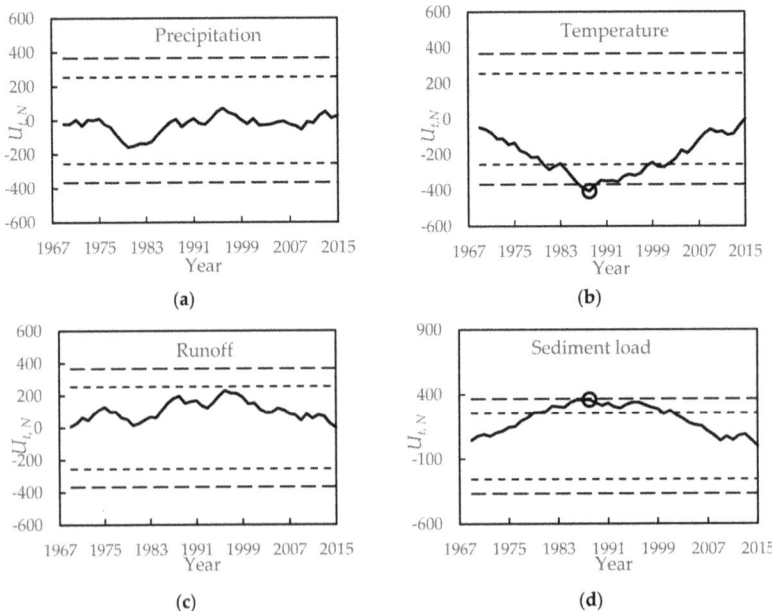

Figure 4. Results of sudden changes in annual (**a**) precipitation, (**b**) temperature, (**c**) runoff and (**d**) sediment load from 1969–2015 in the Xihe River Basin.

3.2. SWAT Model Simulation Results

The results of parameters sensitivity analysis for runoff and sediment load are showed in Table 2. The most sensitive parameters for runoff and sediment load are the base flow alpha factor (ALPHA_BF) and sediment movement linear compensation coefficient (SPCON), respectively. What needs illustration is that the final range of USLE_P is 0.56–1; this corresponds to the farming practices in the Xihe River Basin. In this watershed, the cultivated land is mainly distributed along the river. Contour tillage with conservative farming methods is most common.

Table 2. Parameters sensitivity analysis results in the Xihe River Basin.

Parameter Name and Document Type	Parameter Description	Process	Rank	Beg. Range	End Range
v__ALPHA_BF.gw	Base flow α coefficient	Runoff	1	0.0~1.0	−0.13~0.62
r__CN2.mgt	Runoff curve parameter		6	−0.2~0.2	−0.06~0.23
v__SLSUBBSN.hru	Average grade length		9	10~150	14.03~22.41
v__ESCO.hru	Soil evaporation coefficient		4	0.0~1.0	0.06~0.69
r__SOL_K().sol	Saturated water conductivity coefficient		10	0.0~25	0.88~2.66
r__SOL_Z().sol	Soil depth		8	−0.5~0.5	0.10~0.31
v__CH_K2.rte	River effective water conductivity coefficient		2	−0.01~150	47.97~144.52
v__RCHRG_DP.gw	Water table permeation		3	0.0~1.0	0.26~0.79
v__EPCO.hru	Material transpiration coefficient		5	0.0~1.0	0.37~1.12
v__BIOMIX.mgt	Biological mix efficiency coefficient		7	0.0~1.0	0.03~0.68
v__SPCON.bsn	Sediment movement linear compensation coefficient	Sediment load	1	0.0001~0.01	0.0057~0.0097
v__CH_COV2.rte	River coverage coefficient		2	0.0~1.0	0.63~1.00
v__SPEXP.bsn	Sediment movement index coefficient		4	0.0~1.5	1.20~1.42
v__CH_ERODMO().rte	River erosion coefficient		3	0.0~1.0	0.75~1.00
v__USLE_P.mgt	Water and land conservation measure factors		5	0.0~1.0	0.56~1.00

Notes: Beg. Range shows beginning range; .gw shows groundwater hydrology; .rte shows river hydrology; .hru shows HRU conventional data entry; .mgt shows HRU management; .sol shows soil data entry; .bsn shows conventional basin parameters; v shows parameter changes by the specified value; r shows parameter changes by the original value, 1+ specified value.

According to the results of sudden change analysis, 1969–1987 and 1986–2015 were considered as the calibration and validation periods. The preheating period for both the calibration and validation periods was 2 years, and the runoff was calibrated first, followed by sediment load. The R^2 and Ens of the runoff simulation in the calibration period were 0.92 and 0.91, respectively, while the corresponding values of the validation period were 0.90 and 0.87 with high simulation accuracy. The R^2 and Ens for sediment load simulation in the calibration period were 0.60 and 0.55, respectively, the corresponding values of the validation period were 0.70 and 0.69, which simulation effect are satisfactory. The comparison of the simulated and observed values of the runoff and sediment load is shown in Figure 5a,b.

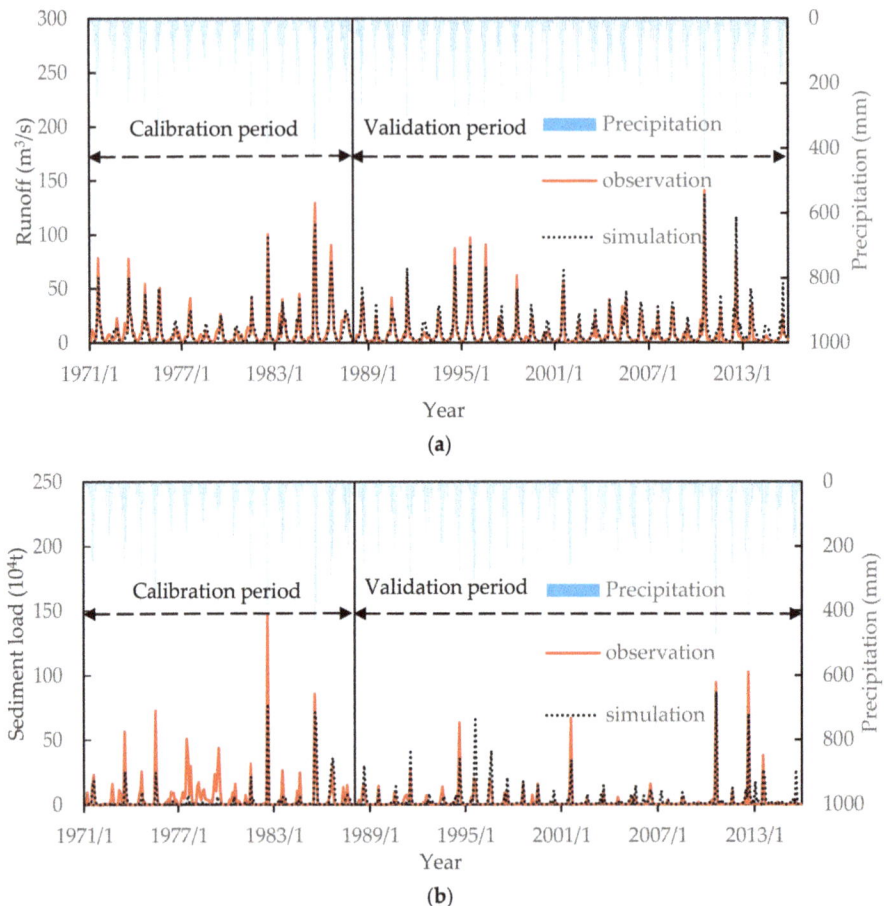

Figure 5. (**a**) runoff and (**b**) sediment load calibration and validation periods simulation results at Xihe River Basin during the period 1971–2015..

3.3. Xihe River Basin Soil Erosion Conditions

Figure 6 shows the spatial distribution of annual average soil erosion modulus in the Xihe River Basin. The study area lies in the northeast black soil region. According to the Classification Standard for Soil Erosion SL 190-2007 [44], the allowable soil loss in the region is 200 t/(km²·a). Soil erosion modulus under 200 t/(km²·a) is classified as micro erosion, and soil erosion modulus between 200 and 2500 t/(km²·a) is classified as mild erosion. As can be seen in Figure 6, only sub-basin regions 1, 2, 5, 7, 9, and 15 are within the allowable range of soil erosion and have an area of 279.063 km², occupying 25.23% of the study area. Other areas of sub-basin experience more erosion, representing an area of 826.937 km², occupying 74.77% of the study area. Furthermore, the most severe soil erosion is in sub-basin region 19 in the Xihe River Basin, which has an average soil erosion modulus of 595.4 t/(km²·a). Overall, the upper part of the basin has a higher erosion rate than the lower part.

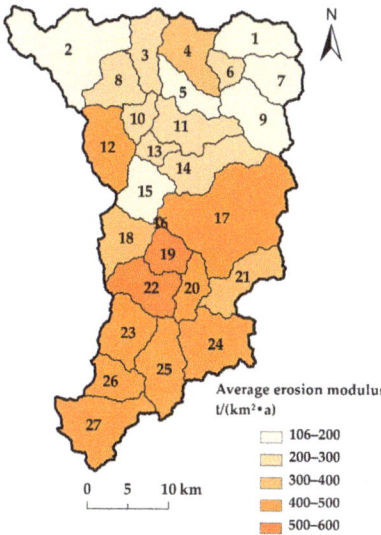

Figure 6. Soil erosion modulus spatial distribution in the Xihe River Basin.

Table 3 shows the annual average soil erosion volume and modulus by types of land use. Cultivated land exhibited the highest levels of erosion, 3.6561×10^5 t, followed by forest land, and lastly grassland. The highest levels of soil erosion modulus are also exhibited by cultivated land, 1856.56 t/(km²·a), followed by grassland, then urban land, unused land and lastly forest land.

Table 3. Xihe River Basin soil erosion by types land use.

Land Use Type	Area/km²	Soil Erosion Volume/10⁴ t	Average Soil Erosion Modulus (t·km^{-2}·a^{-1})
Cultivated land	194.5	36.11	1856.56
Forest land	865.06	1.11	12.83
Grassland	2.78	0.0099	35.61
Unused land	7.3	0.0158	21.64
Urban land	22.04	0.0633	28.72
Water	14.32	-	-

3.4. Effects of Climate Change and Human Activities on Runoff and Sediment Load

Table 4 shows the contributions of climate change and human activities on runoff and sediment load. Compared to the baseline period, annual average runoff decreased by 29.95 m³/s in the impact period. Climate change contributed an increase of 7.95 m³/s accounting for −26.54% of the total change, and human activities contributed a decrease of 37.9 m³/s accounting for 126.54% of the total change. Annual average sediment load decreased by 4.1×10^5 t in the impact period, to which climate change and human activities contributed an increase of 0.36×10^5 t (−8.8% of the total) and a decrease of 4.45×10^5 t (108.8% of the total), respectively. The overall decrease in runoff and sediment load can be attributed to human factors, which exceeded the effect of climate change and may be considered a main factor impacting the hydrological cycle of the Xihe River Basin.

Table 4. Relative contributions of climate change and human activities to runoff and sediment load.

	Period	Q_v	Q_s	ΔQ_t	ΔQ_c	ΔQ_h	$P_c(\%)$	$P_h(\%)$
Runoff (m³/s)	Calibration	136.94	119.62	−29.95	7.95	−37.9	−26.54	126.54
	Validation	106.99	127.57					
Sediment load (10⁵ t)	Calibration	7.18	3.52	−4.1	0.36	−4.45	−8.8	108.8
	Validation	3.08	3.88					

Notes: Q_v shows the calibration and validation period observed values, Q_s shows simulated values.

4. Discussion

Accuracy of runoff simulation of the SWAT model is better than sediment load simulation. The same phenomenon has also been found in other studies [45,46]. In the SWAT model, runoff and peak rate are the main input factors in estimating the sediment discharge equation, but when SWAT model estimates runoff, precipitation intensity, frequency and duration are neglected [47], which would lead to an underestimation of the runoff, finally, the sediment load simulation results in large differences.

The average soil erosion modulus of the Xihe River Basin is 334.06 t/(km²·a), which is similar to the counties and cities in Chaoyang city in the Liaoning Province, the results of which were studied by Yi Kai et al. [48] using the RUSLE (Revised Universal Soil Loss Equation) model. However, our result is lower than that study. The difference is not only because of the difference between climatic conditions and human activities in these two regions, but is also related to the different methods applied. In SWAT, soil erosion is calculated by the USLE (Universal Soil Loss Equation) model. Firstly, the USLE model only considers the changes in rainfall erosivity factor, and does not consider other changes in related to soil erosion, such as the vegetation coverage, the RUSLE model is more complete. Secondly, the RUSLE model obtained parameters through observing or taking similar empirical values, such as soil and water conservation measures factor and vegetation cover factor, without calibrating and validating. The SWAT model selected parameters by sensitivity analysis and calibrated these parameters with observed runoff and sediment load. Then soil erosion was calculated using the best simulated value.

The method of quantifying the contribution rate of climate change and human activities to runoff and sediment load in changes is simple, the disadvantage of it is that it cannot reflect the changes of runoff and sediment load caused by the changes of soil type, topography and geomorphology. During the study period, the contribution rates of human activities to runoff and sediment load reduction were 126.54% and 108.8%, respectively, which were far greater than the impact of climate change. This result is similar to the Jinghe River Basin, an arid inland basin in northwest China [49]. Climate change usually impacts runoff variation periodically and lastingly, whereas human activities impacts runoff variation suddenly and directionally [50]. The results could be useful for understanding the variation and driving factors for runoff and sediment load reduction in the basin.

5. Conclusions

This study analyzed the inter-annual validation trends and sudden changes in hydrological and meteorological variables in the Xihe River Basin from 1969–2015. The SWAT model was used to simulate the hydrological cycle and the spatial distribution characteristics of soil erosion. The study quantified the contributions of climate change and human activities to runoff and sediment load in changes. Which are summarized as follows:

(1) During the period 1969–2015, annual rainfall and temperature exhibited an increasing trend. Whereas annual runoff and sediment exhibited a decreasing trend. A significant trend was observed in the annual temperature and sediment load. In 1988, the annual temperature exhibited a sudden increase, but sediment load showed a sudden decline.
(2) The evaluating indicators showed that the R^2 and Ens of the SWAT model rose to an acceptable level after calibration, which indicated that the hydrological cycle of the Xihe River Basin could

be accurately simulated using the SWAT model. The simulation effect of runoff was better than that of the sediment load.

(3) Soil erosion in the upper part of the Xihe River Basin was more serious than that of the lower part, and the soil erosion modulus for cultivated land was highest. The contribution of human activities to runoff and sediment load changes was greater than that of climate change. Therefore, human activities is the primary factor that affects the hydrological cycle in the Xihe River Basin. In addition, it should be noted that the upstream should be paid more attention. Finally, the specific human activities that affect runoff and sediment load, as well as their modes of action, require further study.

Author Contributions: Conceptualization, S.G. and L.L.; Methodology, S.G. and L.L.; Software, S.G. and L.L.; Validation, L.L. and Z.Z.; Formal Analysis, S.G.; Investigation, S.G.; Resources, L.L. and Z.Z.; Data Curation, S.G.; Writing-Original Draft Preparation, S.G.; Writing-Review & Editing, L.L.; Visualization, S.G.; Supervision, L.L. and Z.Z.; Project Administration, L.L.; Funding Acquisition, L.L.

Funding: This research was funded by National Natural Science Foundation of China grant number 41701208.

Conflicts of Interest: The authors declare no conflicts of interest.

References

1. Pimentel, D. Soil Erosion: A Food and Environmental Threat. *Environ. Dev. Sustain.* **2006**, *8*, 119–137. [CrossRef]
2. Munro, R.N.; Deckers, J.; Haile, M.; Grove, A.T.; Poesen, J.; Nyssen, J. Soil landscapes, land cover change and erosion features of the Central Plateau region of Tigrai, Ethiopia: Photo-monitoring with an interval of 30 years. *Catena* **2008**, *75*, 55–64. [CrossRef]
3. Hu, S.; Cao, M.M.; Liu, Q.; Zhang, T.Q.; Qiu, H.J.; Liu, W.; Song, J.X. Comparative study on the soil conservation function of InVEST model under different perspectives. *Geogr. Res.* **2014**, *33*, 2393–2406. [CrossRef]
4. Qiao, Y.; Qiao, Y. Fast soil erosion investigation and dynamic analysis in the loess plateau of China by using information composite technique. *Adv. Space Res.* **2002**, *29*, 85–88.
5. Dotterweich, M. The history of human-induced soil erosion: Geomorphic legacies, early descriptions and research, and the development of soil conservation—A global synopsis. *Geomorphology* **2013**, *201*, 1–34. [CrossRef]
6. Hamel, P.; Chaplin-Kramer, R.; Sim, S.; Mueller, C. A new approach to modeling the sediment retention service (InVEST 3.0): Case study of the Cape Fear catchment, North Carolina, USA. *Sci. Total Environ.* **2015**, *524–525*, 166–177. [CrossRef] [PubMed]
7. Khanal, S.; Parajuli, P.B. Evaluating the Impacts of Forest Clear Cutting on Water and Sediment Yields Using SWAT in Mississippi. *J. Water Resour. Prot.* **2013**, *5*, 474–483. [CrossRef]
8. Li, T.; Liu, K.; Ma, L.Y.; Bao, Y.B.; Wu, L. Evaluation on Soil Erosion Effects Driven by Land Use Changes over Danjiang River Basin of Qinling Mountain. *J. Nat. Resour.* **2012**, *23*, 2249–2256. [CrossRef]
9. Nicu, I.C. Natural risk assessment and mitigation of cultural heritage sites in North-eastern Romania (Valea Oii river basin). *Area* **2018**. [CrossRef]
10. Chen, S.X.; Yang, X.H.; Xiao, L.L.; Cai, H.Y. Study of Soil Erosion in the Southern Hillside Area of China Based on RUSLE Model. *Resour. Sci.* **2014**, *36*, 1288–1297.
11. Golosov, V.; Koiter, A.; Ivanov, M.; Maltsev, K.; Gusarov, A.; Sharifullin, A.; Radchenko, I. Assessment of soil erosion rate trends in two agricultural regions of European Russia for the last 60 years. *J. Soils Sediments* **2018**. [CrossRef]
12. Panagos, P.; Borrelli, P.; Poesen, J.; Ballabio, C.; Lugato, E.; Meusburger, K.; Montanarella, L.; Alewell, C. The new assessment of soil loss by water erosion in Europe. *Environ. Sci. Policy* **2015**, *54*, 438–447. [CrossRef]
13. Labrière, N.; Locatelli, B.; Laumonier, Y.; Freycon, V.; Bernoux, M. Soil erosion in the humid tropics: A systematic quantitative review. *Agric. Ecosyst. Environ.* **2015**, *203*, 127–139. [CrossRef]
14. Keesstra, S.; Pereira, P.; Novara, A.; Brevik, E.C.; Molina, C.A.; Alcántara, L.P.; Jordán, A.; Cerdà, A. Effects of soil management techniques on soil water erosion in apricot orchards. *Sci. Total Environ.* **2016**, *551–552*, 357–366. [CrossRef] [PubMed]

15. Arnold, J.G.; Allen, P.M.; Bernhardt, G. A comprehensive surface-groundwater flow model. *J. Hydrol.* **1993**, *142*, 47–69. [CrossRef]
16. Arnold, J.G.; Srinivasan, R.; Muttiah, R.S.; Williams, J.R. Large area hydrologic modeling and assessment part I: Model development. *J. Am. Water Resour. Assoc.* **1998**, *34*, 73–89. [CrossRef]
17. Hajigholizadeh, M.; Melesse, A.; Fuentes, H. Erosion and Sediment Transport Modelling in Shallow Waters: A Review on Approaches, Models and Applications. *Int. J. Environ. Res. Public Health* **2018**, *15*, 518. [CrossRef] [PubMed]
18. Cornelissen, T.; Diekkrüger, B.; Giertz, S. A comparison of hydrological models for assessing the impact of land use and climate change on discharge in a tropical catchment. *J. Hydrol.* **2013**, *498*, 221–236. [CrossRef]
19. Abdelwahab, O.M.M.; Ricci, G.F.; De Girolamo, A.M.; Gentile, F. Modelling soil erosion in a Mediterranean watershed: Comparison between SWAT and AnnAGNPS models. *Environ. Res.* **2018**, *166*, 363–376. [CrossRef] [PubMed]
20. Aghakhani Afshar, A.; Hassanzadeh, Y. Determination of Monthly Hydrological Erosion Severity and Runoff in Torogh Dam Watershed Basin Using SWAT and WEPP Models. *Iran. J. Sci. Technol. Trans. Civ. Eng.* **2017**, *41*, 221–228. [CrossRef]
21. Hasan, Z.A.; Hamidon, N.; Yusof, M.S.; Ghani, A.A. Flow and sediment yield simulations for Bukit Merah Reservoir catchment, Malaysia: A case study. *Water Sci. Technol.* **2012**, *66*, 2170–2176. [CrossRef] [PubMed]
22. Xiao, J.C.; Luo, D.G.; Wang, Z.Z. Soil Erosion Simulation in Fuhe Basin Based on SWAT Model. *Res. Soil Water Conserv.* **2013**, *20*, 18–24.
23. Zuo, D.P.; Xu, Z.X.; Yao, W.Y.; Jin, S.Y.; Xiao, P.Q.; Ran, D.C. Assessing the effects of changes in land use and climate on runoff and sediment yields from a watershed in the Loess Plateau of China. *Sci. Total Environ.* **2016**, *544*, 238–250. [CrossRef] [PubMed]
24. Wu, J.W.; Miao, C.Y.; Yang, T.T.; Duan, Q.Y.; Zhang, X.M. Modeling streamflow and sediment responses to climate change and human activities in the Yanhe River, China. *Hydrol. Res.* **2018**, *49*, 150–162. [CrossRef]
25. Alighalehbabakhani, F.; Miller, C.J.; Selegean, J.P.; Barkach, J.; Sadatiyan Abkenar, S.M.; Dahl, T.; Baskaran, M. Estimates of sediment trapping rates for two reservoirs in the Lake Erie watershed: Past and present scenarios. *J. Hydrol.* **2017**, *544*, 147–155. [CrossRef]
26. Yen, H.; Lu, S.L.; Feng, Q.Y.; Wang, R.Y.; Gao, J.G.; Brady, D.; Sharifi, A.; Ahn, J.; Chen, S.T.; Jeong, J.; et al. Assessment of Optional Sediment Transport Functions via the Complex Watershed Simulation Model SWAT. *Water* **2017**, *9*, 76. [CrossRef]
27. Dong, L.H.; Xiong, L.H.; Yu, K.X.; Li, S. Research advances in effects of climate change and human activities on hydrology. *Adv. Water Sci.* **2012**, *23*, 278–285.
28. Yang, C.L.; Ma, X.P.; Hou, W.; Li, F.Y.; Liu, Q.; Li, Y.; Cheng, Z.H.; Kong, W.J. Soil Physical and Chemical Properties of Riparian Zone Along Xi River. *Sci. Technol. Rev.* **2012**, *30*, 61–66. [CrossRef]
29. Pettitt, A.N. A non-parametric approach to the change point problem. *J. R. Stat. Soc.* **1979**, *28*, 126–135. [CrossRef]
30. Zuo, D.P.; Xu, Z.X.; Wu, W.; Zhao, J.; Zhao, F.F. Identification of Streamflow Response to Climate Change and Human Activities in the Wei River Basin, China. *Water Resour. Manag.* **2014**, *28*, 833–851. [CrossRef]
31. Liu, M.F.; Gao, Y.C.; Gan, G.J. Long-Term Trends in Annual Runoff and the Impact of Meteorological Factors in the Baiyangdian Watershed. *Resour. Sci.* **2011**, *33*, 1438–1445.
32. Wang, J.; Gao, Y.; Wang, S. Assessing the response of runoff to climate change and human activities for a typical basin in the Northern Taihang Mountain, China. *J. Earth Syst. Sci.* **2018**, *127*, 1–15. [CrossRef]
33. Li, Y.Y.; Chang, J.X.; Wang, Y.M.; Jin, W.T.; Guo, A.J. Spatiotemporal Impacts of Climate, Land Cover Change and Direct Human Activities on Runoff Variations in the Wei River Basin, China. *Water* **2016**, *8*, 220. [CrossRef]
34. Li, P.F.; Mu, X.M.; Holden, J.; Wu, Y.P.; Irvine, B.; Wang, F.; Gao, P.; Zhao, G.J.; Sun, W.Y. Comparison of soil erosion models used to study the Chinese Loess Plateau. *Earth Sci. Rev.* **2017**, *170*, 17–30. [CrossRef]
35. Yin, J.; He, F.; Xiong, Y.J.; Qiu, G.Y. Effects of land use/land cover and climate changes on surface runoff in a semi-humid and semi-arid transition zone in northwest China. *Hydrol. Earth Syst. Sci.* **2017**, *21*, 183–196. [CrossRef]
36. USDA Soil Conservation Service (SCS). *National Engineering Handbook Section 4 Hydrology*; USDA: Washington, DC, USA, 1972.

37. Monteith, J.L. Evaporation and the environment. In *The State and Movement of Water in Living Organism*; Cambridge University Press: Swansea, UK, 1965.
38. William, J.R. Chapter 25: The EPIC model. In *Computer Models of Watershed Hydrology*; Singh, V.P., Ed.; Water Resources Publications, LLC: Highlands Ranch, CO, USA, 1995; pp. 909–1000.
39. Neitsch, S.L.; Arnold, J.G.; Kiniry, J.R.; Williams, J.R. *Soil and Water Assessment Tool Theoretical Documentation Version 2009*; Technical Report No. 406; Texas Water Resources Institute: College Station, TX, USA, 2011.
40. Che, Z.H. The empirical formulas and curves of soil permeability system number are discussed. *Water Resour. Hydropower Northeast China* **1995**, *19*, 17–19. [CrossRef]
41. Abbaspour, K.C.; Yang, J.; Maximov, I.; Siber, R.; Bogner, K.; Mieleitner, J.; Zobrist, J.; Srinivasan, R. Modelling hydrology and water quality in the pre-alpine/alpine Thur watershed using SWAT. *J. Hydrol.* **2007**, *333*, 413–430. [CrossRef]
42. Nash, J.E.; Sutcliffe, J.V. River flow forecasting through conceptual models part I-a discussion of principles. *J. Hydrol.* **1970**, *10*, 282–290. [CrossRef]
43. Moriasi, D.N.; Arnold, J.G.; Van Liew, M.W.; Bingner, R.L.; Harmel, R.D.; Veith, T.L. Model Evaluation Guidelines for Systematic Quantification of Accuracy in Watershed Simulations. *Am. Soc. Agric. Biol. Eng.* **2007**, *50*, 885–900.
44. SL 190-2007. *Classification Standard for Soil Erosion*; WaterPower Press: Beijing, China, 1977.
45. Zeng, Y.; Wei, L. Runoff and Sediment Simulation in Purple Hilly Area Based on SWAT Model. *J. Geo-Inf. Sci.* **2013**, *15*, 401–407. [CrossRef]
46. Pang, J.P.; Liu, C.M.; Xu, Z.X. Streamflow and Soil Erosion Simulation Based on SWAT Model. *Res. Soil Water Conserv.* **2007**, *14*, 88–93.
47. Assouline, S.; Ben-Hur, M. Effects of rainfall intensity and slope gradient on the dynamics of interrill erosion during soil surface sealing. *Catena* **2006**, *66*, 211–220. [CrossRef]
48. Yi, K.; Wang, S.Y.; Wang, X.; Yao, H.L. The Characteristics of Spatial-temporal Differentiation of Soil Erosion Based on RUSLE Model: A Case Study of Chaoyang City. *Sci. Geogr. Sin.* **2015**, *35*, 365–372.
49. Dong, W.; Cui, B.; Liu, Z.; Zhang, K. Relative effects of human activities and climate change on the river runoff in an arid basin in northwest China. *Hydrol. Process.* **2013**, *28*, 4854–4864. [CrossRef]
50. Miao, C.; Yang, Y.; Liu, B.; Gao, Y.; Li, S. Streamflow changes and its influencing factors in the mainstream of the Songhua River basin, Northeast China over the past 50 years. *Environ. Earth Sci.* **2011**, *63*, 489–499. [CrossRef]

© 2018 by the authors. Licensee MDPI, Basel, Switzerland. This article is an open access article distributed under the terms and conditions of the Creative Commons Attribution (CC BY) license (http://creativecommons.org/licenses/by/4.0/).

Article

The Role of Attenuation and Land Management in Small Catchments to Remove Sediment and Phosphorus: A Modelling Study of Mitigation Options and Impacts

Russell Adams [1,*], Paul Quinn [1], Nick Barber [2] and Sim Reaney [2]

1 School of Engineering, Newcastle University, Newcastle Upon Tyne NE1 7RU, UK; p.f.quinn@ncl.ac.uk
2 Department of Geography, Durham University, Durham DH1 3LE, UK; nick_barbie@hotmail.com (N.B.); sim.reaney@durham.ac.uk (S.R.)
* Correspondence: russell.adams@ncl.ac.uk; Tel.: +44-191-208-5773

Received: 22 August 2018; Accepted: 10 September 2018; Published: 12 September 2018

Abstract: It is well known that soil, hillslopes, and watercourses in small catchments possess a degree of natural attenuation that affects both the shape of the outlet hydrograph and the transport of nutrients and sediments. The widespread adoption of Natural Based Solutions (NBS) practices in the headwaters of these catchments is expected to add additional attenuation primarily through increasing the amount of new storage available to accommodate flood flows. The actual type of NBS features used to add storage could include swales, ditches, and small ponds (acting as sediment traps). Here, recent data collected from monitored features (from the Demonstration Test Catchments project in the Newby Beck catchment (Eden) in northwest England) were used to provide first estimates of the percentages of the suspended sediment (SS) and total phosphorus (TP) loads that could be trapped by additional features. The Catchment Runoff Attenuation Flux Tool (CRAFT) was then used to model this catchment (Newby Beck) to investigate whether adding additional attenuation, along with the ability to trap and retain SS (and attached P), will have any effect on the flood peak and associated peak concentrations of SS and TP. The modelling tested the hypothesis that increasing the amount of new storage (thus adding attenuation capacity) in the catchment will have a beneficial effect. The model results implied that a small decrease of the order of 5–10% in the peak concentrations of SS and TP was observable after adding 2000 m^3 to 8000 m^3 of additional storage to the catchment.

Keywords: runoff; suspended sediment; phosphorus; water quality modelling; mitigation measures; flooding

1. Introduction

It is becoming widely accepted that Nature Based Solutions (NBS—Nature Based Green Instructure Solutions) [1] and "Natural Flood Management" (NFM) (defined by the United Kingdom Environment Agency as part of Working with Natural Processes) [2] can have a positive impact in terms of reducing flooding, most observably by lowering the peak discharge of the flood hydrograph to enable this [3–7]. The construction of different types of "soft engineered" measures (or features) in headwater catchments has become an established part of this strategy [2,8]. Previous studies of the performance of features have concentrated primarily on the attenuation capabilities of these features in terms of reducing flooding, e.g., Belford Burn [9] and Pickering Beck [4,10] (in the U.K.). Moreover, the improvement in water quality (quantifiable by a reduction in concentrations and/or loads of nutrients and sediments) brought about by the construction of features in rural catchments has been studied in the U.K. [11–13], Irish Republic [14], and in New Zealand [15], but this issue has generally received less attention than the mitigating benefits of NFM in terms of reducing storm events.

Avery [16] coined the term "rural sustainable drainage systems" (RSuDS) to reflect the trend to construct wetlands and other types of features in rural catchments, and since the focus of this study is on the use of mitigation features to improve water quality rather than NFM per se, the term RSuDS henceforth is not used. The term "runoff attenuation feature" (RAF) has also been used in the literature [9,11]; essentially, an RSuD is a type of RAF that adds attenuation to a ditch or channel in which it is constructed [8]. However, the design purpose of RSuDs by default leads to sediment trapping and P stripping and associated "buffering" of other chemicals and microbial pollutants. Therefore, their ability to store and hence attenuate larger flood flows is lower. Often RAFs are designed to target flood flows primarily and thus the ideal solution is to both trap sediment and attenuate larger flow. This study explores the key role of adding attenuation to a catchment to target both water quantity and quality issues.

Environmentalists are interested in spatial patterns because they are essential in the scaling-up from localised measurements to larger spatial scales in order to provide assessments of mitigation impacts on pollution at the catchment, regional, or national scale for policy purposes [17]. In terms of addressing the impacts of mitigation features, few studies have attempted to assess the effect of mitigation features at the catchment scale [18]. These impacts have often been monitored by local water sampling and the measurement of runoff entering and leaving the features. For example, water quality, sediments, and nutrients have commonly been measured by automatic water quality samplers in order to collect data on concentrations and (if flow was measured) loads [12]. However, an important research question still remains as to whether impacts measured at the experimental scale are observable downstream, where monitoring points are often located (e.g., U.K. Environment Agency weekly sampling sites). Longer term monitoring programs (e.g., the Irish Agricultural Catchments (IAC) programme [19] and the Demonstration Test Catchments (DTC) programme [20,21]) are required if the larger scale impacts of these features are to be detected, if this is indeed possible with existing monitoring networks, and to address climatic issues (e.g., floods and droughts). Thus, the evidence that shows the rewards of adding attenuation capacity to a catchment are needed by end users, which is one goal of the DTC programme [21].

It is known from measuring water levels to observe runoff events before and after their construction that mitigation features have the potential to add attenuation to ditches and/or headwater streams [4,5]. These features can either: (i) divert water from channels via draw-off structures to separate storage areas or disconnected channels (classified as "off-line"), or (ii) temporarily detain runoff using "in-line" interventions located within low-order streams and ditches [11]. In-line features involve direct intervention in the channel or ditch itself such as the creation of artificial barriers, such as large woody debris (LWD) and engineered log jams (ELJs), and can also be applied in combination with off-line features [22]. Off-line features include riparian buffer strips, swales (vegetated channels), and ponds [5,13,23].

In terms of sediments and nutrients (principally total phosphorus (TP) and suspended sediment (SS)), the evidence from the case studies [11,24] is that these mitigation features can trap significant quantities of particulates with attached, insoluble forms of P. However, Barber [11] stated that few studies [13,23,25] had addressed either the effectiveness of mitigation features at the catchment scale, and suggested that, in order to address the requirement for urgent action with respect to meeting water quality targets in U.K. agricultural catchments, further research was required.

The natural attenuation of SS and bound nutrients, including forms of nitrogen (N) and P in riparian channel systems, is less-widely studied; however, one U.S. study [26] did highlight an important ecosystem function where the channels retained N and P exported from row crop fields during baseflow conditions, thus preventing higher exports into the estuary downstream in their catchment located in South Carolina, USA.

The primary aim of the study is to address the impacts of land management by altering hydrological flow paths and the overall catchment attenuation capacity on flow rates and nutrient losses. The modelling study described below demonstrates whether the impact of adding mitigation features, i.e., RAFs, at the headwater scale can be observable further downstream at a larger measurement scale. This modelling allows an estimate of how much attenuation can be achieved and the corresponding loss of productive land that may be required. A secondary aim is to explore whether the chosen model can simulate improvements to land management in a catchment that are designed to reduce losses of sediment and P in surface runoff. A further important research question poses: "Are there any significant differences in the performance of different types of features?" This can be limited by the available data on the performance of mitigation features [11,24].

A modelling case study to pursue the above aims used data collected from a catchment-based field programme in northern England [11,24] to parameterise the Catchment Runoff Attenuation Flux Tool (CRAFT) [27,28]. Scenarios were considered where: (i) RAFs were simulated by the model (by adding attenuation storage and trapping sediment and associated particulate forms of P) in order to see if the first aim can be achieved by investigating the impact on both the runoff hydrographs and the time series of P and SS concentrations modelled at the catchment outlet and also at the outlet of a mitigated sub-catchment, and (ii) land management options are applied instead to the same sub-catchment in order to reduce surface runoff from fields in the catchment and associated losses of sediment and particulate forms of P in surface runoff.

2. Methods

The methodology underpinning this study can be summarized by the following steps:

1. Develop the CRAFT model (at Newcastle University, Newcastle Upon Tyne, UK) to simulate nutrient and sediment fluxes at the catchment scale and add the capability to attenuate these fluxes in the surface runoff [27,28].
2. Calibrate the CRAFT to the existing runoff, nutrient, and sediment data collected in October 2011–September 2012 and establish a baseline scenario [28].
3. Evaluate the performance of a series of demonstration RAFs in the Mitigation sub-catchment of the NBC (mass of sediment and P trapped) [24] and use the information to inform the future scenarios.
4. Run simulations of the NBC using CRAFT with additional attenuation to represent three scenarios of land management.
5. Interpret the results of these scenarios in terms of (i) local and catchment scale impacts on runoff, TP, and SS; and (ii) compare the modelled changes in TP and SS yield with the measured reductions from step 3.

2.1. Description of the CRAFT

The Catchment Runoff Attenuation Flux Tool (CRAFT) [27,28] was selected for the case study. In this study, use was made of the model's attenuation store, which attenuates the surface runoff generated from rainfall excess. The store is connected, therefore, to only one of the three runoff pathways in CRAFT [27]. The store mimics the physical processes of attenuation using the minimum information requirement [29] (MIR approach) in this case by utilizing a linear storage–discharge relationship [28]. The attenuation storage depth was calculated by the model at each timestep so the total volume of storage in the catchment was obtained by multiplying this by catchment area. It was necessary to specify the maximum storage depth in the model. If the depth in the store exceeded this value, then the excess runoff was added to the outflow from the store. The rate of drainage of the attenuation store was controlled by the K_{LAG} parameter; this is equivalent to the reciprocal of lag time $1/k$ in the storage discharge relationship for a linear routing model ($S = kQ$) [30].

In terms of definitions of various P species, the CRAFT can output the following forms of P: particulate (insoluble, unreactive) P (PUP), particulate (insoluble, reactive) P (PRP), soluble reactive P (SRP) in groundwater, and fast subsurface flow. Therefore, in terms of the modelled concentrations and fluxes the P forms are combined so that: TRP = PRP + SRP and TP = TRP + PUP.

2.2. Description of Case Study

The catchment modelled in the case study (Newby Beck) is one of the three instrumented during the EdenDTC project, located in the River Eden catchment in northwest England [20,31–33]. It contains both control and Mitigation sub-catchments of similar sizes (approximately 2 km^2) out of a total area of 12.5 km^2. Capitals will be used to denote the monitored Mitigation sub-catchment to distinguish it from the modelled one. The main source of sediment in the wider Eden catchment was found to be bank erosion (exacerbated by livestock accessing watercourses and causing poaching), and also caused by streams undercutting the banks leading to instability [34]. Reference [34] found that good vegetation cover ensured that erosion by overland flow entrainment was likely to be a minor source of suspended sediment in the channels; however, field observations made in the EdenDTC project [32] have identified field sources of suspended sediment that become active during runoff events (e.g., tracks and ditches that become transport pathways).

In terms of the potential for future construction of RAFs (primarily as RSuDS), the areas that need to be treated within a sub-catchment can by determined by either: (i) using the current guidelines in terms of what proportion of the catchment area should be treated (this area varies typically from at least 1% (as advocated in New Zealand [15]) for riparian wetland coverage in dairy catchments), and up to 5% (advocated by Quinn [5])); or (ii) calculating the volume of attenuation storage that is required in the catchment to achieve targeted water quality improvements. In Europe, the Water Framework Directive [35] has been responsible for setting water quality targets for rivers and lakes and such targets can be used to examine what improvements are required.

Figure 1 shows a map of the Eden catchment showing the Newby Beck catchment (NBC) and the location of the monitoring sites for rainfall and flow in the catchment (these have been monitored as part of the EdenDTC project). Contours at 20 m intervals depict the topography. The dashed line shows the outline of the Mitigation sub-catchment.

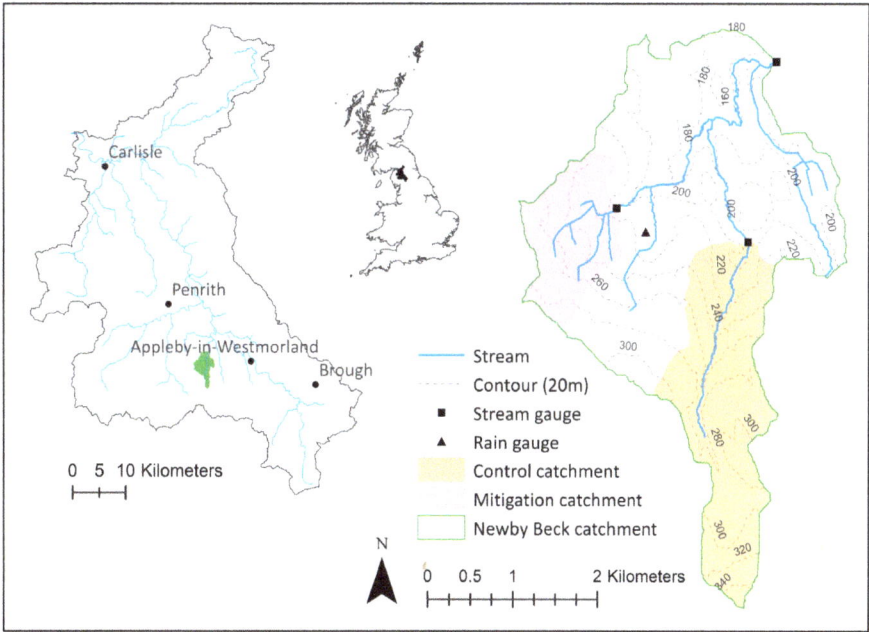

Figure 1. Map of the Eden catchment showing the NBC sub-catchment studied as part of the EdenDTC project, and the location of the monitoring sites for rainfall and flow in the sub-catchments. Contours at 20 m intervals depict the topography.

The monitoring data collected at the NBC outlet (i.e., the entire catchment) at Morland have been summarized previously [20,28,31,32]; the dataset comprises Q, turbidity plus TRP, and TP concentrations measured every 15–30 min but for modelling purposes these values have been converted to hourly values. Observed totals (runoff, rainfall depth and nutrient and sediment loads) are shown here in Table 1 for the time period of interest in this study (April–September 2012). Observed TP concentrations and calculated TP yields were based on samples taken by the bankside equipment and included a small component of dissolved unreactive or organic P (DUP), which was not modelled and assumed to be negligible. Predicted TP export (yields) data are also available from an export coefficient-based modelling study [36], and for the geoclimatic region including the NBC this predicted baseline TP export (year 2000) to be 1.39 kg·ha^{-1}·year^{-1} P, of which 52.5% originated from diffuse sources.

At both catchments, turbidity was measured at 15-min intervals using an YSI 6600 V2 multi-parameter sonde (YSI Incorporated, Yellow Springs, OH, USA). A strong relationship (from regression analysis) between turbidity and SS (the latter was measured from grab samples collected by an autosampler at the NBC outlet during storm events) was identified [31,32], enabling turbidity to act as a proxy for suspended sediment. These data were used for the calculation of an "observed" yield from the catchment for modelling purposes.

Table 1. Observations at the NBC April–September 2012 from monitoring data.

Observation	Value
Catchment Area (km^2)	12.5
Rainfall (mm)	686
Runoff (mm)	303
TP Yield (kg·ha^{-1})	0.73
SS Yield (t·km^{-2})	18.1
TP Load (kg)	908
SS Load (t)	229
TP mean Concentration (mg P·L^{-1})	0.077
SS mean Concentration (mg·L^{-1})	4.3

Nine mitigation features have been constructed in the Mitigation sub-catchment that target surface runoff from two farms located in the headwater area, with a combination of swales, small ponds, and ditches designed to intercept runoff from farm tracks and fields and divert this into the features [24]. The mass of sediment and nutrients trapped by the features has been calculated from the accumulated sediment and then analyzing the removed sediment for nutrient content. These data have been collected on an annual basis since late 2014 and allowed the loads to be estimated from five of the nine features.

In the NBC surface, runoff pathways (including ditches and drains) represent the major runoff pathway for exporting sediment attached P via sediment transport [31,32]. The best fit for a transfer function model for TP load from the NBC was a single store model with a sole quick flow pathway [33]. Therefore, this flow pathway was targeted by the mitigation features in order to develop a strategy that adds attenuation to the outlet hydrograph and load time series of P. Note that no mitigation features were in place in 2012, so the observed data represent baseline conditions prior to any intervention being made.

2.3. Mitigation Modelling Approach

The CRAFT has already been calibrated and validated on the NBC [28]. In this earlier study, several simulations of runoff and P were carried out in order to test different hypotheses of conceptual models for the entire NBC, these differed primarily in whether any attenuation of the surface runoff flow pathway was included in the model. The scenario chosen for use in this study as a baseline was the "lagged" [28] in which the modelled hydrographs have added attenuation representing the natural storage in the catchment during runoff events.

The modelling strategy followed the approach of using simulations where the volume of attenuation storage is selected a priori. Implementing the proposed measures would likely require terrain analysis to identify runoff pathways and Critical Source Areas (CSAs) in order to select suitable sites for constructing off-line features [29,37–39].

A series of curves, such as those shown in Figure 2 below, can be plotted that relate a representative set of outputs from the modelling (e.g., Q_p i.e., peak flow, or the load, or concentration, of TP or SS) to the degree of attenuation or storage in the catchment, where the left axis represents baseline conditions with no additional features present to add storage capacity. Therefore, the origin represents a minimum degree of attenuation and a small amount of storage from the existing floodplain and riparian areas.

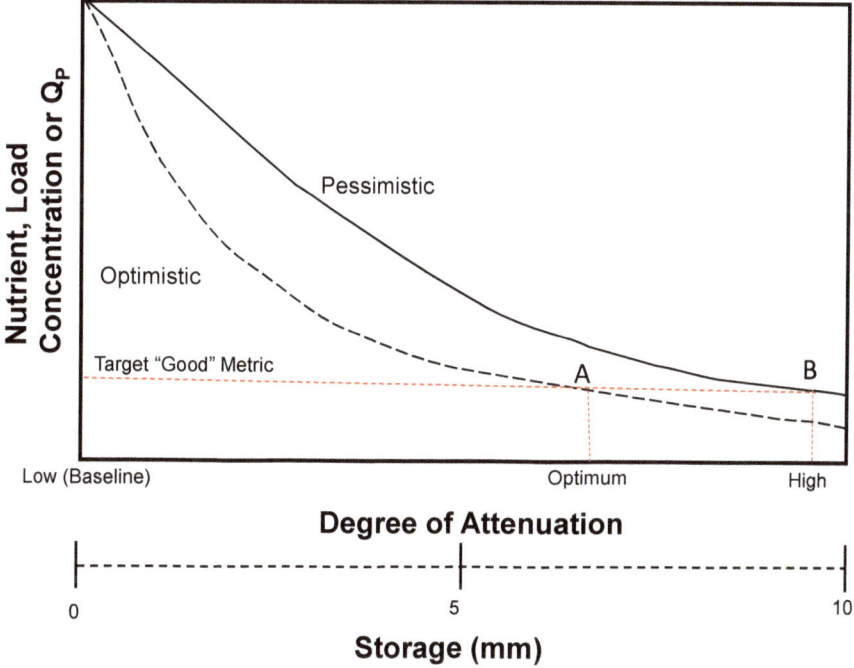

Figure 2. Sketch of relationship between load or concentration and degree of attenuation and storage (both plotted on the x-axis) in a hypothetical catchment indicating how adding storage can achieve a desired target.

Adding features can then be shown to add storage capacity and thus decrease the value of the output(s) until an optimum target is reached. In terms of a conceptual model of adding attenuation storage to a catchment, it may be expedient to consider the UK Water Framework Directive Targets (WFD-UKTAG) [40] for achieving reductions in TRP concentrations in order to improve the ecological status of the catchments (site-specific values were reported by Ockenden [32]). The two curves shown in Figure 2 could represent different types of RAF with a higher or lower optimum design storage capacity that is required to meet the required optima (indicated by the intersections of the curves and the dashed vertical red lines at points A and B). A similar set of curves relating particulate P (PP) load reduction at the catchment outlet to the managed proportion of the catchment for two large catchments in Austria and Hungary were developed [41] using the PhosFate model. The modelled load reductions were achieved by adopting best management practice (BMP) interventions over part of the catchments. In the U.K., a national scale modelling platform based on the export coefficient method simulated several scenarios of nutrient load reductions [36]. The scenarios relevant to the NBC were based on:

(i) on-farm mitigation measures, and (ii) farming practices modified to comply with WFD targets. In (i), the TP export from sheep was reduced by 50% and P fertilizer loads applied to grass and arable crops were reduced by 50%, and in (ii), TP loads from cattle farming were reduced by 25% in addition to the reductions in (i). These scenarios predicted that TP exports would reduce by 22.1% and 32.9% respectively (based on data from the year 2000).

The following model simulations investigated whether the target of reducing P and SS concentrations could be met by adding storage capacity. It could be argued that this added storage capacity is actually offsetting the loss of natural attenuation storage caused by deforestation and agricultural intensification. What the final target should be is debatable but the premise that more storage capacity gives a better status is required as part of a long-term plan to reach WFD-UKTAG [40] status, for example. Even as a basic estimate a target of 10 mm of new storage capacity over 1 km^2 would require 10,000 m^3. This would require a storage pond of 1 m depth with a surface area of 100 m by 100 m. Implementing NBS [1,2] would suggest that this could be spread throughout the catchment in an RAF network, and hopefully soil improvement and buffer zones would add to the storage capacity as well. It is proposed that ditch management is a primary basis for the first two scenarios, hence for 1 m of storage depth, a minimum of 1 km of ditch with an effective width of 10 m would be needed. To gain this storage capacity, a ditch would need to be widened and barriers to flow constructed; Figure 3 is an actual example of how a traditional narrow "V" shaped ditch can be substantially modified (Netherton Burn catchment [11]).

Figure 3. An example of a modified ditch with flow barriers acting as both a sediment trap and flood flow storage in Netherton Burn (photo from Barber 2013, credit to N. Barber).

CRAFT Mitigation Modelling

The three mitigation scenarios were modelled using the CRAFT and discussed below. Figure 4 shows sketches indicating both the flow pathways under each scenario (and the baseline) as well as the design of the RAFs themselves (in Scenarios 2 and 3). For simplicity, the area of the mitigated sub-catchment in the modelling was set to 10% of the NBC area (1.25 km^2), which was slightly smaller than the actual Mitigation sub-catchment (so this will be referred to as the "mitigated sub-catchment").

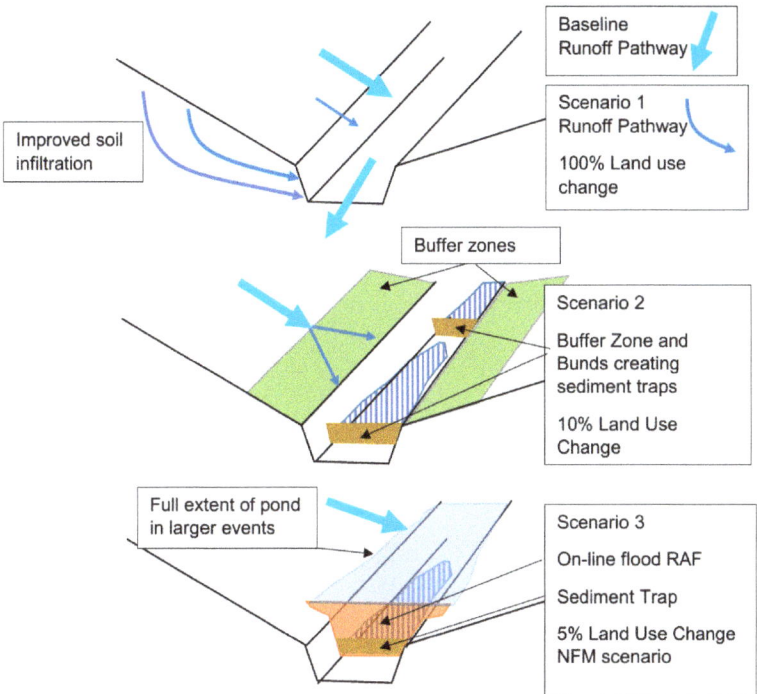

Figure 4. Schematic describing Runoff Pathways and design sketches of RAFs in Scenarios 1–3, and Runoff Pathways in the Baseline simulation. Note that Scenario 3 has a widened ditch eating into the buffer zone.

Scenario 1 (Figure 4 top pane): Managing the topsoil in agricultural areas (primarily grassland) to improve the soil health and reduce SS and P loads transported in surface runoff (indicated by the thick blue lines). A review of crop and soil management techniques [42] designed to reduce the quantities of P and SS lost to watercourses found that firstly in upland areas, one of the main mechanisms responsible for P and SS losses is erosion by overland flow detaching soil particles with attached P during rainfall events. Second, in arable fields, the following options can reduce P losses: reduced tillage strategies, application of soil conditioners, crop rotation, crop management, and planting either catch or cover crops. In grasslands, reducing stocking densities and preventing highly trafficked areas from developing (e.g., around feed lots or water troughs) in order to avoid poaching of the soil are the main options available to reduce P losses from the surface runoff. This scenario assumed that the entire mitigated sub-catchment was treated in this way by increasing the maximum infiltration capacity of the soil in the model and evaluated the results in terms of changes to P fluxes. It was assumed that some "pollution swapping" could take place between particulate and soluble forms of P, hence the total P loads may not reduce as much as expected before the "experiment" was carried out. This was essentially a 100% intervention to the land use management across the mitigated sub-catchment. The green lines indicate the major flow pathways under this scenario, which was dominated by infiltration and subsurface flow.

Scenario 2 (Figure 4 middle pane): Treating the mitigated sub-catchment's watercourses with riparian buffer strips by adding 2000 m^3 of additional storage to the sub-catchment and trapping 40% of the SS and attached P transported via surface runoff (overland flow). As a first estimate, it was assumed that up to 10% of the sub-catchment area (i.e., 12.5 ha) was set aside for these measures. This was a 10% land use change intervention option idealising the green/ecological corridor with minimum management other than sediment removal. As the soil was still degraded, overland flow was still dominant during events.

Scenario 3 (Figure 4 lower pane): Adding engineered RAFs to address food flows and to trap and remove 80% of the SS and attached P transported via the surface runoff (overland flow). These features were assumed to be a combination of offline storages and inline ditches that added 8000 m³ of additional storage to the sub-catchment. As a first estimate, it was assumed that 5% of the mitigated sub-catchment area (i.e., 6.25 ha) was set aside and modified for these measures [5]. This was a reduced area, but it required a widening of the ditch to give the increased volume of storage needed to cope with larger storm events. Essentially, the NFM component was being optimised here, and thus a higher level of maintenance may be needed to keep the total infrastructure operating at its optimum.

Scenarios 2 and 3 were modelled as follows: considering that there are two parameters that can be varied in the CRAFT to represent the attenuation and trapping, these are added by the features in the mitigation sub-catchment only, namely (i) K_{LAG} and (ii) a removal or trapping "efficiency" K_{RE}, that apply to the modelled pathways of (a) TP (i.e., PRP and PUP) and (b) SS transported by the surface runoff pathway. Their values were pre-selected for Scenarios 2 and 3 (see Table 2) based on expert judgement and evidence from the field experiments conducted in the mitigation sub-catchment.

Table 2. Details of the scenarios modelled in the NBC: parameter Values and storages (S). Baseline values are shown for comparison.

Scenario	K_{LAG} (h^{-1})	S (Total) (mm)	S (Total) (m³)	S (Added) (m³)	K_{RE} (-)	Other Parameters
Baseline	0.75	4.86	6075	0	0	No changes
1	0.75	4.86	6075	0	0	Increased Infiltration Capacity
2	0.83	6.46	8075	2000	0.4	No changes
3	0.93	11.3	14,075	8000	0.8	No changes

The parameter values used for the baseline were:

$K_{LAG} = 0.75$ h^{-1} (representing natural attenuation, from the "lagged" simulation [28]).
$K_{RE} = 0$ (for the baseline with no trapping).

The unmitigated portion of the catchment was modelled with CRAFT using the calibrated baseline parameter set [28]. There was no additional attenuation of flow, sediments, and nutrients, or removal simulated in this portion of the catchment, or in the fast subsurface and deep groundwater flow pathways in any of the scenarios anywhere.

In the CRAFT, nutrients (in this case PUP and PRP fluxes) and SS were modelled slightly differently from runoff in that their removal was permitted from the attenuation store, which represents the ability of the modelled features to remove (i.e., trap) particulates. Therefore, the mass balance Equations (1) to (3) for the stores become (for the components of PRP, PUP and SS transported by the surface runoff (SR) pathway, which is indicated by "SR" in parentheses):

$$PRPL(SR)_{out} = PRPL(SR)_{in} (1 - K_{LAG})(1 - K_{RE}) \tag{1}$$

$$PUPL(SR)_{out} = PUPL(SR)_{in} (1 - K_{LAG})(1 - K_{RE}) \tag{2}$$

$$SSL(SR)_{out} = SSL(SR)_{in} (1 - K_{LAG})(1 - K_{RE}) \tag{3}$$

where PRPL$_{out}$, PUPL$_{out}$, and SSL$_{out}$ were the loads per time step of PRP, PUP, and SS from the mitigated sub-catchments, respectively. CRAFT outputs were the specific discharge (i.e., runoff depth) and specific yield (i.e., load/unit area). Therefore, the model outputs were scaled up by the areas of the mitigated (A_{mit}) and unmitigated areas of the catchments (A_{umit}) to obtain the total flow and loads from the entire catchment. SRP loads from the fast subsurface and slow groundwater flow components were added to the loads from the surface runoff pathways in a mass balance to compute a total TP load. Concentrations of TP and SS at the catchment's or sub-catchment's outlet were calculated by dividing the total loads by the total flow. The flow from the catchment Q_{catch} was the sum of the flow components from both the mitigated and unmitigated areas, where the suffixes "mit" and "umit" denote these respectively

$$Q_{catch} = Q_{mit} + Q_{umit} \tag{4}$$

Nutrient (TP) and SS concentrations were calculated in the same way, where TPL, TRPL, and SSL were the loads with suffixes (as above) indicating which part of the catchment (suffix "catch") the load originated from

$$TPL_{catch} = TPL_{mit} + TPL_{umit} \tag{5}$$

where

$$TPL = PUPL(SR) + TRPL \tag{6}$$

$$SSL_{catch} = SSL_{mit} + SSL_{umit} \tag{7}$$

A simplified, additive mixing model was used to calculate the flows and loads at the outlet, with no additional attenuation added to represent the in-stream reaches (clearly this was a simplification of reality where in-stream routing could further attenuate the outlet hydrograph [43]). This assumption holds for small catchments where the main stem channel length is less than 10 km and the travel time (lag) between hillslope and outlet is of the order of a few hours. The functional unit defined by a CRAFT sub-catchment could equate to a representative elementary area (REA) [44].

Lastly, the maximum volumes of added storage (V_{add}) required in the mitigated sub-catchment were supplied by the user for each scenario. The area (A_{mit}) of the mitigated sub-catchment is set to 10% of the total catchment area (A_t). The model calculated the depth in the attenuation store at each timestep (D_{add}), as it works with depths rather than volumes. The attenuation store was empty at the start of the simulation. The maximum storage volume required in the attenuation store V_{att} could therefore be calculated using Equations (8)–(10).

$$A_{mit} = 0.1 A_t \tag{8}$$

$$V_{att} = V_{nat} + V_{add} \tag{9}$$

where, in general terms, at each timestep t:

$$V(t) = A_{mit} D(t) \tag{10}$$

Thus, the required model parameter, D_{max} was calculated using Equation (11):

$$D_{max} = V_{att} / A_{mit} \tag{11}$$

where D_{max} was the maximum depth of water in the attenuation store (m), V_{nat} was the modelled volume of natural storage in the catchment, which comes from the results of calibrated baseline model results (by extracting D_{max} from these).

Since the outflow Q_{mit} from the attenuation store was a function of K_{LAG}, the required value of this parameter could be back-calculated from Equation (12). In practice this was obtained by increasing the value of the parameter until the desired value of D_{max} was achieved in the attenuation store.

$$Q_{mit} = V_{att}(1 - K_{LAG}) \tag{12}$$

The results from these simulations provided the flows and loads of TP and SS from the mitigated sub-catchment (i.e., Q_{mit}, TPL_{mit} and SSL_{mit} in Equations (4)–(7)).

3. Results

The CRAFT had already been calibrated to baseline conditions during October 2011–September 2012 in the NBC [28]. In this study the parameter values from the "lagged" scenario, where natural attenuation were included, were used for the baseline simulation. The modelled runoff is shown in Figure 5 for comparison against the observed runoff, for the period January 1 to September 30, 2012 at the NBC outlet. Table 1 summarizes the observations recorded during the entire period between April and September 2012 including the three selected events. The SS yield of 18 t km^{-2} over a six-month period appears to fit into the middle range of estimates for the larger Eden catchment made at 14 monitoring points, which was 4–73 t km^{-2} year [34]. The lowest monitoring point measured yield from a 1373 km^2 sub-catchment of the River Eden [34].

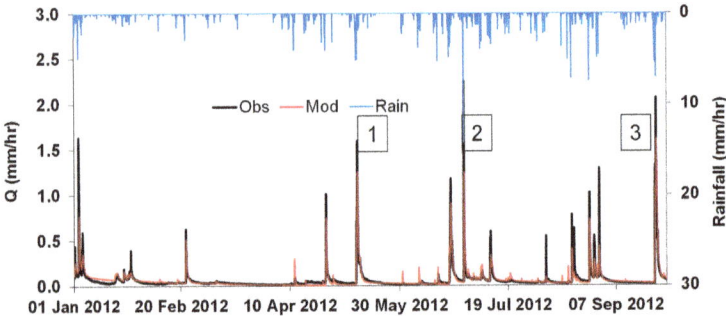

Figure 5. Time series plots of observed ("Obs") and modelled ("Mod") (baseline) runoff plus rainfall at the NBC outlet from January–September 2012.

3.1. Results from the 2012 Events (Baseline Simulation)

In terms of investigating the temporal scaling of the results, three large events during a wet 5-month period in 2012 were analysed, and these are denoted by the numbered boxes in Figure 5. According to the observations, the annual (2011–2012) exports of TP and SS from the NBC were 1762 kg and 477.5 tonnes respectively [28,32], and from the three events during 2012, the modelled TP load was predicted to be 725 kg and the modelled SS load was 322 tonnes from the baseline simulation, which indicates that these three events alone contributed a significant proportion of the annual P and sediment exports.

3.2. Results from Scenarios 1–3

Table 2 shows the model parameter values used in each of the scenarios along with the values used in the baseline simulation (for comparison). It also shows the total and added storage in the attenuation component of the model in Scenarios 1–3.

The left-hand pane of Figure 6 shows the modelled and observed runoff at the NBC outlet for Event 3. The modelled attenuation storage per unit time and area (expressed as a depth) is also shown by the solid black line, and both modelled runoff and storage are shown for the baseline case only. The right-hand pane of Figure 6 shows the modelled runoff from the mitigated sub-catchment for the (unmitigated) baseline (solid red line) and Scenarios 2 and 3 (dotted and dashed red lines). The black lines show the attenuation storage during the event for the three simulations (same line styles as runoff). Its value is shown on the right-hand axis and Scenario 3 had a far greater effect on the shape of the hydrograph than Scenario 2 due to the much greater additional added storage (8000 m^3 vs. 2000 m^3, which is the equivalent of storing an additional 6.4 mm vs. 1.6 mm of runoff during events in the attenuation store). The effect was to both flatten and delay the hydrograph peak due to added attenuation. Both runoff and storage in Scenario 1 were identical to the baseline values as no attenuation storage was added to the model in this scenario, so these are not shown for clarity.

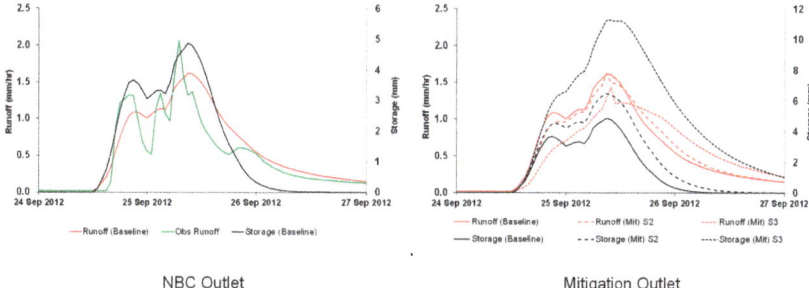

NBC Outlet   Mitigation Outlet

Figure 6. Time series plots of runoff (Q) and storage (S) during Event 3: Left hand (LH) pane shows modelled and observed Q (red) and modelled S (black) for baseline only at the NBC outlet. Right hand (RH) pane shows modelled Q (red) and S (black) at the mitigated sub-catchment outlet with different line types representing the baseline and Scenarios 2 (S2) and 3 (S3).

Figure 7 shows the results from Scenarios 1–3 alongside the baseline (blue) results for the SS (Figure 7a) and TP (Figure 7b) concentrations in event 3. These were extracted at the NBC outlet (left hand panes) and mitigated sub-catchment outlet (right hand panes) and are shown as different coloured lines for each scenario. The effect on TP concentrations during event 3 can be seen in Figure 7b. The two scenarios that added attenuation storage and trapped particulate P (PUP) (Scenario 2 (green) and Scenario 3 (red)) reduced the peak TP concentrations, but Scenario 1 actually increased the TP concentrations during the falling limb event due to an increase in SRP in the fast subsurface flow pathway; however, the maximum TP concentration was reduced by eliminating the surface runoff pathway as a major source of PUP through improved soil management. Hence, the overall outcome for the event was better even though some instantaneous values worsened. Scenarios 1 (purple) and 3 (red) had the greatest effect on TP concentrations during event 3 in terms of reducing them; in the case of Scenarios 2 and 3, this was due to reducing the amount of PUP exported from the mitigated sub-catchment through trapping sediment with attached P.

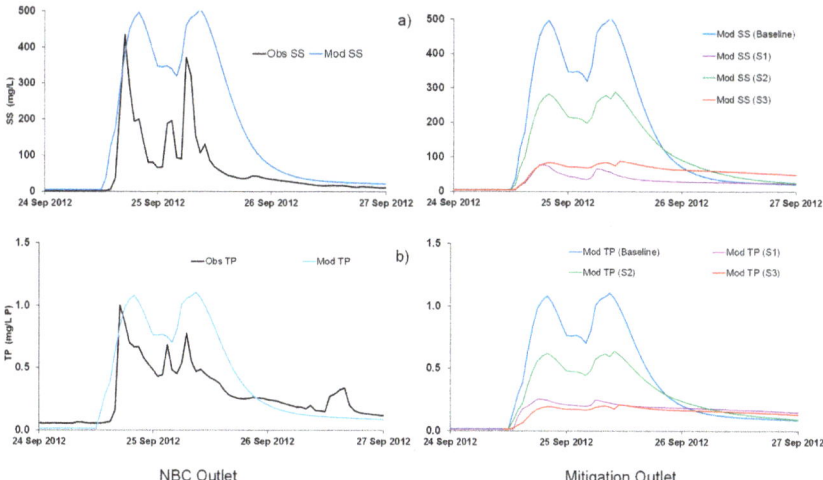

NBC Outlet   Mitigation Outlet

Figure 7. (**a**) LH pane shows the modelled ("Mod") and observed ("Obs") SS concentrations at the NBC outlet. RH pane shows the SS concentrations at the mitigated sub-catchment outlet (Scenarios 1–3 shown by colored lines, with baseline in blue). (**b**) LH pane shows modelled (black) and observed (blue) TP concentrations at the NBC outlet. RH pane shows the TP concentrations at the mitigated sub-catchment outlet (Scenarios 1–3 shown by colored lines, with baseline in blue). Only Event 3 is shown.

In the left-hand panes of Figure 7a,b, the black lines represent observed concentrations and the blue lines represent the modelled concentrations (at the NBC outlet) from the baseline simulation. The model performed reasonably well in capturing the peak concentrations, although there are some errors in timing of the peaks (compared with the blue line). The model underpredicted the TP concentrations slightly.

In terms of P yields, Figure 8 shows the modelled yields of TP and SS from the baseline simulation alongside the yields from Scenarios 1–3 from the mitigated sub-catchment. The bars are split into three coloured segments, each representing the yield transported by each of the three flow pathways (SR = surface runoff, "Fast S/S" = fast subsurface, "Slow G/W" = slow groundwater). The sum of the three segments was thus the total yield from all three pathways added together. The results covered a six-month period in 2012. Therefore, these yields included periods of low flows in addition to events; however, based on field observations and the fluxes transported by the modelled flow pathways, surface runoff during events contributed to the vast majority of the total SS and P losses from this catchment.

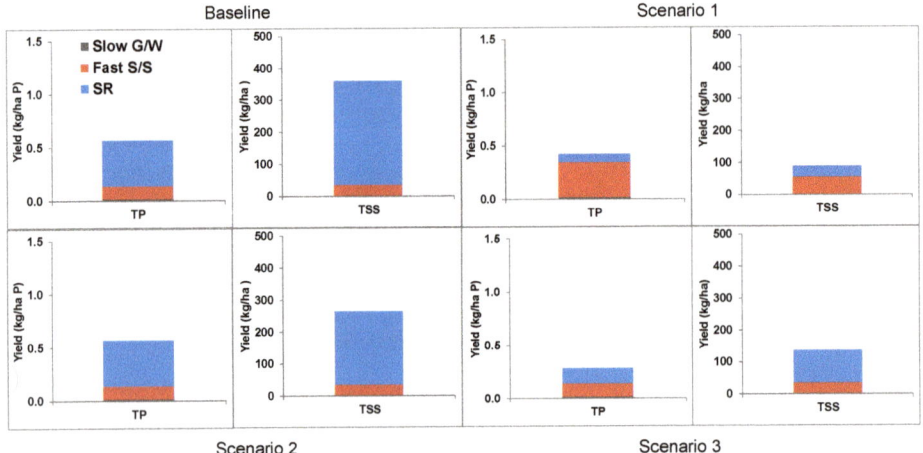

Figure 8. Plots showing modelled yields of TP and SS exported from the mitigated sub-catchment under all three scenarios in comparison with the modelled yields from the baseline.

In Scenario 1, the total yield of SS and TP reduced quite drastically due to the reduction in transport via the fast surface flow pathway. There was some pollution swapping into SRP (transported by the fast subsurface flow pathway) as a result of these interventions increasing the amount of P that could infiltrate through the upper soil layer.

The results from Scenarios 2 and 3, where there has been simulation of the removal of SS and P from the surface runoff pathway by mitigation RAFs, show that there was a considerable reduction in SS TP yields according to the model. This reduction was achieved by trapping SS and attached particulate P, hence also reducing the components of PUP and PRP transported by the surface runoff pathway by 40–80%. These results were for the mitigated sub-catchment (1.25 km^2) only and the results scaled up to the NBC outlet are discussed below.

3.3. Comparison of Recent Sampling Campaign and Model Results

The most recent available results from the mitigation features were as follows [24]. The mass of TP and SS trapped by the five features ranged from 1.5 to 2.5 kg and 0.5 to 2.8 tonnes, respectively (total mass 9.8 kg of P and 6.5 tonnes of sediment). These totals represent the masses of TP and SS collected over a 9-month period ending in summer 2015. Expressed as a yield of TP (loads per unit area over the Mitigation sub-catchment), the reductions were 0.06 kg·ha^{-1} of TP and 0.04 t·ha^{-1} of SS. At present, only 24% of the sub-catchment area is treated by these features (3% of the NBC).

Model simulations using the parameter combination that was described above in the NBC over a 6-month period in 2012 that included the three events in April–September estimated that the fluxes of P trapped by mitigation features would be 0.23 kg·ha^{-1} of PUP and 0.06 kg·ha^{-1} of PRP in Scenario 2, and 0.46 kg·ha^{-1} of PUP and 0.11 kg·ha^{-1} of PRP in Scenario 3. The ratio between these yields corresponds closely to the ratios between the values of the parameter K_{RE} in Scenarios 2 and 3 (40% and 80%, respectively).

4. Discussion

It is important to remember that European agricultural catchments have been heavily modified by centuries of intensive farming practices such that the degree of attenuation and storage have been reduced from prehistoric times when the catchments were in pristine (forested) condition prior to anthropogenic modification.

One pertinent research question relates to the detectability of mitigation features at the catchment scale, which in this case was ten times larger than the sub-catchment where the features were located. The results in terms of reducing the mean and maximum concentrations of SS and TP during Event 3 were analysed and were shown in the upper pane of Figure 9 for the three scenarios. Note that the effects of adding the features on the modelled Q_p at the NBC outlet were not evaluated for reasons given above.

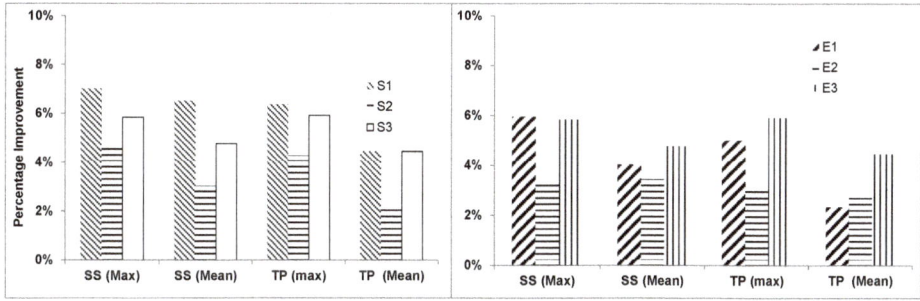

Figure 9. Plots showing the percentage reductions in TP and SS concentrations modelled at the NBC outlet achieved under: LH pane shows the three scenarios (legend indicates the scenario number) relative to the baseline in Event 3 only, and the RH pane shows the improvements under Scenario 3 relative to the baseline only, with Events 1–3 shown for comparison (improvements in Events 1–3 shown with different fill patterns).

Scenario 1 evaluated the soil management improvements covering 10% of the total catchment to reduce surface runoff and particulate P loads transported by this pathway. The event mean and maximum TP concentrations were reduced by 4.7% and 6.4%, respectively, due to reductions in the PUP concentrations transported by the surface runoff pathway. The reductions in the event maximum TP concentrations were 4.3–5.9% during Event 3 at the NBC outlet when both attenuation storage and the removal of P were simulated in Scenarios 2 and 3. The reductions in Events 1 and 2 were similar under the three scenarios to Event 3 and are shown in the lower pane of Figure 9 for Scenario 3 only. These reductions are likely to be conservative estimates since the effect of the in-stream routing and dispersion along the channel reach that connected the mitigated sub-catchment outlet to the main channel outlet were not incorporated into the model structure.

In Scenario 1, there is considerable scope however for reducing the maximum TP concentration and TP yields by this method, as shown by the reduction in the maximum TP concentration in Event 1 of over 6% (Figure 9). In Scenario 2, the value of K_{RE} of 40% was quite conservative compared to some of the recent findings from U.K. field studies [11,12,24,45], but the load reduction was still predicted to be up to 4% at the catchment outlet with a smaller predicted decrease in the mean and maximum event TP concentration (less than those decreases achieved by Scenarios 1 (TP) and 3 (TP)). The model results showed that adding additional attenuation and removing P from 10% of the catchment reduced event loads of TP by up to 8% at the outlet under Scenario 3 (where K_{RE} was 80%), which was the best performing of the three scenarios.

In terms of locating mitigation features of size ca. 2000–8000 m^3 in one of the headwater sub-catchments of the NBC, the volumes required in Scenarios 2 and 3 do not seem excessively large if one assumes that a maximum water depth of 1 m is permissible under the relevant legislation (thus requiring 0.2 and 0.8 ha of the catchments, respectively, to be set aside for storage). In the U.K., stringent government guidelines [21] apply to all features greater than 10,000 m^3 capacity, which may make constructing larger features expensive and probably uneconomic in terms of a cost–benefit ratio. For comparison, the five recent pilot features constructed in the NBC Mitigation sub-catchment had a total storage capacity of circa 400 m^3 and treated only 38.4 ha (less than 0.031% of the total NBC catchment by area). It is important to consider that RAFs should be constructed in other parts of the NBC in order to achieve the desired reduction in P concentrations and loads.

It should be stressed that further research is required to evaluate the performance of these mitigation features when constructed at a larger scale than previously, i.e., up to 4–8 times greater than at the scale (17–34 ha) of Belford Burn [11,12,45] and up to 3 times greater than the 38 ha of the NBC Mitigation sub-catchment [24]. It is vital that trapped material is recovered by removal of sediment in well-designed traps or by removing vegetation.

Field studies are required to ascertain whether Scenario 1 could deliver the results shown here in reality; however, subsurface field drains are known to be a source of high SRP loads from agricultural catchments. Under Scenario 2, the strategic location of ditches and small channels in the landscape may play a crucial role in future management option for surface runoff driven agricultural pollution. The space in and around ditches afforded by buffer zones and fenced off channels could allow the construction of long riparian mitigation zones without significant impact on farming. The ability to address both NFM and pollution targets and create new ecological habitats may justify the investment and maintenance of RAFs [6]. The implication from all three scenarios is that an extensive network of RAFs is needed in farmed landscapes, and this could require a considerable shift in standard farming and environmental payments for schemes to be taken up and maintenance to be ensured. In reality, a mixture of all three scenarios being delivered would be attractive and that will very much depend on the local conditions and the farming community.

It is possible that unforeseen constraints, such as planning regulations, infrastructure restrictions and land ownership issues, may further restrict the widespread adoption of mitigation measures [4,22]. Therefore, the exact location of these flow pathways and identification of suitable sites for constructing features is beyond the scope of this study.

Further reductions in loads could be achieved in any case by targeting a larger percentage of the catchment for the construction of mitigation features, which in rural, pastoral farming-based areas like Cumbria (where the EdenDTC project is based) may be achievable, especially on a seasonal basis under the right conditions and government policies to compensate farmers. However, this policy might not be so attractive if the catchment contains arable land of higher value to the farmers than the rough and improved pasture in the upper Eden [22].

5. Conclusions

An attenuation component has been added to the CRAFT model to represent the storage and attenuation of surface runoff during events and also the trapping of particulate forms of P and suspended sediment, which provides a methodology for modelling NBS.

Currently, the removal efficiency of the trapping process has to be supplied by the user based on expert knowledge. The effect of adding attenuation and trapping SS plus particulate forms of P has been modelled in a small (1.25 km^2) sub-catchment. Due to the variability in the removal efficiencies measured in the field from previous studies of NBS, two modelling scenarios with removal efficiencies of 40% and 80% have been evaluated here with the latter clearly providing the best results in terms of improvements in reducing sediment and nutrient (P) fluxes and concentrations.

According to the model results, at the outlet of the entire Newby Beck catchment (ca. 10× larger), a small reduction in the peak SS and TP concentrations would be observable during an event where surface runoff is the dominant pathway. A third scenario evaluated the impacts of reducing surface runoff to represent improved land management. This scenario reduced the amount of SS and P transported by the surface runoff pathway but increased the amount of SRP transported by the fast subsurface flow pathway. The reduction in both SS fluxes and maximum event concentration was more pronounced since the majority of the modelled flux was predicted to occur via the surface runoff pathway.

This modelling study represents a first step towards developing and testing a fully integrated model that can simulate both natural and added attenuation storage. Mitigation features are represented in the model as an aggregated storage effect rather than through utilizing a physically-based model of each feature individually. Research into the implications of adding new attenuation features to catchments with lighter soils and groundwater dominated runoff as well as adoption and maintenance issues is also needed.

One final conclusion arising from this study is the need to provide guidance to end users. The hypothetical attenuation versus mitigation curve can inform a policymaker on how to set annual targets over time that dictates the total amount of attenuation added to a catchment over a longer period. Hence, these guidelines would indicate how long it would take to reach the environmental target(s) based on an annual financial budget.

Author Contributions: Conceptualization, R.A. and P.Q.; Data curation, N.B.; Formal analysis, R.A.; Investigation, R.A.; Methodology, R.A., P.Q., and N.B.; Software, R.A.; Writing—original draft, R.A.; Writing—review and editing, R.A., P.Q., N.B., and S.R.

Funding: The Eden Demonstration Test Catchment (Eden DTC) research platform that funded this research and supplied the monitoring data used in this paper is funded by the Department for Environment, Food and Rural Affairs (DEFRA) (project WQ0210) and is further supported by the Welsh Assembly Government and the Environment Agency.

Acknowledgments: The Eden DTC (www.edendtc.org.uk) team includes research staff from Lancaster University, University of Durham, Newcastle University (PROACTIVE team), Eden Rivers Trust, Centre for Ecology and Hydrology, British Geological Survey, and Newton Rigg College, and their assistance in undertaking this research is gratefully acknowledged.

Conflicts of Interest: The authors declare no conflict of interest.

References

1. NWRM—Natural Water Retention Measures. European Union. Available online: www.nwrm.eu (accessed on 9 February 2018).
2. Environment Agency Working with Natural Processes to Reduce Flood Risk. Available online: www.gov.uk/government/publications/working-with-natural-processes-to-reduce-flood-riskPublished31/10/2017 (accessed on 9 February 2018).
3. O'Connell, P.E.; Ewen, J.; O'Donnell, G.; Quinn, P. Is there a link between agricultural land-use management and flooding? *Hydrol. Earth Syst. Sci.* **2007**, *11*, 96–107. [CrossRef]
4. Nisbet, T.R.; Marrington, S.; Thomas, H.; Broadmeadow, S.B.; Valatin, G. *Slowing the Flow at Pickering. Final Report for the Department of Environment, Food and Rural Affairs*; Project RMP5455; DEFRA: London, UK, 2011.
5. Quinn, P.; O'Donnell, G.; Nicholson, A.; Wilkinson, M.; Owen, G.; Jonczyk, J.; Barber, N.; Hardwick, M.; Davies, G. *Potential Use of Runoff Attenuation Features in Small Rural Catchments for Flood Mitigation*; Newcastle University, Environment Agency, Royal Haskoning DHV: Newcastle Upon Tyne, UK, 2013.
6. Charlesworth, S.M.; Booth, C.A. *Sustainable Surface Water Management: A Handbook for SUDS*; John Wiley & Sons: Chichester, UK, 2016.

7. SEPA. *Natural Flood Management Handbook*, 2nd ed.; Scottish Environment Protection Agency: Stirling, UK, 2015; Available online: https://www.sepa.org.uk/media/163560/sepa-natural-flood-management-handbook1.pdf (accessed on 9 February 2018).
8. Duffy, A.; Moir, S.; Berwick, N.; Shabashow, J.; D'Arcy, B.; Wade, R. *Rural Sustainable Drainage Systems: A Practical Design and Build. Guide for Scotland's Farmers and Landowner*; CREW: Scotland, UK, 2015; Available online: http://crew.ac.uk/publications (accessed on 6 June 2018).
9. Wilkinson, M.E.; Quinn, P.F.; Welton, P. Runoff management during the September 2008 floods in the Belford catchment, Northumberland. *J. Flood Risk Manag.* **2010**, *3*, 285–295. [CrossRef]
10. Nisbet, T.R.; Thomas, H. *Restoring Floodplain Woodland for Flood Alleviation. Final Report for the Department of Environment, Food and Rural Affairs*; Project SLD2316; DEFRA: London, UK, 2008.
11. Barber, N. Sediment, Nutrient and Runoff Management and Mitigation in Rural Catchments. Ph.D. Thesis, Newcastle University, Newcastle Upon Tyne, UK, 2014.
12. Barber, N.J.; Quinn, P.F. Mitigating diffuse water pollution from agriculture using soft engineered runoff attenuation features. *Area* **2012**, *44*, 454–462. [CrossRef]
13. Deasy, C.; Quinton, J.N.; Silgram, M.; Bailey, A.P.; Jackson, B.; Stevens, C.J. Contributing understanding of mitigation options for phosphorus and sediment to a review of the efficacy of contemporary agricultural stewardship measures. *Agric. Syst.* **2010**, *103*, 105–109. [CrossRef]
14. Doody, D.G.; Archbold, M.; Foy, R.H.; Flynn, R. Approaches to the implementation of the Water Framework Directive: Targeting mitigation measures at critical source areas of diffuse phosphorus in Irish catchments. *J. Environ. Manag.* **2012**, *93*, 225–234. [CrossRef] [PubMed]
15. Wilcock, R.J.; Müller, K.; van Assema, G.B.; Bellingham, M.A.; Ovenden, R. Attenuation of nitrogen, phosphorus and E. coli inputs from pasture runoff to surface waters by a farm wetland: The importance of wetland shape and residence time. *Water Air Soil Pollut.* **2012**, *223*, 499–509. [CrossRef]
16. Avery, L.M. *Rural Sustainable Drainage Systems (RSuDS)*; Environment Agency: Bristol, UK, 2012.
17. Blöschl, G. Scaling in hydrology. *Hydrol. Process.* **2001**, *15*, 709–711. [CrossRef]
18. Biggs, J.; Stoate, C.; Williams, P.; Brown, C.; Casey, A.; Davies, S.; Diego, I.G.; Hawczak, A.; Kizuka, T.; McGoff, E.; et al. *Water Friendly Farming Autumn 2016 Update*; Freshwater Habitats Trust: Oxford, UK; Game & Wildlife Conservation Trust: Fordingbridge, UK, 2016; Available online: https://freshwaterhabitats.org.uk/wp-content/uploads/2016/11/Water-Friendly-Farming-update-2016.pdf (accessed on 18 April 2018).
19. Fealy, R.M.; Buckley, C.; Mechan, S.; Melland, A.; Mellander, P.E.; Shortle, G.; Wall, D.; Jordan, P. The Irish Agricultural Catchments Programme: Catchment selection using spatial multi-criteria decision analysis. *Soil Use Manag.* **2010**, *26*, 225–236. [CrossRef]
20. Owen, G.J.; Perks, M.T.; Benskin, C.M.H.; Wilkinson, M.E.; Jonczyk, J.; Quinn, P.F. Monitoring agricultural diffuse pollution through a dense monitoring network in the River Eden Demonstration Test Catchment, Cumbria, UK. *Area* **2012**, *44*, 443–453. [CrossRef]
21. McGonigle, D.F.; Burke, S.P.; Collins, A.L.; Gartner, R.; Haft, M.R.; Harris, R.C.; Haygarth, P.M.; Hedges, M.C.; Hiscock, K.M.; Lovett, A.A. Developing demonstration test catchments as a platform for transdisciplinary land management research in England and Wales. *Environ. Sci. Process. Impacts* **2014**, *16*, 1618–1628. [CrossRef] [PubMed]
22. Metcalfe, P.; Beven, K.; Hankin, B.; Lamb, R. A modelling framework for evaluation of the hydrological impacts of nature-based approaches to flood risk management, with application to in-channel interventions across a 29 km^2 scale catchment in the United Kingdom. *Hydrol. Process.* **2017**, *31*, 1734–1748. [CrossRef]
23. Bechmann, M.; Deelstra, J.; Stålnacke, P.; Eggestad, H.O.; Øygarden, L.; Pengerud, A. Monitoring catchment scale agricultural pollution in Norway: Policy instruments, implementation of mitigation methods and trends in nutrient and sediment losses. *Environ. Sci. Policy* **2008**, *11*, 102–114. [CrossRef]
24. Barber, N.J.; Reaney, S.N.; Barker, P.A.; Benskin, C.; Burke, S.; Cleasby, W.; Haygarth, P.; Jonczyk, J.C.; Owen, G.J.; Snell, M.A.; et al. The treatment train approach to reducing nonpoint source pollution from agriculture. In Proceedings of the AGU Fall Meeting, San Francisco, CA, USA, 12–16 December 2016.
25. Kay, P.; Edwards, A.C.; Foulger, M. A review of the efficacy of contemporary agricultural stewardship measures for ameliorating water pollution problems of key concern to the UK water industry. *Agric. Syst.* **2009**, *99*, 67–75. [CrossRef]

26. Ensign, S.H.; McMillan, S.K.; Thompson, S.P.; Piehler, M.F. Nitrogen and phosphorus attenuation within the stream network of a coastal, agricultural watershed. *J. Environ. Qual.* **2006**, *35*, 1237–1247. [CrossRef] [PubMed]
27. Adams, R.; Quinn, P.F.; Bowes, M.J. The Catchment Runoff Attenuation Flux Tool, A minimum information requirement nutrient pollution model. *Hydrol. Earth Syst. Sci.* **2015**, *19*, 1641–1657. [CrossRef]
28. Adams, R.; Quinn, P.F.; Perks, M.; Barber, N.J.; Jonczyk, J.; Owen, G.J. Simulating high frequency water quality monitoring data using a catchment runoff attenuation flux tool (CRAFT). *Sci. Total Environ.* **2016**, *572*, 1622–1635. [CrossRef] [PubMed]
29. Heathwaite, A.L.; Quinn, P.F.; Hewett, C.J.M. Modelling and managing critical source areas of diffuse pollution from agricultural land using flow connectivity simulation. *J. Hydrol.* **2005**, *304*, 446–461. [CrossRef]
30. Singh, V.P. Is Hydrology kinematic? *Hydrol. Process.* **2002**, *16*, 667–716. [CrossRef]
31. Perks, M.T.; Owen, G.J.; Benskin, C.M.H.; Jonczyk, J.; Deasy, C.; Burke, S.; Haygarth, P.M. Dominant mechanisms for the delivery of fine sediment and phosphorus to fluvial networks draining grassland dominated headwater catchments. *Sci. Total Environ.* **2015**, *523*, 178–190. [CrossRef] [PubMed]
32. Ockenden, M.C.; Deasy, C.E.; Benskin, C.M.; Beven, K.J.; Burke, S.; Collins, A.L.; Evans, R.; Falloon, P.D.; Forber, K.J.; Hiscock, K.M.; et al. Potential effects of changing climate on catchment processes and nutrient transfers: Evidence from high temporal concentration-flow dynamics in headwater catchments. *Sci. Total Environ.* **2016**, *548*, 325–339. [CrossRef] [PubMed]
33. Ockenden, M.C.; Tych, W.; Beven, K.; Collins, A.; Evans, R.; Falloon, P.; Forber, K.; Hiscock, K.; Hollaway, M.; Kahana, R.; et al. Prediction of storm transfers and annual loads with data-based mechanistic models using high-frequency data. *Hydrol. Earth Syst. Sci.* **2017**, *18*, 6425–6444. [CrossRef]
34. Mills, C.F.; Bathurst, J.C. Spatial variability of suspended sediment yield in a gravel-bed river across four orders of magnitude of catchment area. *Catena* **2015**, *133*, 14–24. [CrossRef]
35. European Commission. *Directive 2000/60/EC: Establishing a Framework for Community Action in the Field of Water Policy. (The Water Framework Directive)*; European Commission: Brussels, Belgium, 2000.
36. Greene, S.; Johnes, P.J.; Bloomfield, J.P.; Reaney, S.M.; Lawley, R.; Elkhatib, Y.; Freer, J.; Odoni, N.; Macleod, C.J.; Percy, B. A geospatial framework to support integrated biogeochemical modelling in the United Kingdom. *Environ. Model. Softw.* **2015**, *68*, 219–232. [CrossRef]
37. Hewett, C.J.; Quinn, P.F.; Heathwaite, A.L.; Doyle, A.; Burke, S.; Whitehead, P.G.; Lerner, D.N. A multi-scale framework for strategic management of diffuse pollution. *Environ. Model. Softw.* **2009**, *24*, 74–85. [CrossRef]
38. Lane, S.N.; Reaney, S.M.; Heathwaite, A.L. Representation of landscape hydrological connectivity using a topographically driven surface flow index. *Water Resour. Res.* **2009**, *45*. [CrossRef]
39. Reaney, S.M.; Lane, S.N.; Heathwaite, A.L.; Dugdale, L.J. Risk-based modelling of diffuse land use impacts from rural landscapes upon salmonid fry abundance. *Ecol. Model.* **2011**, *222*, 1016–1029. [CrossRef]
40. WFD-UKTAG. *UKTAG River Assessment Method—Phosphorus: River Phosphorus Standards*; Water Framework Directive-United Kingdom Technical Advisory Group (WFD-UKTAG): Stirling, UK, 2014.
41. Kovacs, A.; Honti, M.; Zessner, M.; Eder, A.; Clement, A.; Blöschl, G. Identification of phosphorus emission hotspots in agricultural catchments. *Sci. Total Environ.* **2012**, *433*, 74–88. [CrossRef] [PubMed]
42. Schoumans, O.F.; Chardon, W.J.; Bechmann, M.E.; Gascuel-Odoux, C.; Hofman, G.; Kronvang, B.; Dorioz, J.M. Mitigation options to reduce phosphorus losses from the agricultural sector and improve surface water quality: A review. *Sci. Total Environ.* **2014**, *468*, 1255–1266. [CrossRef] [PubMed]
43. Beven, K. On the generalized kinematic routing method. *Water Resour. Res.* **1979**, *15*, 1238–1242. [CrossRef]
44. Wood, E.F.; Sivapalan, M.; Beven, K.; Band, L. Effects of spatial variability and scale with implications to hydrologic modeling. *J. Hydrol.* **1988**, *102*, 29–47. [CrossRef]
45. Wilkinson, M.E.; Quinn, P.F.; Barber, N.J.; Jonczyk, J. A framework for managing runoff and pollution in the rural landscape using a Catchment Systems Engineering approach. *Sci. Total Environ.* **2014**, *468*, 1245–1254. [CrossRef] [PubMed]

 © 2018 by the authors. Licensee MDPI, Basel, Switzerland. This article is an open access article distributed under the terms and conditions of the Creative Commons Attribution (CC BY) license (http://creativecommons.org/licenses/by/4.0/).

Article

Shear Stress-Based Analysis of Sediment Incipient Deposition in Rigid Boundary Open Channels

Necati Erdem Unal

Department of Civil Engineering, Hydraulics Division, Istanbul Technical University, Maslak, 34469 Istanbul, Turkey; neu@itu.edu.tr; Tel.: +90-212-285-3727

Received: 24 September 2018; Accepted: 4 October 2018; Published: 9 October 2018

Abstract: Urban drainage and sewer systems, and channels in general, are treated by the deposition of sediment that comes from water collecting systems, such as roads, parking lots, land, cultivation areas, and so forth, which are all under gradual or sudden change. The carrying capacity of urban area channels is reduced heavily by sediment transport that might even totally block the channel. In order to solve the sedimentation problem, it is therefore important that the channel is designed by considering self-cleansing criteria. Incipient deposition is proposed as a conservative method for channel design and is the subject of this study. With this aim, an experimental study carried out in trapezoidal, rectangular, circular, U-shape, and V-bottom channels is presented. Four different sizes of sand were used as sediment in the experiments performed in a tilting flume under nine different longitudinal channel bed slopes. A shear stress approach is considered, with the Shields and Yalin methods used in the analysis. Using the experimental data, functionals are developed for both methods. It is seen that the bed shear stress changes with the shape of the channel cross-section. Incipient deposition in rectangular and V-bottom channels starts under the lowest and the highest shear stress, respectively, due mainly to the shape of the channel cross-section that affects the distribution of shear stress on the channel bed.

Keywords: incipient deposition; sediment transport; self-cleansing; sewer systems; shear stress; urban drainage system

1. Introduction

The sediment transport issue has always been an important scientific and practical problem [1,2], and kept its importance with the changes in hydrology [3–5] emerging with the change in the sediment load of urban watersheds and alluvial streams. As the outlets of urban watersheds, the sewer and urban drainage systems are heavily affected by any change in their watersheds. Therefore, research on sediment transport is continuously needed for sustainable practice in drainage systems.

Sediment deposition is avoided as it causes numerous unwanted problems in urban drainage and sewer systems. It reduces the hydraulic capacity of the channel by decreasing the flow cross-sectional area or blocking the channel. The performance and efficiency of drainage systems is heavily affected by deposition. Additional funds should be invested to keep the system working. Furthermore, sedimentation creates environmental problems. Such problems in drainage and sewer systems could be prevented or minimized by the use of self-cleansing criteria, with which sediment particles deposited at the channel bed start to move [6], or sediment particles suspended within the flow are transported without being deposited [7–9]. Transportation of sediment particles without deposition is preferred in drainage system design to keep the channel bed clean [10–14]. In this regard, incipient deposition is a concept linked to the channel design, and identified as the sediment transport mode in which sediment particles are clustered visibly in certain areas at the channel bed [15]. In other words, sediment particles are transported as bed load or accumulated at the channel bed, but without making a permanent deposited bed layer [15–17]. Flow velocity is sufficiently low for incipient deposition of sediment. At the incipient deposition, sediment particles in suspension within flow start moving downward to reach the channel bed.

Incipient deposition has been studied in several fixed bed channels by Loveless [15] who assumed that incipient motion and incipient deposition were similar concepts but with a slight difference [18]. The experimental data of Loveless [15] fit the sediment transport models of May [19] and Ackers [20]. Safari et al. [16] studied the incipient deposition concept using the experimental data of Loveless [15]. As a conclusion, velocity at incipient deposition was found to be higher than velocity at incipient motion for non-cohesive sediments. Incipient deposition is observed when flow velocity decreases gradually to a level that allows sediment particles to deposit. In the opposite case, when flow velocity increases gradually, sediment particles with no motion start moving when flow velocity reaches a level high enough for incipient motion. The former is called the incipient deposition velocity, which is higher than the latter, the incipient motion velocity. This is a hysteretic curve with a higher threshold velocity for incipient deposition and lower threshold velocity for the incipient motion. In this paradigm, there is a shear stress (or velocity) threshold below which erosion does not occur, and a lower threshold above which deposition does not occur; erosion and deposition occur simultaneously between the two thresholds [21]. Therefore, results denied the common assumption that incipient deposition and incipient motion are the same.

In order to explain the exact difference between the incipient deposition and incipient motion, Aksoy and Safari [22] performed a preliminary study on incipient motion and incipient deposition in a trapezoidal cross-section channel, and found that the flow has higher shear stress at incipient deposition than at the incipient motion. For the sake of achieving more conclusive results, a new set of experiments with a wider range of sediment size in different channel cross-sections seemed important. Not only because of its importance, but also for the sake of getting experimental data with a wider range in terms of sediment size, channel cross-section, channel slope, discharge, and so forth, Unal et al. [23] constructed a laboratory experimental setup to study the incipient motion and incipient deposition in trapezoidal, rectangular, circular, U-shape, and V-bottom channels, and performed experiments for the self-cleansing design of fixed bed systems.

In this study, experimental data from an indoor laboratory flume was analyzed to understand the incipient deposition of sediment particles within flow. The shear stress approach was considered for the analysis in which the Shields [24] and Yalin [25] methods were used.

2. Mechanism of Particle Motion and Methodology

Sediment particles in flow move under the influence of two types of hydrodynamic forces; the first of which has a positive impact through the drag force and the lift force, while the second discourages motion through the buoyed weight of sediment and the resisting force against motion. The drag force should be equal to the resistance force in the sediment threshold condition.

The shear stress- and velocity-based approaches were commonly used in the analysis of sediment threshold and incipient deposition in this study. The velocity-based approach has been applied on the experimental data existing in the literature [26]. The shear stress approach was used in this study for which Shields [24] and Yalin [25] methods are considered.

2.1. Shields Method

The shear stress approach used in the incipient deposition of sediment is based on the shear velocity (u_*), defined as

$$u_* = \sqrt{\frac{\tau_{id}}{\rho}} \quad (1)$$

in which τ_{id} is the bed shear stress under the incipient deposition condition and ρ is the specific mass of water. The dimensionless shear stress is calculated by

$$\tau_{id}^* = \frac{\tau_{id}}{\rho g d(s-1)}. \tag{2}$$

in which g is the acceleration due to gravity, d is the median size of sediment particles, and s is the sediment relative mass density. The dimensionless shear stress (τ_{id}^*) is indicated as

$$\tau_{id}^* = f(Re^*) \tag{3}$$

where Re^* is the particle Reynolds number (Re^*) defined by

$$Re^* = \frac{u_* d}{\nu} \tag{4}$$

in which ν is the kinematic viscosity of water.

Experimental data has been used by many researchers [9,15–17,22,23,26] to determine the functional between τ_{id}^* and Re^* for practical problems of incipient motion [27,28]. In this study, the same methodology was adopted for the incipient deposition.

Average flow velocity (V_{id}) calculated from the incipient deposition experimental data is used to calculate the incipient deposition shear stress by

$$\tau_{id} = \frac{\lambda \rho V_{id}^2}{8} \tag{5}$$

in which λ is the channel friction factor to be calculated by the Colebrook–White equation [29] as

$$\lambda = \frac{1}{4\left[\log\left(\frac{k_b}{14.8R} + \frac{0.22\nu}{R\sqrt{gRS}}\right)\right]^2} \tag{6}$$

in which k_b is the roughness height of the bed taken to be the same as the median size of sediment particles (d), R is the hydraulic radius of the channel, and S is the slope of the channel bed.

2.2. Yalin Method

Yalin [25] suggested a combination of the dimensionless parameters initially proposed by Shields [24] as

$$\tau_{id}^* = f(D_{gr}) \tag{7}$$

in which D_{gr} is the dimensionless grain size parameter defined by

$$D_{gr} = \left[\frac{(s-1)gd^3}{\nu^2}\right]^{1/3} \tag{8}$$

The shear velocity is eliminated, and only fluid and sediment characteristics are retained in the formulation.

3. Experiments

An experimental setup was configured as in Figure 1 [22,23]. An iron-made support structure was constructed. Twelve meter-long transparent acrylic glass (plexiglass) channels were mounted on the support structure. Five different cross-sections were considered for the channels; they were trapezoidal, rectangular, circular, U-shape, and V-bottom. The surface width of the rectangular, U-shape, and V-bottom channels was 300 mm, while the trapezoidal channel had the same width at the bottom, and outer angles of 60° at the 30 cm-long side walls. The U-shape and V-bottom channels had a cross-fall of 50 mm longitudinally along the centerline of the bottom. The inner diameter of the circular channel was 290 mm.

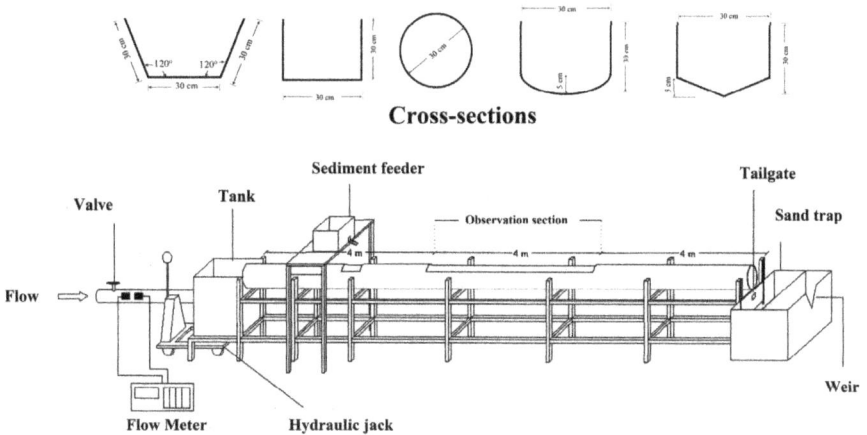

Figure 1. Experimental setup and cross-sections of the channels.

In the experiments, the bed slope was changed between 0.00147 and 0.01106. Four non-cohesive sands were poured into the channel from the sediment feeder placed 3 m upstream of the observation section of the channel. The granulometric curve and the characteristics of the sands are shown in Figure 2 and Table 1, from which it can be seen that they have uniform size distribution. The discharge was measured by an ultrasonic flowmeter (BSUF-TTCL, Bass Instruments, Istanbul, Turkey) with an accuracy better than 1.0% of read. Sediment motion was observed in each experiment in the 4 m-long observation section of the channel, 4 m from the inlet and 4 m from the outlet of the channel (Figure 1). Uniform flow conditions were satisfied in the channel before observations and measurements were done.

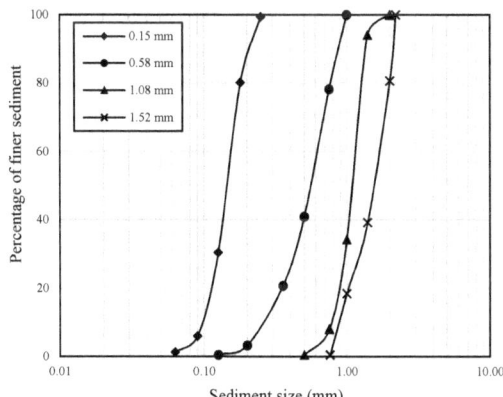

Figure 2. Granulometric curve of sediment.

Table 1. Sediment characteristics. d: median diameter, s: relative mass density, σ_g: geometric standard deviation of sediment particles.

d (mm)	0.15	0.58	1.08	1.52
s	2.60	2.63	2.56	2.60
σ_g	1.3	1.6	1.3	1.4

Experiments started with a flow velocity high enough to achieve the non-deposition condition. The average velocity was adjusted by increasing or decreasing flow discharge into the channel. In the non-deposition condition, sediment particles are prevented from being deposited; that is, sediment particles within flow are in motion. Flow velocity was gradually decreased until incipient deposition was achieved; that is, flow switches from non-deposition to incipient deposition. It was assumed that incipient deposition was satisfied when sediment particles were clustered visibly in certain areas at the channel bottom [15]. In this case, flow velocity is sufficiently low for incipient deposition of sediment. Incipient deposition was observed in the same form in trapezoidal and rectangular channels due to their flat bed. However, in the channels with U-shape and V-bottom, sediment particles were deposited in the center line of the channel with the same cross-fall. The form of the deposition depends on the channel bed. Sediment particles were accumulated on each other along the narrow centerline in the V-bottom channel, while accumulation in the circular and U-shape channels was not that narrow, as due to the wider bed along the centerline width, sediment particles spread over the bed width to make a deposited sediment layer instead.

4. Results

4.1. Shields Method

The incipient deposition experimental data of the channels are plotted on the Shields diagram (Figure 3), with the upper and lower limits as proposed by Paphitis [30]. Using the experimental data, τ_{id}^* and Re^* were calculated using Equations (2) and (4), respectively, and functional relationships were developed by curve fitting to the measured data as

$$\tau_{id}^* = 0.74(Re^*)^{-0.86} \qquad 3.13 < Re^* < 47.61 \qquad r^2 = 0.913 \tag{9}$$

$$\tau_{id}^* = 0.32(Re^*)^{-0.97} \qquad 2.36 < Re^* < 29.89 \qquad r^2 = 0.953 \tag{10}$$

$$\tau_{id}^* = 0.57(Re^*)^{-0.83} \qquad 3.02 < Re^* < 41.11 \qquad r^2 = 0.972 \tag{11}$$

$$\tau_{id}^* = 0.55(Re^*)^{-0.72} \qquad 3.19 < Re^* < 48.11 \qquad r^2 = 0.960 \tag{12}$$

$$\tau_{id}^* = 0.79(Re^*)^{-0.78} \qquad 3.46 < Re^* < 52.62 \qquad r^2 = 0.907 \tag{13}$$

in the range $2.36 < Re^* < 52.62$ for the trapezoidal, rectangular, circular, U-shape, and V-bottom channels, respectively. In Equations (9)–(13), r^2, the determination coefficient, shows the goodness of fit of the curves. The data of the five channels are close to each other and partially overlap in some of the cases. It is seen from Figure 3 that, for the 0.15 mm- and 0.58 mm-particle size sands, incipient deposition shear stress remained above the upper limit of the Shields curve for the non-rectangular channels; however, in the rectangular channel, it is on the upper limit curve for the finest sand (the 0.15 mm-particle size sand) and on the average curve for the 0.58 mm-particle size sand. It should be kept in mind that the Shields curve has been developed for the incipient motion of sediment in loose boundary channels. Therefore, Figure 3 indicates, for sand finer than 0.58 mm, that the incipient deposition shear stress in rigid boundary channels is higher than the incipient motion shear stress in loose boundary channels. For sand with a 1.08 mm diameter, the incipient deposition shear stress remains between the upper limit and the average curve in the non-rectangular channels. It is on the lower limit curve in the rectangular channel case. It is also shown in Figure 3 that the incipient deposition shear stress for the coarsest sand with a 1.52 mm diameter remains between the average and lower limit curves in all the channels other than the rectangular cross-section. It is below the lower limit curve for the rectangular channel. Generally, the incipient deposition shear stress of coarse sediment (1.08 mm and 1.52 mm) is lower than incipient motion shear stress in loose boundary channels. The coarser the sediment, the lower the shear stress under which the sediment particles initiate deposition within flow.

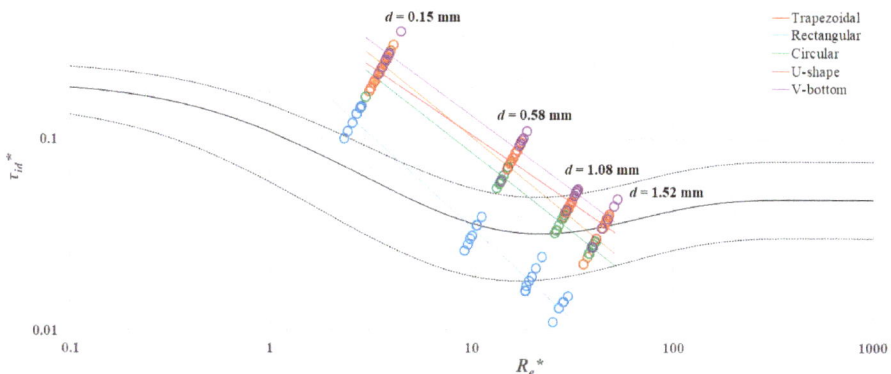

Figure 3. Relationship between the incipient deposition shear stress and Reynolds number based on the Shields method (circles show values calculated from the measurements, lines are the fitted equations).

It can also be seen that incipient deposition starts under lower shear stress in the rectangular channel (Figure 3). The incipient deposition shear stress of the trapezoidal, circular, U-shape, and V-bottom channels are close to each other. However, sediment particles in the V-bottom and U-shape channels initiate deposition under higher shear stress. The rectangular channel on the other hand obviously has different performance than the other channels; it allows sediment to move within flow until a lower shear stress is approached.

4.2. Yalin Method

Incipient deposition experimental data are plotted on the Yalin diagram in Figure 4. For the five channels, the dimensionless incipient deposition shear stress (τ_{id}^*) and grain size (D_{gr}) were calculated by Equations (2) and (8), respectively. Utilizing the incipient deposition experimental data,

$$\tau_{id}^* = 0.84 D_{gr}^{-0.92} \qquad r^2 = 0.913 \qquad (14)$$

$$\tau_{id}^* = 0.47 D_{gr}^{-0.99} \qquad r^2 = 0.953 \qquad (15)$$

$$\tau_{id}^* = 0.68 D_{gr}^{-0.89} \qquad r^2 = 0.972 \qquad (16)$$

$$\tau_{id}^* = 0.65 D_{gr}^{-0.80} \qquad r^2 = 0.960 \qquad (17)$$

$$\tau_{id}^* = 0.87 D_{gr}^{-0.86} \qquad r^2 = 0.907 \qquad (18)$$

are proposed in the range of $3.76 < D_{gr} < 38.06$ for the trapezoidal, rectangular, circular, U-shape, and V-bottom channels, respectively. It can be seen that the incipient deposition shear stress remains above the upper limit of the Yalin curve for the trapezoidal, U-shape, and V-bottom channels. It is on the upper limit for the fine sand ($d = 0.15$ mm) in the rectangular channel. For the medium sand ($d = 0.58$ mm) it is on and below the average curve in the circular and rectangular channels, respectively. For the coarser sand ($d = 1.08$ mm), the incipient deposition shear stress remains between the upper limit and the average curve in the trapezoidal, U-shape, and V-bottom channels. It is below the lower limit for the rectangular channel, and on the average curve for the circular channel. For the coarsest sand (d = 1.52 mm), the incipient deposition shear stress is between the average curve and the lower limit in the channels with no rectangular cross-section. It is below the lower limit of the curve for the rectangular channel.

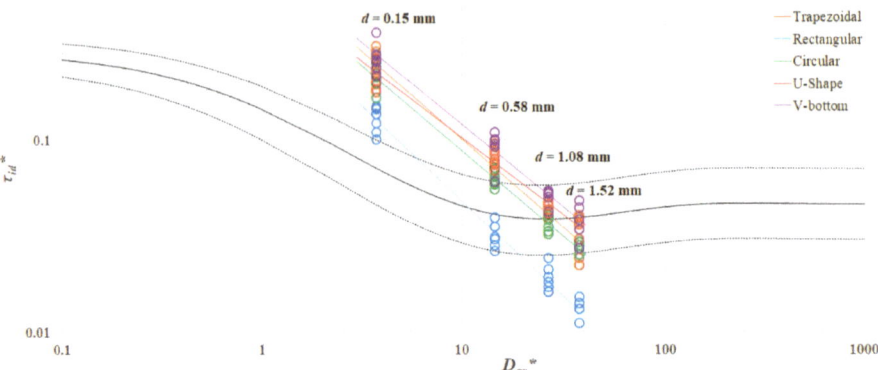

Figure 4. Relationship between the incipient deposition shear stress and grain size based on the Yalin method (circles show values calculated from the measurements, lines are the fitted equations).

The incipient deposition shear stress calculated from the experimental data (Figure 4) shows that sediment particles within flow in a rectangular channel start deposition under lower shear stress compared to the other channels. In other words, sediment particles within flow in the rectangular channel among the tested channels could be kept moving within flow under the lowest shear stress. The non-rectangular channels all have similar performance. Among the non-rectangular channels, the incipient deposition shear stress is lowest in the circular channel and the highest in the V-bottom channel. As a general result, it is clear that the coarser the sediment, the lower the shear stress under which sediment particles initiate deposition within flow. It can be said that the shape of the channel cross-section significantly affects the incipient deposition shear stress. Consequently, the incipient deposition shear stress is lower in the rectangular channel compared to the other channels.

5. Discussion

The results of the shear stress approach (Shields and Yalin methods) show that the channel cross-sectional shape significantly affects the incipient deposition shear stress. This is due to the change in the wall-normal component of the gravitational forces from which the friction force stems [31]. It is important to stress that the concept of incipient deposition is quite different to that of incipient motion. Not only the driving forces for the motion, but also the resisting forces against the motion are radically different in these two concepts. For the incipient motion, the friction coefficient is the static friction coefficient, which is accepted to be equal to the tangent of the friction angle between the bed and the grain (which becomes the angle of repose for a bed composed of identical particles). On the other hand, for incipient deposition, the friction coefficient is the dynamic friction coefficient, which is the tangent of the dynamic friction angle between the bed and the grain. Fredsoe and Deigaard [31] pointed out this important difference, and carried out an analysis for the correction of the transport rate of sediment particles moving on a transverse slope.

It should additionally be emphasized that, since the bed conditions in this study are smooth (i.e., the bed is the so-called "starving bed" where sediment particles travel on an otherwise hydraulically-smooth bed), one could expect that the changes with respect to the bed shape become even more pronounced compared to a regular bed composed of identical particles. Finally, it is possible to state that the critical shear stress curves for the incipient deposition for each of the given channel cross-sections (trapezoidal, rectangular, circular, U-shape, and V-bottom) are expected to be different from each other, unless the channel is very wide, or unless a separate sophisticated correction for the critical shear stress for the deposition is carried out for each and every case. Such a correction, which was out of the scope of this study, should be expected to involve an averaging of the critical shear stress, possibly by integration across the channel cross-section.

The Shields and Yalin methods provide similar results. The behavior of the channels in terms of the magnitude of the incipient deposition shear stress does not change with the selected method. In both methods, the sediment particles initiate deposition under the lowest shear stress in the rectangular channel, and under the highest shear stress in the V-bottom channel. This is an expected result because the incipient deposition shear stress is calculated from the channel, sediment, and flow characteristics.

Another point worth discussing is the trend in the dimensionless incipient deposition shear stress against the particle Reynolds number in the Shields curve, and the grain size in the Yalin curve. The range of the experiments is $2.36 < Re^* < 52.62$ for the Shields curve and $3.76 < D_{gr} < 38.06$ for the Yalin curve. Within these ranges, both curves gradually decrease with increasing Re^* and D_{gr}. Experimental observation fits the general character of the Shields and Yalin curves within the range of the experiments.

It should also be mentioned that the incipient deposition shear stress decreases with increasing channel bed slope. In other words, sediment particles move within flow without being deposited until the shear stress becomes low enough to initiate deposition. Deposition starts under lower shear stress when the channel has a steeper slope. This is an observation for all sediment sizes and channel cross-sections.

One more point to discuss is the similarity with the velocity-based analysis of the incipient deposition performed by Aksoy et al. [26] under which the Novak-Nalluri [32] and Yang [33] methods were used. The rectangular cross-section channel was marked with the lowest velocity at the incipient deposition. Therefore, based on either the shear stress or the velocity approach, rectangular channels seem to be preferable.

Finally, as a general discussion, when the shear stress approach is compared with the velocity-based approach [26], there is a dilemma in the selection of the appropriate approach—shear stress or velocity. In the shear stress approach, two dimensionless parameters are used, namely the shear stress and the particle Reynolds number, both of which are dependent on shear stress; its critical value cannot be determined directly, but can only be calculated through a trial-and-error technique implicitly. In the velocity approach, critical velocity is calculated by an explicit solution. This gives an advantage to the velocity approach in terms of computation. However, as Vanoni [34] demonstrated, in using the velocity approach, flow depth or hydraulic radius, and in general channel cross-section, must be specified.

6. Conclusions

Incipient deposition is a different concept as opposed to the common assumption that the incipient deposition and incipient motion are the same. Therefore, the incipient deposition was considered solely in this study. An experimental analysis was performed for fixed bed channels with trapezoidal, rectangular, circular, U-shape, and V-bottom cross-sections. Experimental data was analyzed using the shear stress approach, under which the Shields and Yalin methods were considered to calculate the incipient deposition shear stress. Both methods showed that the incipient deposition starts under lower shear stress in the rectangular channel compared to the non-rectangular channels. This indicates that rectangular channels have higher efficiency of sediment transport as sediment particles deposit at lower velocities. This gives an advantage to rectangular channels in the design of urban drainage and sewer systems or irrigation canals. The trapezoidal, circular, U-shape, and V-bottom channels have similar performance, among which the circular channel has the lowest incipient deposition shear stress and the V-bottom channel has the highest. This makes the circular channel the second most preferable channel after the rectangular channel, and the V-bottom channel the least. Analysis brings the conclusion that the cross-section of the channel significantly affects the shear stress at incipient deposition. The general trends in the Shields and Yalin curves were traced with the experimental data within the data range tested. One observation is that the incipient deposition shear stress decreases with the increasing channel bed slope; that is, sediment deposition in the channel is delayed when the channel has a steeper slope. The outputs of this study are expected to be considered together with the available literature and be employed for practical use in rigid boundary channel design. Further experiments are encouraged to extend the range of sediment size such that the validity range of the developed equations is increased.

Funding: The Scientific and Technological Research Council of Turkey (TUBITAK) Project number 114M283, The Scientific Research Projects (BAP) Unit of Istanbul Technical University Project number 37973.

Acknowledgments: The author is thankful to the Reviewers, Guest Editors and Academic Editor for their important comments with which the paper has been well improved.

Conflicts of Interest: The author declares no conflict of interest.

References

1. Graf, W.H.; Acaroglu, E.R. Sediment transport in conveyance systems (Part 1)/A physical model for sediment transport in conveyance systems. *Hydrol. Sci. J.* **1968**, *13*, 20–39. [CrossRef]
2. Bogardi, J. *Sediment Transport in Alluvial Streams*; Akademiai Kiado: Budapest, Hungary, 1974.
3. Montanari, A.; Young, G.; Savenije, H.H.G.; Hughes, D.; Wagener, T.; Ren, L.L.; Koutsoyiannis, D.; Cudennec, C.; Toth, E.; Grimaldi, S.; et al. "Panta Rhei—Everything flows": Change in hydrology and society—The IAHS scientific decade 2013–2022. *Hydrol. Sci. J.* **2013**, *58*, 1256–1275. [CrossRef]
4. Ceola, S.; Montanari, A.; Krueger, T.; Dyer, F.; Kreibich, H.; Westerberg, I.; Carr, G.; Cudennec, C.; Elshorbagy, A.; Savenije, H.; et al. Adaptation of water resources systems to changing society and environment: A statement by the International Association of Hydrological Sciences. *Hydrol. Sci. J.* **2016**, *61*, 2803–2817. [CrossRef]
5. McMillan, H.; Montanari, A.; Cudennec, C.; Savenije, H.; Kreibich, H.; Krüger, T.; Liu, J.; Meija, A.; van Loon, A.; Aksoy, H.; et al. Panta Rhei 2013–2015: Global perspectives on hydrology, society and change. *Hydrol. Sci. J.* **2016**, *61*, 1174–1191. [CrossRef]
6. Bong, C.H.J.; Lau, T.L.; Ghani, A.A.; Chan, N.W. Sediment deposit thickness and its effect on critical velocity for incipient motion. *Water Sci. Technol.* **2016**, *74*, 1876–1884. [CrossRef] [PubMed]
7. May, R.W.; Ackers, J.C.; Butler, D.; John, S. Development of design methodology for self-cleansing sewers. *Water Sci. Technol.* **1996**, *33*, 195–205. [CrossRef]
8. Butler, D.; May, R.; Ackers, J. Self-cleansing sewer design based on sediment transport principles. *J. Hydraul. Eng.* **2003**, *129*, 276–282. [CrossRef]
9. Safari, M.J.S.; Aksoy, H.; Unal, N.E.; Mohammadi, M. Non-deposition self-cleansing design criteria for drainage systems. *J. Hydro-Environ. Res.* **2017**, *14*, 76–84. [CrossRef]
10. Mayerle, R.; Nalluri, C.; Novak, P. Sediment transport in rigid bed conveyances. *J. Hydraul. Res.* **1991**, *29*, 475–495. [CrossRef]
11. May, R.W.P. *Sediment Transport in Pipes and Sewers with Deposited Beds*; Report SR 320; HR Wallingford: Wallingford, UK, 1993.
12. Ota, J.J.; Nalluri, C. Urban storm sewer design: Approach in consideration of sediments. *J. Hydraul. Eng.* **2003**, *129*, 291–297. [CrossRef]
13. Vongvisessomjai, N.; Tingsanchali, T.; Babel, M.S. Non-deposition design criteria for sewers with part-full flow. *Urban Water J.* **2010**, *7*, 61–77. [CrossRef]
14. Ota, J.J.; Perrusquia, G.S. Particle velocity and sediment transport at the limit of deposition in sewers. *Water Sci. Technol.* **2013**, *67*, 959–967. [CrossRef] [PubMed]
15. Loveless, J.H. Sediment Transport in Rigid Boundary Channels with Particular Reference to the Condition of Incipient Deposition. Ph.D. Thesis, University of London, London, UK, 1992.
16. Safari, M.J.S.; Aksoy, H.; Mohammadi, M. Incipient deposition of sediment in rigid boundary open channels. *Environ. Fluid Mech.* **2015**, *15*, 1053–1068. [CrossRef]

17. Safari, M.J.S.; Aksoy, H.; Mohammadi, M. Artificial neural network and regression models for flow velocity at sediment incipient deposition. *J. Hydrol.* **2016**, *541*, 1420–1429. [CrossRef]
18. Task Committee. Sediment transportation mechanics: Initiation of motion. *J. Hydraul. Div. ASCE* **1966**, *92(HY2)*, 291–314.
19. May, R.W.P. *Sediment Transport in Sewers*; Report No. IT 222; Hydraulic Research Station: Wallingford, UK, 1982.
20. Ackers, P. Sediment transport in sewers and the design implications. In Proceedings of the International Conference on Planning, Construction, Maintenance, and Operation of Sewerage Systems, Reading, UK, 12–14 September 1984; pp. 215–230.
21. Sanford, L.P.; Halka, J.P. Assessing the paradigm of mutually exclusive erosion and deposition of mud, with examples from upper Chesapeake Bay. *Mar. Geol.* **1993**, *114*, 37–57. [CrossRef]
22. Aksoy, H.; Safari, M.J.S. *Incipient Deposition of Sediment Particles in Rigid Boundary Channels*; Technical Report 113M062; Scientific and Technological Research Council of Turkey (TÜBİTAK): Istanbul, Turkey, 2014. (In Turkish)
23. Unal, N.E.; Aksoy, H.; Safari, M.J.S. *Self-Cleansing Drainage System Design by Incipient Motion and Incipient Deposition-Based Models*; Technical Report 114M283; Scientific and Technological Research Council of Turkey (TÜBİTAK): Istanbul, Turkey, 2016. (In Turkish)
24. Shields, A. *Application of Similarity Principles and Turbulence Research to Bed-Load Movement*; Translation from German; California Institute of Technology: Pasadena, CA, USA, 1936.
25. Yalin, M.S. *Mechanics of Sediment Transport*; Pergamon Press: Oxford, NY, USA, 1972.
26. Aksoy, H.; Safari, M.J.S.; Unal, N.E.; Mohammadi, M. Velocity-based analysis of incipient deposition in rigid boundary channels. *Water Sci. Technol.* **2017**, *76*, 2535–2543. [CrossRef] [PubMed]
27. Ippen, A.T.; Verma, R.P. *The Motion of Discrete Particles along the Bed of a Turbulent Stream: International Hydraulics Convention*; St. Anthony Falls Laboratory: Minneapolis, MN, USA, 1953; pp. 7–20.
28. Novak, P.; Nalluri, C. Sediment transport in smooth fixed bed channels. *J. Hydraul. Div. ASCE* **1975**, *101*, 1139–1154.
29. Butler, D.; May, R.W.P.; Ackers, J.C. Sediment transport in sewers Part 1: Background. *Proc. Inst. Civ. Eng.-Water Marit. Energy* **1996**, *118*, 103–112. [CrossRef]
30. Paphitis, D. Sediment movement under unidirectional flows: An assessment of empirical threshold curves. *Coas. Eng.* **2001**, *43*, 227–245. [CrossRef]
31. Fredsoe, J.; Deigaard, R. *Mechanics of Coastal Sediment Transport*; World Scientific: Singapore, 1992.
32. Novak, P.; Nalluri, C. Incipient motion of sediment particles over fixed beds. *J. Hydraul. Res.* **1984**, *22*, 181–197. [CrossRef]
33. Yang, C.T. Incipient motion and sediment transport. *J. Hydraul. Div. ASCE* **1973**, *99*, 1679–1704.
34. Vanoni, V.A. *Sedimentation Engineering*; American Society of Civil Engineers (ASCE): Reston, VA, USA, 2006.

© 2018 by the author. Licensee MDPI, Basel, Switzerland. This article is an open access article distributed under the terms and conditions of the Creative Commons Attribution (CC BY) license (http://creativecommons.org/licenses/by/4.0/).

Article

Riverbed Migrations in Western Taiwan under Climate Change

Yi-Chiung Chao [1], Chi-Wen Chen [1,2,*], Hsin-Chi Li [1] and Yung-Ming Chen [1]

1. National Science and Technology Center for Disaster Reduction, No. 200, Sec. 3, Beixin Road, Xindian District, New Taipei City 23143, Taiwan; ycchao@ncdr.nat.gov.tw (Y.-C.C.); hsinchi@ncdr.nat.gov.tw (H.-C.L.); ymchen@ncdr.nat.gov.tw (Y.-M.C.)
2. Center for Spatial Information Science, The University of Tokyo, 5-1-5 Kashiwanoha Kashiwa, Chiba 277-8568, Japan
* Correspondence: kevin4919@gmail.com; Tel.: +81-4-7136-4306

Received: 2 November 2018; Accepted: 8 November 2018; Published: 12 November 2018

Abstract: In recent years, extreme weather phenomena have occurred worldwide, resulting in many catastrophic disasters. Under the impact of climate change, the frequency of extreme rainfall events in Taiwan will increase, according to a report on climate change in Taiwan. This study analyzed riverbed migrations, such as degradation and aggradation, caused by extreme rainfall events under climate change for the Choshui River, Taiwan. We used the CCHE1D model to simulate changes in flow discharge and riverbed caused by typhoon events for the base period (1979–2003) and the end of the 21st century (2075–2099) according to the climate change scenario of representative concentration pathways 8.5 (RCP8.5) and dynamical downscaling of rainfall data in Taiwan. According to the results on flow discharge, at the end of the 21st century, the average peak flow during extreme rainfall events will increase by 20% relative to the base period, but the time required to reach the peak will be 8 h shorter than that in the base period. In terms of the results of degradation and aggradation of the riverbed, at the end of the 21st century, the amount of aggradation will increase by 33% over that of the base period. In the future, upstream sediment will be blocked by the Chichi weir, increasing the severity of scouring downstream. In addition, due to the increased peak flow discharge in the future, the scouring of the pier may be more serious than it is currently. More detailed 2D or 3D hydrological models are necessary in future works, which could adequately address the erosive phenomena created by bridge piers. Our results indicate that not only will flood disasters occur within a shorter time duration, but the catchment will also face more severe degradation and aggradation in the future.

Keywords: aggradation; CCHE1D; climate change; degradation; dynamical downscaling; flow discharge; migration; riverbed; sediment

1. Introduction

Global climate change has led to changes in the spatial and temporal distribution of precipitation, evaporation, and runoff, which in turn contribute to environmental issues such as frequent droughts, increased flooding, waterlogging disasters, and aggravated soil erosion [1]. At the present time, one of the main problems in soil erosion research around the world is to assess the impact of climate change on sediment cycling [2]. It is expected that the increase in global precipitation will affect the extent, frequency, and magnitude of soil erosion and sediment redistribution [3,4]. This could lead to more severe hydrological cycles [5] and increased rainfall erosivity [6]. Nevertheless, contrasting impacts have recently been found in different parts of the world. Zhao et al. [7] observed that owing to anthropogenic causes, the sediment export from the Loess Plateau of China has increased significantly, but with the recent climate change, stream flow and sediment load have decreased enormously. Foster et al. [8] analyzed changes in sediment transport for the Karoo uplands, South Africa, and found that the sediment yield has generally increased over the past decades, which is related to factors such as the increase in the frequency of high magnitude rainfall events.

Numerical models have been employed in engineering studies to predict the flow discharge and riverbed migrations along rivers. Several 1-D models have been proposed, and each model has its advantages and limitations [9]. The use of these models requires the proper simulation of hydraulic conditions [10]. The development of 1-D models is mainly focused on numerical aspects of the advection–dispersion equation [11–13] and exchange with dead zones [14–16]. Some 1-D models, such as the SIMCAT [17], QUAL2KW [18], and Multiphysics software COMSOL [19], are limited to steady flow conditions, while other models, such as the OTIS [20], CCHE1D [21], MIKE 11 [22], SD model [23], HEC-RAS [24], and ADISTS [25], can simulate unsteady flows and solute transport. In general, the validation testing focuses on theoretical situations or simplified river geometries in limited spatial and temporal scales [26].

Climate change is an important issue that has attracted growing attention in recent years. The report of the International Panel on Climate Change (IPCC) noted that rises in temperature and sea level are ongoing and that appropriate adaptations to reduce disaster risk are necessary [1]. In the current changing environment, both the probability of the occurrence of a strong typhoon and the rainfall intensity during a typhoon event will increase [27–29]. Many studies have explored the potential effects of different climate change scenarios on runoff and sediment yield in catchments [30–34]. In addition, there are also several studies focused on the impact of a changing climate on fluvial hydro-morphodynamics, as well as on the uncertainties related to future scenarios [35–41]. According to the Taiwan climate change report, the frequency of extreme rainfall events has also increased in the past 20 years in Taiwan [42]. Under the impact of climate change, extreme rainfall events will become more frequent, and riverbed migrations, such as degradation and aggradation, will become more serious, possibly affecting existing constructions and land use, and even human life and property. Therefore, the objectives of this study were to calibrate and validate a general one-dimensional channel network model for the Choshui River in western Taiwan and to assess changes in flow discharge and riverbed caused by typhoon events for the base period (1979–2003) and the end of the 21st century (2075–2099) according to the climate change scenario and dynamical downscaling of rainfall data in Taiwan.

2. Study Area

The Choshui River, located in western Taiwan, originates from the saddle between the main peak and east peak of Mount Hehuan. It flows through Nantou County, Chiayi County, Yunlin County, and Changhua County before reaching the Taiwan Strait. Having a mainstream about 187 km long, it is the longest river in Taiwan, and it has an average slope of 0.53%. The main tributaries of Choshui River are the Danda River, Jyunda River, Chenyoulan River, Tonpuze River, and Chingshui River. The Choshui River catchment covers an area of about 3157 km^2, which is the second largest catchment area in Taiwan. The topography slopes downward from east to west and the area of elevation >1000 m accounts for 64% of the catchment area. The slope decreases from east to west, and the area of slope >29% accounts for 62.5%, with an average slope of 54.5% (Figure 1).

The Choshui River catchment is characterized by friable Miocene and Pliocene sedimentary, metasedimentary, and low-grade metamorphic rocks and Quaternary alluvium deposits [43]. The soil cover in this area is mainly composed of colluvial soil, alluvial soil, and lithosol, with some yellow soil and laterite [44]. The Choshui River catchment can be divided into upstream and downstream areas at the Chunyun Bridge. The upstream area is well-covered with vegetation; the total area of natural forest and artificial forest is about 470 km^2, accounting for 92% of all land use. In the downstream area, agriculture occupies 1530 km^2, accounting for 85.5% of the land use, and the construction area is about 56 km^2, accounting for 3.14%.

The annual average rainfall in the Choshui River catchment is 2200 mm, but the spatial distribution of rainfall varies greatly. The annual rainfall in the upstream area can average 2500 mm, in contrast to 1100 mm in the downstream area and only 830 mm in the coastal area. The flow discharge of the Choshui River varies enormously between wet and dry seasons. About 80% of the total flow discharge is discharged during the wet season (May–October) [45]. According to [46], from the convergence of the Choshui River and the Chingshui River to the estuary, the flow discharges of the five-, 10-, 50-, and 100-year recurrence intervals are 0.65×10^4, 1.11×10^4, 1.44×10^4, 2.27×10^4, and 2.66×10^4 m^3/s, respectively.

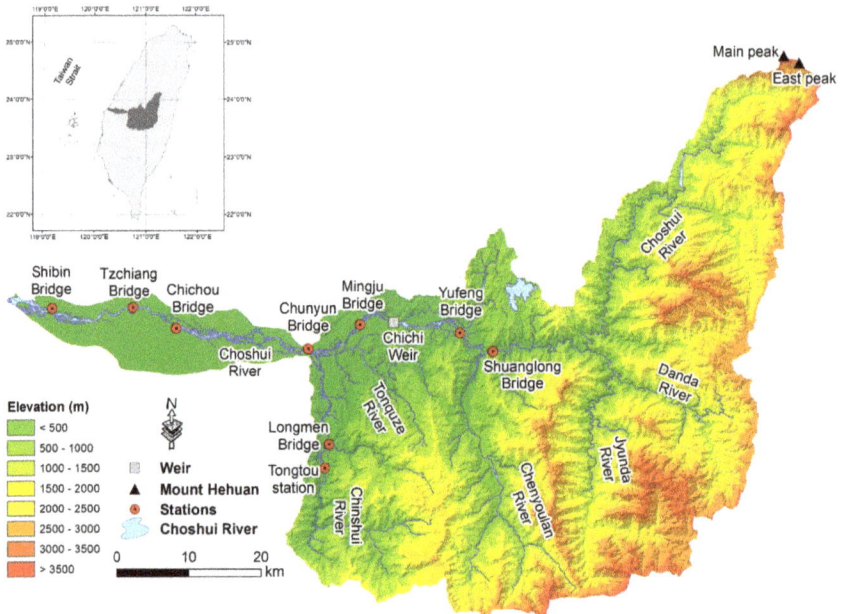

Figure 1. Choshui River catchment of Taiwan.

3. Data and Methods

3.1. CCHE1D Model

The National Center for Computational Hydroscience and Engineering of the University of Mississippi has developed a general one-dimensional channel network model that can simulate steady and unsteady flows and sedimentation processes in dendritic channel networks, including fractional sediment transport, riverbed aggradation and degradation, riverbed materials, bank erosion, and the resulting channel morphologic changes [21], which is called the CCHE1D model. The CCHE1D model solves the Saint–Venant equation and applies the implicit Preissmann four-point finite-difference and discrete-governing equation. A linear iterative method was developed in the model for the discrete-governing equation, which is solved by the double-sweep method. The one-dimensional hydraulic governing equations are as follows [47,48]:

Continuity equation:

$$\frac{\partial A}{\partial t} + \frac{\partial Q}{\partial x} = q \tag{1}$$

Momentum equation:

$$\frac{\partial}{\partial t}\left(\frac{Q}{A}\right) + \frac{\partial}{\partial x}\left(\frac{\beta Q^2}{2A^2}\right) + g\frac{\partial h}{\partial x} + g\left(S_f - S_0\right) = 0 \qquad (2)$$

where x and t are the spatial (m) and temporal (s) axes, A is the flow area (m^2), Q is the flow discharge (m^3/s), q is the side discharge per unit channel length (m^2/s), h is the water surface elevation (m), S_f and S_0 are the friction slope and riverbed slope (m/m), β is the correction factor due to the non-uniformity of velocity distribution in the cross section, and g is the gravitational acceleration (m/s^2).

In addition, when the inertial force is small, the diffusion wave model can be used for the hydraulic routing. In general, when the Froude number is less than 0.5, the first two items on the left side of Equation (2) can be ignored. Then the momentum equation can be rewritten into the diffusion wave model, and the equation is as follows:

$$\frac{\partial h}{\partial x} + S_f - S_0 = 0 \qquad (3)$$

Equation (3) can still be used when the Froude number is larger than 0.5 if the simulation case is close to a uniform flow, because the first two items on the left side of Equation (2) will be sufficiently small to be ignored.

In terms of the sediment transport model, the CCHE1D model uses the non-equilibrium sediment transport concept to calculate the non-uniform sediment transport in the river. The equation for the non-equilibrium and non-uniform sediment transport control of total load is expressed as follows:

$$\frac{\partial (AC_{tk})}{\partial t} + \frac{\partial Q_{tk}}{\partial x} + \frac{1}{L}(Q_{tk} - Q_{t*k}) = q_{lk} \qquad (4)$$

where C_{tk} is the section-averaged sediment concentration of size class k (kg/m^3), Q_{tk} is the actual sediment transport rate (kg/s), Q_{t*k} is the sediment transport capacity or the so-called equilibrium transport rate (kg/s), L is the non-equilibrium adaptation length of sediment transport (m), and q_{lk} is the side inflow or outflow sediment discharge from bank boundaries or tributary streams per unit channel length (kg/m/s).

The CCHE1D model offers the following four sediment discharge formulas for calculation: (1) SEDTRA module [49]; (2) Wu et al. formula [50]; (3) Modified Ackers and White's 1973 formula [51]; and (4) Modified Engelund and Hansen's 1967 formula [47]. In this study, we used the Wu et al. formula for our analyses because it is the most suitable formula for Taiwan, according to previous studies [52,53].

3.2. Parameter Setting for the Model

This study used the measured profile data of the Choshui River (2004, 2008, and 2012), provided by the 4th River Management Office of Taiwan, to build the model and set parameters. We used the measured profile data of 2004 as the initial topography for flood event simulations after 2004. Then the measured profile data of 2008 were used for calibrating the model, and those of 2012 were used for validation. We simulated the mainstream of the Choshui River for about 76 km from the Shuanglong Bridge (5 km upstream of the Yufeng Bridge) to the estuary. The tributary of the Choshui River, the Chingshui River, for about 13 km from the Longmen Bridge to the convergence with the mainstream was also simulated, and the results were used as the input data for the Choshui River.

This study used flow discharge during the wet seasons from 2004 to 2007 as the model calibration and continued to use flow discharge during the wet seasons from 2008 to 2011 as the validation. We used the change of time series in flow discharge during the period of the wet season for simulations. Flow discharge observed at the Yufeng Bridge station was used as the upstream boundary condition of the Choshui River, while flow discharge observed at the Tongtou station (2.5 km upstream of the

Longmen Bridge) was used as the upstream boundary condition of the Chingshui River. However, the Tongtou station was abandoned on 31 August 2009, so for points later than that, we used the flow discharge observed at the Longmen Bridge station (Figure 1). For the downstream, the open boundary method was applied for simulations.

Data of the grain size distribution of the riverbed of the Choshui River were measured in 2007, while those of the Chingshui River were measured in 2008 [46,54]. Manning's roughness coefficients for a riverbed of 0.029–0.044 were used, as recommended in government reports [46,54]. We applied the rating curves of flow discharge and sediment discharge to calculate the input of sediment, and also considered wash load in the calculations (Figure 2). Correlation coefficients (r^2) of rating curves for the Choshui River and the Chingshui River are 0.58 and 0.47 respectively. Although the r^2 is not very high, this study is mainly focused on riverbed migrations caused by heavy rainfall, which show a better correlation between flow discharge and sediment concentration when flow discharge is high (Figure 2). For the other general parameters for sediment transport simulation, this study set the suspended load adaptation coefficient (α) as 1.0, bed load adaptation length (L) as 1000 m, and mixing layer as 2.0 m. According to [46], nine groups of representative particle sizes were selected, ranging from 0.001 to 1000 m, and a porosity of 0.4 was used. Yeh et al. [50] have made a sensitivity analysis on parameters of CCHE1D model in the Choshui River and found that the sensitivity of most parameters is not high. Only the thickness of mixing layer is sensitive to the degradation and aggradation of riverbed. This study followed the model parameters suggested by them for the parameter setting to get the most accurate results.

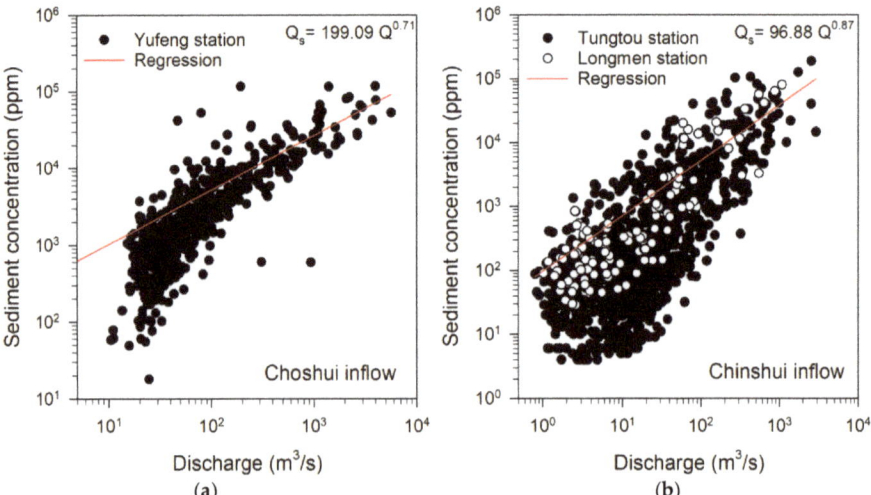

Figure 2. Rating curves of flow discharge and sediment concentration for (**a**) the Yufeng Bridge station (1994/2–2013/5), and (**b**) the Tongtou station (1994/6–2009/8) and the Longmen Bridge station (2009/10–2013/5).

3.3. Assessing Degradation and Aggradation Conditions of Rivers under Climate Change

The resolution of the general circulation model (GCM) used for climate estimation and simulation is usually 100–200 km. Consequently, it is difficult to identify the structure of a typhoon with a radius of less than several hundred kilometers. Owing to the low spatial resolution of the model, it is impossible to reasonably simulate the extreme weather system with a physical model, and it is impossible to carry out extreme weather estimation under climate change [55,56]. Therefore, this study used a climate projection simulated from the high-resolution (approximately 20 km horizontal resolution) atmospheric general circulation model (AGCM) developed by the Meteorological Research Institute (MRI) under the Japan Meteorological Agency (JMA) [57]. The low boundary drive by ensemble mean sea surface temperature (SST) of the climate change scenario of representative concentration pathways 8.5 (RCP8.5) in Coupled Model Intercomparison Project Phase 5 (CMIP5) [58] is called MRI-AGCM. This model has been validated for the simulation of typhoon generation [59,60]. However, the 20-km horizontal resolution still could not necessarily reflect the precipitation intensity caused by topographical effects in Taiwan. Therefore, the MRI-AGCM data were dynamically downscaled using the Weather Research and Forecasting Model (WRF), proposed by the US National Center for Atmospheric Research (NCAR), to obtain more realistic rainfall data for Taiwan with spatial and temporal resolutions of 5 km per hour.

Based on the downscaled rainfall data, there were 87 typhoons during the base period (1979–2003) and will be 43 typhoons in the last quarter of the 21st century. We selected eight typhoons in the base period and four typhoons at the end of the century, which accounted for the top 10% of the total rainfall for the Choshui River catchment among all typhoons in the two periods, to discuss the impact of climate change on riverbed migrations.

4. Results and Discussion

4.1. Model Calibration and Validation

This study used flow discharge during the wet seasons from 2004 to 2007 as the model calibration and continued to use flow discharge during the wet seasons from 2008 to 2011 as the validation. The total times for both periods of calibration and validation were 18,790 and 16,590 h, respectively. Figure 3 shows the comparison result between calibration and validation of the measured and simulated flow discharges at the Chunyun Bridge. We can find that the simulated peak flow discharge during typhoons is underestimated both in calibration and validation. Even so, trends of simulated and measured flow discharge are similar. The maximum percent errors of peak discharge (EQ_p) are −1.98% in the calibration and −9.50% in the validation. Overall, the coefficient of efficiency (CE) of the CCHE1D model in the calibration and validation are 0.85 and 0.83, respectively, indicating that the CCHE1D model can accurately simulate the rise and decline of peak discharge; therefore, we can use this model for simulating sediment transport under climate change scenarios.

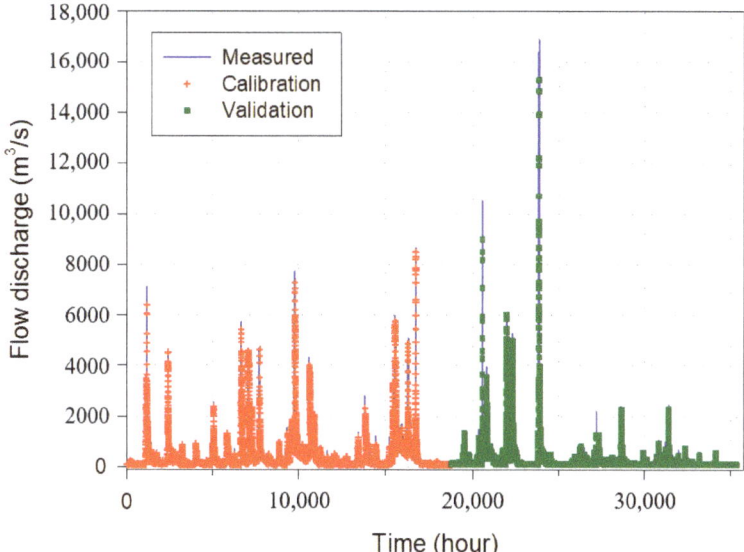

Figure 3. Calibration and validation of the measured and simulated flow discharge at the Chunyun Bridge.

In terms of degradation and aggradation in the river, this study used the measured profile data of 2004 as the initial topography for flood event simulations after 2004. Then the measured profile data of 2008 were used for calibrating the model and the measured profile data of 2012 were used for validation. Figure 4a shows the calibration result of the riverbed changes during 2004–2008 using the lowest point of each section along the river. The CCHE1D model can reasonably reproduce the process of degradation and aggradation in the river. Figure 4b shows the validation results of the riverbed changes for 2008–2012. It can be found that if this set of parameters is used, although the aggradation in downstream was underestimated, the riverbed changes in degradation and aggradation for most parts could be well validated. This represents that the CCHE1D model can truly grasp characteristics of degradation and aggradation in the Choshui River, so this model can also be used to simulate the degradation and aggradation in the river under climate change scenarios.

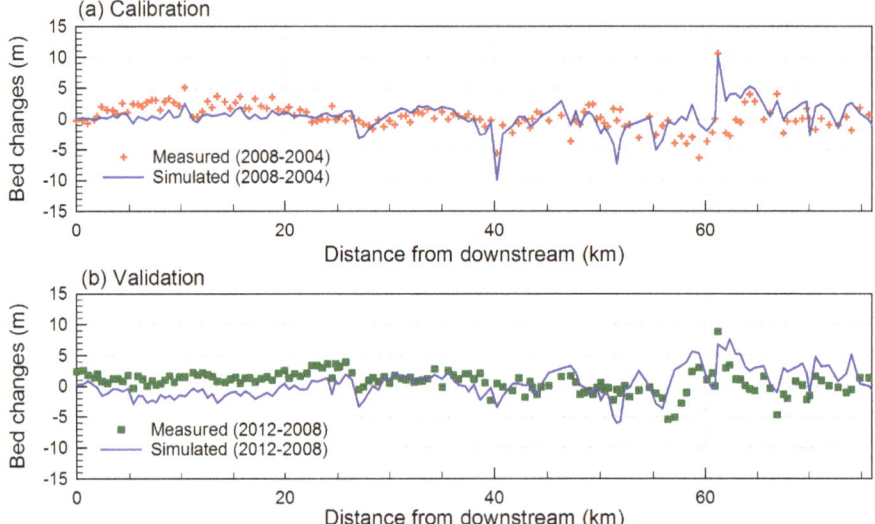

Figure 4. (a) Calibration result of the riverbed changes during 2004–2008 and (b) the validation result of the riverbed changes during 2008–2012 using the lowest point of each section along the river.

4.2. Peak Flow Discharge during Typhoons under Climate Change

Based on the downscaled rainfall data, there were 87 typhoons during the base period (1979–2003) and will be 43 typhoons in the last quarter of the 21st century (2075–2099). Table 1 shows the simulation results of peak flow discharge of each of the top 20 typhoons during the two periods, accounting for the top 20 of the total rainfall amounts for the Choshui River catchment among all typhoons, at the Chunyun Bridge. According to the government report [44], the peak flow discharge of the 100-year recurrence interval (Q_{100}) in the Choshui River before the Chingshui River merges with it is 25,394 m^3/s, which was used as the basis for the design of levees. We can find from the simulation results that the peak flow discharge during two typhoon events in the base period is greater than Q_{100}, while the peak flow discharge during three typhoon events at the end of the 21st century is greater than Q_{100}. This result indicates that, relative to the base period, the Choshui River is more likely to have a flood overflow at the end of the 21st century.

In addition, the peak flow discharge of the two-year recurrence interval (Q_2) in the Choshui River before the Chingshui River merge point is 5829 m^3/s [44]. The peak flow discharge caused by two typhoon events in the base period is less than Q_2, and the peak flow discharge of six typhoon events at the end of the 21st century is less than Q_2. This result suggests that the flood disasters caused by extreme rainfall events will be less frequent at the end of the 21st century than in the base period. We could assess from the above results that the sediment transport in the Choshui River was more frequent in the base period than it will be at the end of the 21st century; however, at the end of the 21st century, the severity of sediment migration will be more obvious than that in the base period.

Table 1. Simulation results of peak flow discharge of each of the top 20 typhoons during the two periods at the Chunyun Bridge.

Event Ranking	Base Period (m^3/s)	End of the 21st Century (m^3/s)
1	16,325	14,000
2	27,005 **	26,873 **
3	20,897	29,513 **
4	10,015	13,278
5	25,577 **	27,600 **
6	7422	15,706
7	23,515	23,062
8	8788	11,561
9	23,195	14,170
10	11,830	22,305
11	11,620	17,313
12	14,842	11,857
13	5071 *	9660
14	8931	8662
15	6519	5045 *
16	9048	2794 *
17	7128	2175 *
18	7977	3716 *
19	9437	3730 *
20	5341 *	1068 *

* represents that the peak flow discharge is less than that of the two-year recurrence interval (Q_2 = 5829 m^3/s).
** represents that the peak flow discharge is greater than that of the 100-year recurrence interval (Q_{100} = 25,394 m^3/s).

Here we selected the top 10% typhoons during the base period (1979–2003; n = 8) and during the end of the 21st century (2075–2099; n = 4), respectively, according to the total rainfall for the Choshui River catchment, for the following discussion on peak flow discharge along the whole reach of the Choshui River. The average peak flow discharges at the Yufeng Bridge in the base period and at the end of the 21st century are 17,443 and 20,916 m^3/s, respectively. On average, it takes about 48 and 40 h to reach the peak values. We can notice that the average peak flow caused by extreme rainfall events at the end of the 21st century will increase by 20% from the base period, but the time required to reach the peak at the end of the 21st century will be 8 h shorter than that in the base period. It means that we will face greater flood events and need more urgent response time in the future. Figure 5a shows the simulation result of the most severe flood event at the end of the 21st century (peak flow discharge = 29,513 m^3/s at the Chunyun Bridge). During the most severe flood event at the end of the 21st century, the locations where the overflow may occur are the estuary and the confluence of the Choshui River and Chingshui River (Figure 5b,c). Generally, in the case of normal energy transfer, energy transfer from upstream to downstream will be affected by topography and other factors, and energy will be lost, resulting in a lower water level downstream than upstream. The above water level is the result of quantitative simulations; however, the real situation is a dynamic flow. Therefore, the real downstream water level may be lower than the simulation result of the quantitative flow owing to energy losses, but it is still necessary to pay attention to flood control safety.

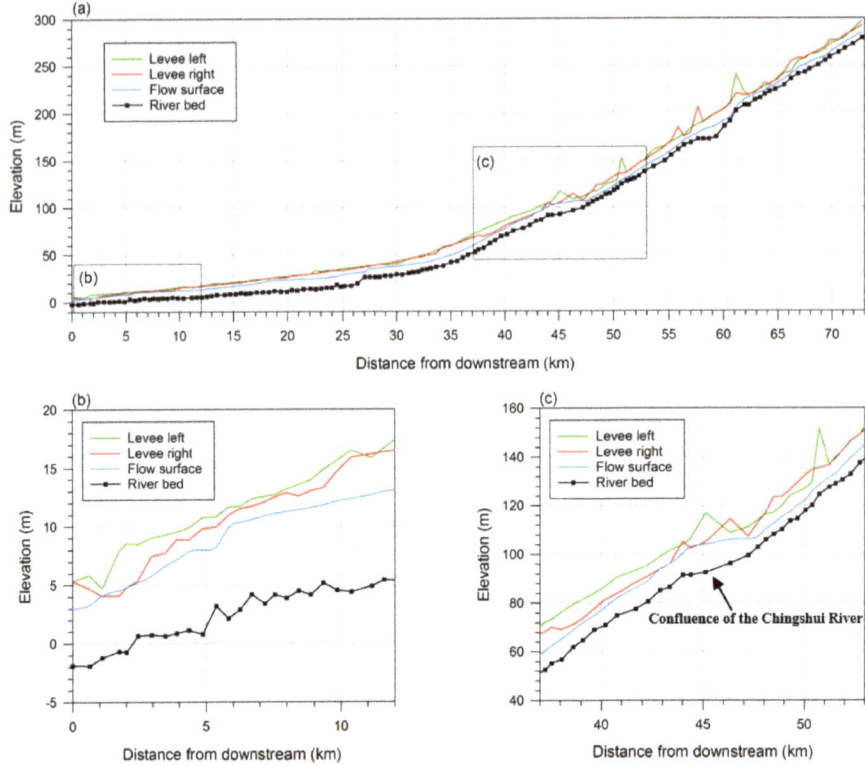

Figure 5. Simulation result of the most severe flood event at the end of the 21st century.

4.3. Changes in Degradation and Aggradation of the Riverbed under Climate Change

Changes in degradation and aggradation of the riverbed of the Choshui River under climate change scenarios are shown in Figure 6. The Choshui River can be divided into three simulation zones according to the artificial weir and the intersection of the main stream and tributary: (I) the estuary to the Chunyun Bridge; (II) the Chunyun Bridge to the Chichi weir; (III) the Chichi weir to the Shuanglong Bridge. It can be found from the results that zone I presents a slight aggradation from the base period to the end of the 21st century; however, the problem of pier scouring still requires attention. Zone II is located at lower reaches of the Chichi weir, so the riverbed suffers from strong changes in degradation and aggradation. At the end of the 21st century, the degree of degradation will be more serious than that in the base period. Because of the effect of the Chichi weir, zone III has obvious sediment accumulation in the upper part of the weir, resulting in the aggradation of the riverbed in the area. However, the location of the Yufeng Bridge shows clear degradation. Overall, the simulation results show that the entire area of simulated reaches is aggraded and that the amount of aggradation increases by about 33% at the end of the 21st century as compared with the base period.

Li et al. [61] pointed out that the landslide volume in the Choshui River catchment will increase by 14% at the end of the 21st century, relative to that of the base period. This should also be reflected in the results of degradation and aggradation of the riverbed. Landslides in the Choshui River catchment will increase in the future, and landslide materials will accumulate in rivers. On the other hand, the increase of peak flow discharge in the future will severely flush the surrounding structures and downstream areas, which may cause crises such as damaged bridges and dam dumping.

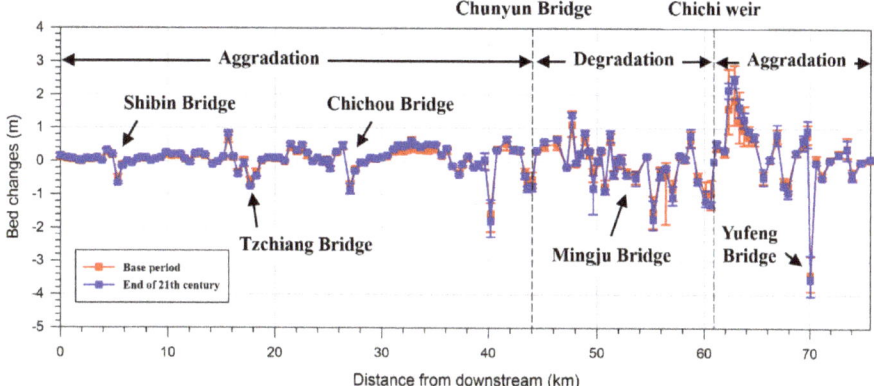

Figure 6. Changes in erosion and aggradation of the riverbed in the Choshui River under climate change scenarios.

Although degradation and aggradation of the riverbed are natural phenomena of a river, under the effect of climate change, the rainfall pattern will change to short duration-high intensity rainfall in the future [27–29]. This change will result in more serious degradation and aggradation of the riverbed in a catchment than occurs in the present situation. It could be imagined that not only will the response time during flood events be shortened, but also more serious riverbed migrations will occur in the future.

5. Conclusions

This study used the flow discharge during the wet seasons from 2004 to 2012 for calibration and validation of the CCHE1D model, a general one-dimensional channel network model that can simulate steady and unsteady flows and sedimentation processes in dendritic channel networks, along the Choshui River in western Taiwan. The CCHE1D model was used for simulating changes in flow discharge and riverbed caused by typhoon events for the base period (1979–2003) and the end of the 21st century (2075–2099). According to the climate change scenario of RCP8.5 and dynamically downscaled rainfall data in Taiwan, from the base period to the end of the 21st century, average peak flow caused by extreme rainfall events will increase by about 20%, but the time required to reach the peak at the end of the 21st century will be 8 h shorter than that in the base period. For the same time period, the amount of aggradation along the Choshui River will increase by about 33% from the base period to the end of the 21st century. Upper and lower parts of the Chichi weir will show large changes in aggradation and degradation, respectively, and the pier may suffer from serious scouring owing to the increase in peak flow discharge in the future. Our results could provide useful information to related authorities for the decision making on plans of public construction and land use in a changing climate. In future works, it is essential to use 2D or 3D hydrological models to adequately address problems such as the erosive phenomena created by bridge piers.

Author Contributions: Y.-C.C. collected the data, performed the statistical and simulation analysis, and drafted the manuscript. C.-W.C. conceived of the study, participated in its design and coordination, and drafted the manuscript. H.-C.L. and Y.-M.C. participated in the design of the study and provided methods. All authors read and approved the final manuscript.

Funding: This research was supported by the Taiwan Climate Change Projection and Information Platform (TCCIP; MOST106-2621-M-865-001).

Acknowledgments: We would like to thank the theme C of the Program for Risk Information on Climate Change of Japan (SOUSEI-C) for providing the MRI-AGCM data.

Conflicts of Interest: The authors declare no conflict of interest.

References

1. IPCC. *Climate Change 2007: The Physical Science Basis. Summary for Policymakers*; Contribution of working group I to the fourth assessment report; The Intergovernmental Panel on Climate Change: Geneva, Switzerland, 2007.
2. Mullan, D.; Favis-Mortlock, D.; Fealy, R. Addressing key limitations associated with modelling soil erosion under the impacts of future climate change. *Agric. Meteorol.* **2012**, *156*, 18–30. [CrossRef]
3. Peizhen, Z.; Molnar, P.; Downs, W.R. Increased sedimentation rates and grain sizes 2–4 Myr ago due to the influence of climate change on erosion rates. *Nature* **2001**, *410*, 891–897. [CrossRef] [PubMed]
4. Pruski, F.F.; Nearing, M.A. Climate-induced changes in erosion during the 21st century for eight US locations. *Water Resour. Res.* **2002**, *38*. [CrossRef]
5. Nearing, M.A.; Jetten, V.; Baffaut, C.; Cerdan, O.; Couturier, A.; Hernandez, M.; Le Bissonnais, Y.; Nichols, M.H.; Nunes, J.P.; Renschler, C.S.; et al. Modeling response of soil erosion and runoff to changes in precipitation and cover. *Catena* **2005**, *61*, 131–154. [CrossRef]
6. Nearing, M.A.; Pruski, F.F.; O'Neill, M.R. Expected climate change impacts on soil erosion rates: A review. *J. Soil Water Conserv. USA* **2004**, *59*, 43–50.
7. Zhao, G.; Mu, X.; Wen, Z.; Wang, F.; Gao, P. Soil erosion, conservation and eco-environment changes in the Loess Plateau of China. *Land Degrad. Dev.* **2013**, *24*, 499–510. [CrossRef]
8. Foster, I.D.L.; Rowntree, K.M.; Boardman, J.; Mighall, T.M. Changing sediment yield and sediment dynamics in the Karoo uplands, South Africa; post-European impacts. *Land Degrad. Dev.* **2012**, *23*, 508–522. [CrossRef]
9. Cox, B.A. A review of currently available in-stream water-quality models and their applicability for simulating dissolved oxygen in lowland rivers. *Sci. Total. Environ.* **2003**, *314*, 335–377. [CrossRef]
10. Leibundgut, C.; Maloszewski, P.; Küülls, C. *Tracers in Hydrology*; Wiley-Blackwell: Hoboken, NJ, USA, 2009.
11. Russell, T.F.; Celia, M.A. An overview of research on Eulerian–Lagrangian Localised Adjoint Methods (ELLAM). *Adv. Water Resour.* **2002**, *25*, 1215–1231. [CrossRef]
12. Rubio, A.D.; Zalts, A.; El Hasi, C.D. Numerical solution of the advection–reaction–diffusion equation at different scales. *Environ. Model. Softw.* **2008**, *23*, 90–95. [CrossRef]
13. Shen, C.; Phanikumar, M.S. An efficient space-fractional dispersion approximation for stream solute transport modeling. *Adv. Water Resour.* **2009**, *32*, 1482–1494. [CrossRef]
14. Bencala, K.E.; Walters, R.A. Simulation of solute transport in a mountain pool-and-riffle stream: A transient storage model. *Water Resour. Res.* **1983**, *19*, 718–724. [CrossRef]
15. Wörman, A.; Packman, A.I.; Johansson, H.; Jonsson, K. Effect of flow-induced exchange in hyporheic zones on longitudinal transport of solutes in streams and rivers. *Water Resour. Res.* **2002**, *38*, 61. [CrossRef]
16. Anderson, E.J.; Phanikumar, M.S. Surface storage dynamics in large rivers: Comparing three-dimensional particle transport, one-dimensional fractional derivative, and multirate transient storage models. *Water Resour. Res.* **2011**, *47*. [CrossRef]
17. UK Environment Agency. *SIMCAT 7.6: A Guide and Reference for Users*; UK Environment Agency: Bristol, UK, 2001.
18. Pelletier, G.J.; Chapra, C.S.; Tao, H. QUAL2Kw, A framework for modeling water quality in streams and rivers using a genetic algorithm for calibration. *Environ. Model. Softw.* **2006**, *21*, 419–425. [CrossRef]
19. Ani, E.C.; Wallis, S.; Kraslawski, A.; Agachi, P.S. Development, calibration and evaluation of two mathematical models for pollutant transport in a small river. *Environ. Model. Softw.* **2009**, *24*, 1139–1152. [CrossRef]
20. Runkel, R.L. *One-Dimensional Transport with Inflow and Storage (OTIS): A Solute Transport Model for Streams and Rivers*; Water-Resources Investigations Report; United States Geological Survey: Reston, VA, USA, 1998; 80p.
21. Vieira, D.A.N. Integrated Modeling of Watershed and Channel Processes. Ph.D. Thesis, University of Mississippi, Oxford, MS, USA, 2004.
22. Danish Hydraulic Institute (DHI). *MIKE 11-A Modelling System for Rivers and Channels*; Danish Hydraulic Institute: Horsholm, Denmark, 2007.
23. Deng, Z.Q.; Jung, H.S. Scaling dispersion model for pollutant transport in rivers. *Environ. Model. Softw.* **2009**, *24*, 627–631. [CrossRef]
24. US Army Corps of Engineers (USACE). *HEC-RAS River Analysis System*; User's Manual—Version 4.1; USACE: Washington, DC, USA, 2010.

25. Launay, M.; Le Coz, J.; Camenen, B.; Walter, C.; Angot, H.; Dramaisa, G.; Faure, J.F.; Coquery, M. Calibrating pollutant dispersion in 1-D hydraulic models of river networks. *J. Hydro-Environ. Res.* **2015**, *9*, 120–132. [CrossRef]
26. Zerihun, D.; Furman, A.; Warrick, A.W.; Sanchez, C.A. Coupled surface–subsurface solute transport model for irrigation borders and basins. I. model development. *J. Irrig. Drain. Eng.* **2005**, *131*, 396–406. [CrossRef]
27. Emanuel, K. Increasing destructiveness of tropical cyclones over the past 30 years. *Nature* **2005**, *436*, 686–688. [CrossRef] [PubMed]
28. Webster, P.J.; Holland, G.J.; Curry, J.A.; Chang, H.R. Changes in tropical cyclone number, duration, and intensity in a warming environment. *Science* **2005**, *309*, 1844–1846. [CrossRef] [PubMed]
29. Tu, J.Y.; Chou, C.; Chu, P.S. The abrupt shift of typhoon activity in the vicinity of Taiwan and its association with western North Pacific–East Asian climate change. *J. Clim.* **2009**, *22*, 3617–3628. [CrossRef]
30. Bussi, G.; Francés, F.; Horel, E.; López-Tarazón, J.A.; Batalla, R.J. Modelling the impact of climate change on sediment yield in a highly erodible Mediterranean catchment. *J. Soils Sediments* **2014**, *14*, 1921–1937. [CrossRef]
31. Lee, S.Y.; Hamlet, A.F.; Grossman, E.E. Impacts of climate change on regulated streamflow, hydrologic extremes, hydropower production, and sediment discharge in the Skagit river basin. *Northwest Sci.* **2016**, *90*, 23–43. [CrossRef]
32. Laura, M.; Tartari, G.; Salerno, F.; Valsecchi, L.; Bravi, C.; Lorenzi, E.; Genoni, P.; Guzzella, L. Climate change impacts on sediment quality of Subalpine reservoirs: Implications on management. *Water* **2017**, *9*, 680. [CrossRef]
33. Ren, Z.; Feng, Z.; Li, P.; Wang, D.; Cheng, S.; Gong, J. Response of runoff and sediment yield from climate change in the Yanhe watershed, China. *J. Coast. Res.* **2017**, *80*, 30–35. [CrossRef]
34. Dahl, T.A.; Kendall, A.D.; Hyndman, D.W. Impacts of projected climate change on sediment yield and dredging costs. *Hydrol. Process.* **2018**, *32*, 1223–1234. [CrossRef]
35. Ashmore, P.; Church, M. *The Impact of Climate Change on Rivers and River Processes in Canada*; Geological Survey of Canada: Ottawa, ON, Canada, 2001; Volume 555.
36. Bronstert, A. Floods and climate change: Interactions and impacts. *Risk Anal. Int. J.* **2003**, *23*, 545–557. [CrossRef]
37. Pfister, L.; Kwadijk, J.; Musy, A.; Bronstert, A.; Hoffmann, L. Climate change, land use change and runoff prediction in the Rhine–Meuse basins. *River Res. Appl.* **2004**, *20*, 229–241. [CrossRef]
38. Döll, P.; Zhang, J. Impact of climate change on freshwater ecosystems: A global-scale analysis of ecologically relevant river flow alterations. *Hydrol. Earth Syst. Sci.* **2010**, *14*, 783–799. [CrossRef]
39. Guerrero, M.; Nones, M.; Saurral, R.; Montroull, N.; Szupiany, R.N. Parana River morphodynamics in the context of climate change. *Int. J. River Basin Manag.* **2013**, *11*, 423–437. [CrossRef]
40. Piniewski, M.; Meresa, H.K.; Romanowicz, R.; Osuch, M.; Szcześniak, M.; Kardel, I.; Okruszko, T.; Mezghani, A.; Kundzewicz, Z.W. What can we learn from the projections of changes of flow patterns? Results from Polish case studies. *Acta Geophys.* **2017**, *65*, 809–827. [CrossRef]
41. Garijo, C.; Mediero, L. Influence of climate change on flood magnitude and seasonality in the Arga River catchment in Spain. *Acta Geophys.* **2018**, *66*, 769–790. [CrossRef]
42. Hsu, H.H.; Chou, C.; Wu, Y.C.; Lu, M.M.; Chen, C.T.; Chen, Y.M. *Climate Change in Taiwan: Scientific Report 2011 (Summary)*; National Science Council: Taipei, Taiwan, 2011.
43. Ho, C.S. *An Introduction to the Geology of Taiwan: Explanatory Text of the Geologic Map of Taiwan*; Central Geological Survey, Ministry of Economic Affairs: Taipei, Taiwan, 1988.
44. Water Resources Planning Institute (WRPI). *Review Report on Jhuo-Shuei River Basin Integrated Regulation Plan*; Water Resources Agency, Ministry of Economic Affairs: Taichung, Taiwan, 2011.
45. Water Resources Planning Institute (WRPI). *A Study of Flood Control and Sediment Management Due to Climate Change of Jhuoshuei River*; Water Resources Agency, Ministry of Economic Affairs: Taichung, Taiwan, 2013.
46. Water Resources Planning Institute (WRPI). *Review Report on River Planning of Jhuo-Shuei River*; Water Resources Agency, Ministry of Economic Affairs: Taichung, Taiwan, 2007.
47. Wu, W.; Vieira, D.A. *One-Dimensional Channel Network Model CCHE1D Version 3.0—Technical Manual*; The University of Mississippi, National Center for Computational Hydroscience and Engineering: Oxford, MS, USA, 2002.

48. Wu, W.; Vieira, D.A.; Wang, S.S.Y. One-Dimensional Numerical Model for Nonuniform Sediment Transport under Unsteady Flows in Channel Networks. *J. Hydraul. Eng.* **2004**, *130*, 914–923. [CrossRef]
49. Garbrecht, J.; Kuhnle, R.; Alonso, C. A Sediment Transport Capacity Formulation for Application to Large Channel Networks. *J. Soil Water Conserv.* **1995**, *50*, 527–529.
50. Wu, W.; Wang, S.S.Y.; Jia, Y. Nonuniform sediment transport in alluvial rivers. *J. Hydraul. Res.* **2000**, *38*, 427–434. [CrossRef]
51. Proffitt, G.T.; Sutherland, A.J. Transport of non-uniform sediments. *J. Hydraul. Res.* **1983**, *21*, 33–43. [CrossRef]
52. Yeh, K.C.; Wang, S.Y.; Wu, W.M.; Lien, H.C.; Liao, C.T. *Implement and Application Study of NCCHE's River Migration Models*; Water Resources Planning Institute, WRA: Taichung, Taiwan, 2007.
53. Liao, C.T.; Yeh, K.C.; Huang, M.W. Development and application of 2-D mobile-bed model with bedrock river evolution mechanism. *J. Hydro-Environ. Res.* **2014**, *8*, 210–222. [CrossRef]
54. Water Resources Planning Institute (WRPI). *Report on River Planning of Cing-Shuei River, Tributary of Jhuo-Shuei River*; Water Resources Agency, Ministry of Economic Affairs: Taichung, Taiwan, 2008.
55. Roble, R.G.; Ridley, E.C. A thermosphere-ionosphere-mesosphere-electrodynamics general circulation model (TIME-GCM): Equinox solar cycle minimum simulations (30–500 km). *Geophys. Res. Lett.* **1994**, *21*, 417–420. [CrossRef]
56. Johns, T.C.; Carnell, R.E.; Crossley, J.F.; Gregory, J.M.; Mitchell, J.F.; Senior, C.A.; Tett, S.F.B.; Wood, R.A. The second Hadley Centre coupled ocean-atmosphere GCM: Model description, spinup and validation. *Clim. Dynam.* **1997**, *13*, 103–134. [CrossRef]
57. Mizuta, R.; Yoshimura, H.; Murakami, H.; Matsueda, M.; Endo, H.; Ose, T.; Kamiguchi, K.; Hosaka, M.; Sugi, M.; Yukimoto, S.; et al. Climate simulations using MRI-AGCM3.2 with 20-km grid. *J. Meteor. Soc. Jpn.* **2012**, *90A*, 233–258. [CrossRef]
58. Mizuta, R.; Arakawa, O.; Ose, T.; Kusunoki, S.; Endo, H.; Kitoh, A. Classification of CMIP5 future climate responses by the tropical sea surface temperature changes. *SOLA* **2014**, *10*, 167–171. [CrossRef]
59. Murakami, H.; Wang, Y.; Yoshimura, H.; Mizuta, R.; Sugi, M.; Shindo, E.; Adachi, Y.; Yukimoto, S.; Hosaka, M.; Kusunoki, S.; et al. Future changes in tropical cyclone activity projected by the new high-resolution MRI-AGCM. *J. Clim.* **2012**, *25*, 3237–3260. [CrossRef]
60. Murakami, H.; Hsu, P.C.; Arakawa, O.; Li, T. Influence of model biases on projected future changes in tropical cyclone frequency of occurrence. *J. Clim.* **2014**, *27*, 2159–2181. [CrossRef]
61. Li, H.C.; Shih, H.J.; Wu, T.; Chao, Y.C.; Chen, W.B.; Cheng, C.T.; Chen, D.R.; Jang, J.H.; Chen, Y.M. *The Loss Assessment for Cho-Shui River Basin in the End of Century under Climate Change*; NCDR 104-T09; National Science and Technology Center for Disaster Reduction: Taipei, Taiwan, 2015.

© 2018 by the authors. Licensee MDPI, Basel, Switzerland. This article is an open access article distributed under the terms and conditions of the Creative Commons Attribution (CC BY) license (http://creativecommons.org/licenses/by/4.0/).

Article

A Shear Reynolds Number-Based Classification Method of the Nonuniform Bed Load Transport

Gergely T. Török [1,2,*], János Józsa [1,2] and Sándor Baranya [2]

1   MTA-BME Water Management Research Group, Hungarian Academy of Science—Budapest University of Technology and Economics, Műegyetem rakpart 3, H-1111 Budapest, Hungary; jozsa.janos@epito.bme.hu
2   Department of Hydraulic and Water Resources Engineering, Faculty of Civil Engineering, Budapest University of Technology and Economics, Műegyetem rakpart 3, H-1111 Budapest, Hungary; baranya.sandor@epito.bme.hu
*   Correspondence: torok.gergely@epito.bme.hu; Tel.: +36-1-463-2248

Received: 30 November 2018; Accepted: 24 December 2018; Published: 3 January 2019

Abstract: The aim of this study is to introduce a novel method which can separate sand- or gravel-dominated bed load transport in rivers with mixed-size bed material. When dealing with large rivers with complex hydrodynamics and morphodynamics, the bed load transport modes can indicate strong variation even locally, which requires a suitable approach to estimate the locally unique behavior of the sediment transport. However, the literature offers only few studies regarding this issue, and they are concerned with uniform bed load. In order to partly fill this gap, we suggest here a decision criteria which utilizes the shear Reynolds number. The method was verified with data from field and laboratory measurements, both performed at nonuniform bed material compositions. The comparative assessment of the results show that the shear Reynolds number-based method operates more reliably than the Shields–Parker diagram and it is expected to predict the sand or gravel transport domination with a <5% uncertainty. The results contribute to the improvement of numerical sediment transport modeling as well as to the field implementation of bed load transport measurements.

Keywords: bed load transport; shear Reynolds number; mixed-size bed material; complex morphodynamics

1. Introduction

In terms of typical geomorphological features, rivers can be divided into three main section types: upper, middle and lower course [1]. One of the decisive differences among these river features is the erosion capacity of the flow, which constantly decreases along the river. In fact, it can be stated that the erosion prevails along the upper course rivers, yielding coarser gravel bed material and significant bed load transport. As to the middle course, all the three characteristic sediment transport processes, such as the local erosion of the bed, the transport of coarse and fine particles together with river bed aggradation take place. Finally, the lower course type rivers can be generally characterized with the deposition of fine sediments transported from upstream. Due to the spatially and temporally varying bed load transport capacity of the flow and the changing river planforms (straight, meandering, or braided river channel patterns [2]), the morphological features can also show significant variability.

Many classification methods can be found in the literature to describe the morphological properties of river sections (e.g., [3–7]). The main goal of these tools is to typify and predict the dominant morphological properties and processes, such as defining the predominant sediment transport mode (bed load, or suspended load) [8], the channel pattern type [1,9], the specific bed material grain size [7], the erosion capacity [10], or the bed armor measure [6]. According to these methods, the morphological properties and processes can be estimated as the function of a few hydro-morphological variables, e.g., the water depth, the channel width, the mean flow velocity, the mean grain size of the bed material, or the longitudinal bed slope.

For instance, there are classification methods, suggesting that the dominant grain size of the bed load (sand or gravel) can be obtained based on the d_{50} grain size of the bed material [8]. In most cases, such a classification is straightforward due to the fact that a dominant bed material fraction can be chosen, considering, e.g., sand-bed (d_{50} < 2 mm) or gravel-bed (d_{50} > 2 mm) streams [8]. However, there are several situations, when such a clear distinction cannot be made as both sand- and gravel-dominated zones appear even in shorter river reaches. Moreover, the locally dominant bed load fraction can alter according to the current flow conditions. So, only the motion of finer particles of a nonuniform bed material can occur during low-water regime. In turn, the higher flood waves increase the sediment transport capacity of the flow, resulting in probably gravel-dominated bed load. That is, because the bed load is mainly fed by the bed itself [11], very different grain sizes can characterize the bed load transport within river sections with mixed-size bed material. These sections are typical to middle-course rivers, at the transition sections in terms of bed material.

Case Study

Such a river reach is the upper Hungarian Danube reach between rkm 1798 and rkm 1795 (Figure 1). At this section, the middle course resulted in a meandering pattern type, with nonuniform bed material (d_{50} ranges between 0.32 and 70.5 mm) [9,12]. Furthermore, conventional river regulation measures (e.g., groin fields) were installed, which further enhanced the diversity of the morphological properties and processes [13–17]. The following parameters characterize the river: the main channel width at mean water regime ranges between 150 m and 350 m; the average water surface gradient is around 0.0002–0.00025 [17], whereas the characteristic flow discharges are Q_m = 2000 m^3/s (mean flow), Q_{bf} = 4300–4500 m^3/s (range of bankfull discharge), Q_{100} = 10,400 m^3/s (100 year flood event) [18] and Q_{200} = ~10,800 m^3/s (200 year flood event) [19]. Here, the dominant fraction in the bed material, and hence in the bed load, shows a strongly varying spatial distribution. That is, bed armoring and gravel bed load takes place in the main channel, while clear sand transport can be observed in the near-bank zones and in the groin fields. Finally, the accumulation and erosion of the gravel–sand mixture is detectable at some places, resulting in the formation of side bars (Figure 1).

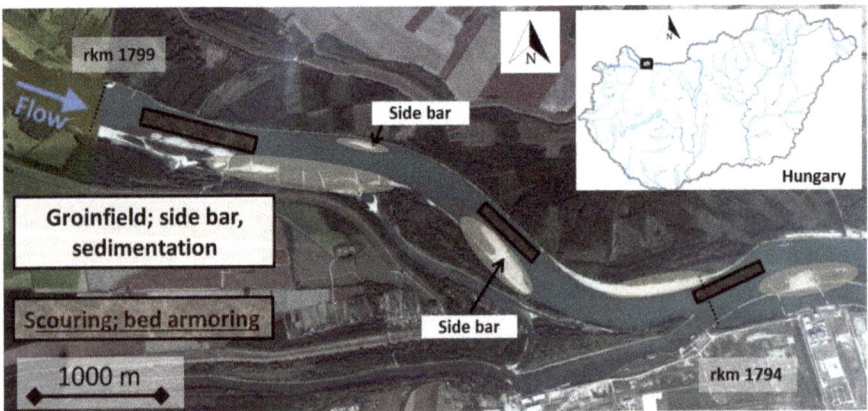

Figure 1. The sketch of the investigated section of the Danube River.

The herein-presented study focuses on this issue: how the dominant fraction of the bed load transport can be determined, when both the river hydrodynamics and the morphological features show strong heterogeneity within the studied reach. When local or reach-scale morphodynamic analysis is carried out, it is far from evident what sort of empirical sediment transport formula to use. There can be found formulas for finer fractions (e.g., [20]), as well as for coarser fractions (e.g., [21]), both having limitations when considering mixed-size bed material. However, no such straightforward method can be found in the literature, which suggests the locally dominant bed load transport mode and thus the more suitable formula for nonuniform bed material. An attempt is made in this study to fill this gap, providing a methodology that works more accurately compared to earlier published methods.

2. Materials and Methods

2.1. Problem Statement

Significant variability of the dominant grain size in the bed load can take place in nonuniform bed material, e.g., in the case of the investigated Danube reach. Figure 2 shows the density function of a bimodal bed material sample, collected from the lower side bar (see Figure 1, green columns, left vertical axis). The black line (and the right vertical axis) represents the critical dimensionless bed shear stress values (or critical Shields number, Equation (1)), as the function of the grain size, calculated by the Shields curve [22]. If the dimensionless bed shear stress exceeds the critical value, the given particle is in motion, otherwise it stays calm.

$$\tau^* = \frac{\tau}{g(\rho_s - \rho_w)d}, \quad (1)$$

where τ^* is the Shields number, τ is the bed shear stress, g is the acceleration due to gravity, ρ_s is the sediment density, ρ_w is the water density and d is the grain diameter.

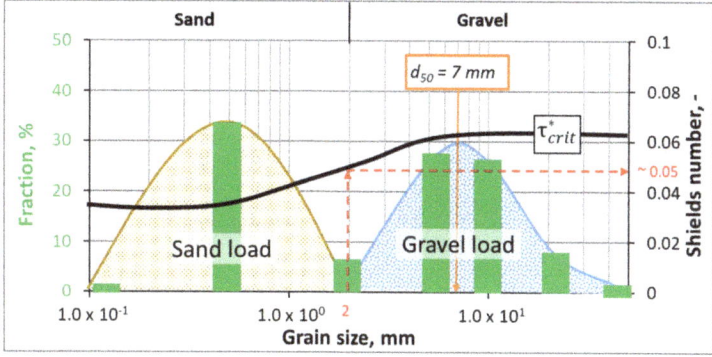

Figure 2. The density function of a bed material sample collected from a side bar (green curve). The Shields curve-related critical bed shear stress values are depicted by the black line.

As the fraction rates indicate, sand and gravel also appear well separately, both with significant amount (>30%). The grain size distribution indicates a d_{50} of 7 mm, which would refer to gravel transport. However, examining the critical Shields numbers, it turns out that the dominant bed load grain size seems to be inconstant, varying depending on the flow conditions. If the dimensionless bed shear stress is in a lower range ($\tau^* <$ ~0.05, or equivalently $\tau <$ ~2 N/m^2), only sand (<2 mm) movement can be expected. In turn, increasing dimensionless bed shear stress ($\tau^* >$ 0.05, or equivalently $\tau >$ ~2 N/m^2) results in the motion of even the coarser gravels (>2 mm), while the motion of the finer sand fraction tends to be rather suspended, like rolling [8], resulting in even coarser bed load. Concluding the above, the d_{50} grain size of the bed material itself is not able to indicate the current dominant grain size in the bed load, especially not in the case of nonuniform bed material under varying flow conditions (for further similar study cases see, e.g., [23,24]).

When implementing computational morphodynamic models in cases like the above-presented mixed-size bed material, the variable dominant fraction in the bed load can lead to inappropriate application of sediment transport models. Nevertheless, the numerical simulation of morphodynamic processes can contribute to the better performance of river engineering activities, e.g., when planning restoration measures, and it is considered a widely applied investigation tool with many benefits. However, the reliable numerical description of the interaction between river hydrodynamics and the morphological features is a challenging task in rivers with uniform bed material composition, but shows even higher difficulties in the case of mixed-size bed material conditions (e.g., [25–29]). When the grain sizes cover a smaller range in the bed material, the potential erosion results also in a quite uniform grain size range of the bed load transport. For such conditions, many bed load sediment transport models have already been developed and successfully validated [22,30–33]. On the other hand, the choice of the bed load sediment transport formula is still a challenge when the characteristic grain sizes can significantly change both in time and space [23,25,34–36]. Accordingly, the most suitable numerical formula can change in time and space. For instance, as many studies present [34,37–44], the van Rijn model [20] could provide better estimation of sediment transport in sand-dominated zones, whereas the Wilcock and Crowe model [21] suits gravel-dominated mixed-size cases better. Török et al. [34] also proved that the combined and parallel application of these two models provides better results when simulating sediment transport in mixed-size bed material. In the referred study, a physical indicator was defined as a parameter to decide which formula is to be applied. It is, however, not evident what this parameter should be. In such cases, the prediction precision of the dominant fraction in the bed load transport plays an emphatic role in the accuracy of the sediment transport calculation possibilities.

2.2. Existing Method for Bed Load Classification

One applicable method, the Shields–Parker river sedimentation diagram, can be used as a comprehensive bed-material and sediment transport nature classification tool for a particular river reach, with uniform bed content [7]. The diagram indicates the Shields number (τ^*, or equivalently τ^*_{bf}, Equation (1)) based on the d_{50} specific mean grain size versus the so-called explicit particle Reynolds number (Re_p, Equation (2), where $d = d_{50}$ in the case of the Shields–Parker diagram). The diagram includes the well-known Shields curve (continuous black curve), which shows if the given particle is in motion (points above the curve), or remains still on the bed surface (points below the curve) [22].

$$Re_p = \frac{\sqrt{gRd}\,d}{\nu}, \qquad (2)$$

where $R = (\rho_s - \rho_W)/\rho_W$, ρ_W is the water, ρ_s is the sediment density, and ν is the kinematic viscosity of the fluid.

Parker stated [8] that alluvial rivers can usually be divided into two types based on the function of particle Reynolds number (Re_p, Equation (2)): sand-bed and gravel-bed rivers. According to Parker, a distinction can be defined based on Re_p, which actually depends only on d_{50} (therefore it can, in fact, be considered as the dimensionless substitution of the grain size). That is, considering that $d = 2$ mm is the threshold value between the sand and gravel fractions, the critical value for Re_p also can be calculated according to Equation (1), where $d = 2$ mm results in $Re_p = 360$ ($Re_p > 360 \rightarrow$ gravel bed, $Re_p < 360 \rightarrow$ sand-bed), which critical value is depicted by a vertical narrow in Figure 3, within the Re_p range of 100–1000.

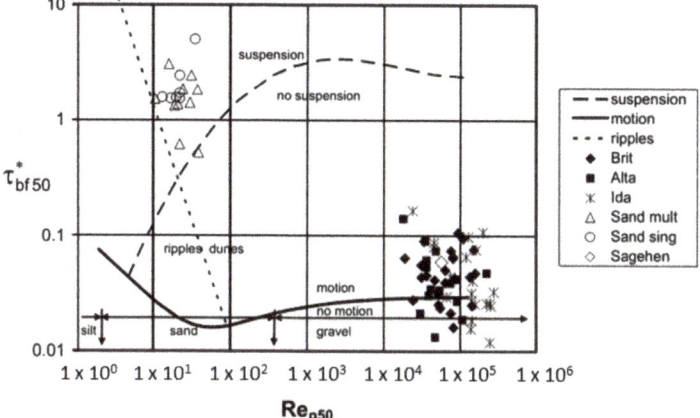

Figure 3. Shields–Parker diagram [7].

Consequently, the method basically yields a grain size-based classification method for near-uniform bed materials. This simplified approach is not capable of classifying characteristic local morphodynamic processes for complex situations in terms of river bed morphology and inhomogeneous bed material composition [7] (e.g., in the case of bed material in Figure 2).

The basic idea here is to keep the simplicity of the classification method shown by Parker, but at the same time to involve a new indicator, which accounts for the local hydrodynamic effects that better characterize the interaction between the water flow and the river bed, i.e., the bed shear stress. Thus, based on literature review, field and laboratory measurements, a conjecture is introduced in this paper. For a better understanding of the herein-proposed methodology, the most relevant findings of the related publications are summarized in the following.

2.3. Introducing the Novel Method for Bed Load Classification

Shields published a well-known study regarding sediment transport investigation in 1936 [22]. The research work investigated the critical state of the bed-forming particles, applying a mathematical model of forces acting on a solitary particle. Shields pointed out that the rate of the acting forces is not constant, but the relation of the effective force of the flow to the resistance of a grain can be expressed with a universal function of the shear Reynolds number (Re^*, Equation (3)). The interpretation of the shear Reynolds number is expressed in Equation (3): the ratio of grain size to the thickness of the laminar boundary layer.

$$Re^* = \frac{u_* d}{\nu} = \frac{d}{\nu/u_*} = \frac{Grain\ diameter}{Thickness\ of\ viscous\ layer}, \quad (3)$$

where u^* is the shear velocity ($u_* = (\frac{\tau}{\rho})^{0.5}$, where τ is the local bed shear stress). The following table demonstrates the difference between the shear Reynolds number and the explicit particle Reynolds number. The constant and the variable parameters in Equations (2) and (3) are depicted in Table 1. The last column indicates the ratio of Re^* to Re_p. The main outcome of the table is that Re^* implicates not only the grain size (d), but also a term regarding the acting forces of the flow (u^*). This remark confirms the Re^*-based investigation of Shields, instead of the Re_p-based method.

Table 1. The constant and the variable parameters in the equations of the shear Reynolds number (Re^*) and the explicit particle Reynolds number (Re_p) (Equations (2) and (3)).

	Constants [1]	Variables	Re^*/Re_p
Re^*	ν	d, u_*	$u_*/\sqrt{d}$
Re_p	ν, g, R	d	

[1] The values of the constant terms are: $\nu = 10^{-6}\,\frac{m^2}{s}$, $g = 9.81\,\frac{m}{s^2}$, $R = 1.65$.

The Shields curve, actually, shows the relation between the shear Reynolds number and the acting forces [22,45]. The investigation proves that the acting forces and the stability of a grain depend not on the grain sizes, but on the hydraulic regime formed directly above the grains [7,46,47] (Considering the possible development of different kinds of turbulent flow pattern, e.g., in sand-bed rivers bed-form-induced turbulent flow, while in gravel-bed rivers, friction-induced turbulent pattern can occur [20,32,48–52]). It leads to two important consequences: (i) as the effective acting force depends not only on the grain size, but also on the acting forces of the flow, it is suggested that the sediment transport natures need to be classified based on the shear Reynolds number, too; (ii) due to this, the validation limits of the sediment transport-related mathematical models (that is, the sediment transport models) should be defined as the function of the shear Reynolds number.

The investigation of the applicability of the van Rijn bed load transport formula [20] also enhances the benefits of the shear Reynolds number-based investigation, against the grain-size-based (or characteristic grain-size-based). As one of the most widely tested and applied sediment transport formulas for river engineering applications, the van Rijn bed load transport equation [20] considers the application of the Shields curve. The transport model was developed and validated for sand transport measurements, and, therefore, the application limit was defined accordingly ($d < 2$ mm) [7]. However, the validity range of the base mathematical model can be defined by the shear Reynolds number, and the grain size is not a suitable variable for this reason. Namely, the drag coefficient (C_D), the lift coefficient (α_L), or the effect of the spinning motion depends on the shear Reynolds number [53–56], and the recommended values of these variables in the van Rijn model were calibrated and validated for lower shear Reynolds numbers ($Re^* < 400$) [20,57,58]. Therefore, the determination of the applicability of the van Rijn model is reasonable as the function of the shear Reynolds number, which is the hydraulic smoother ($Re^* < 400$) regime.

At the same time, the lower Re^* range implies lower bed shear velocity (u^*) and smaller particles (d) (see Equation (3)), assuming the domination of the less-resistant finer fractions in the bed load and stable coarser grains. This is proved by the about 600 measurement data [20,57,58]. Also, Parker mentions [8] that, at rougher hydraulic regime ($Re^* > 1000$), the motion of the sand fractions is rather suspended like rolling. Consequently, finer fractions disappear from the bed load and the coarser gravels remain and become dominant. Accordingly, the van Rijn bed load model is not applicable in rougher hydraulic ($Re^* > 1000$) regime.

The laboratory experiments, where coarser ($d_{50} > 5.3$ mm) and nonuniform bed material conditions were investigated by Wilcock et al. [26], are also taken into consideration here. Figure 4 shows the rate of the gravel load to the sand load as the function of the shear Reynolds number. Within any mixtures, the higher the shear Reynolds number, the coarser the composition of the bed load. It is visible that the gravel load is equal to the sand load when the Re^* is between 300 and 400, around 350 (Figure 4). Above this, the gravel transport is dominant. In turn, below 350 the sand transport prevails, which underlines the earlier assumption, i.e., the fine sediment bed load transport takes place and dominates in the smoother hydraulic range. Consequently, despite that the d_{50} characteristic grain size indicates only gravel transport, sand-dominated sediment transport was also found.

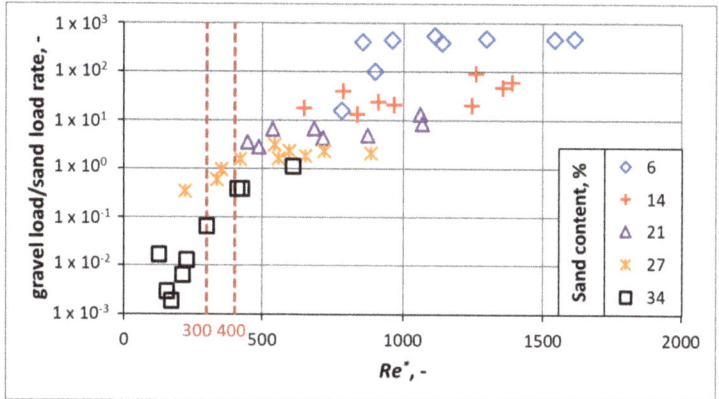

Figure 4. Calculated rates of the gravel load to the sand load as the function of the shear Reynolds number, based on the laboratory experiments of Wilcock et al. [26].

Based on the above-described summary, the following findings are emphasized

- Shields has already presented [22] that the shear Reynolds number (Re^*) is a suitable variable (instead of the grain sizes) to define the relation of the acting forces and thus to classify the sediment transport in the near-surface zone. In other words, the effective force of the flow (together with the resistance of the grain) must be taken into account for the proper investigation of the stability and mobility of bed-forming particles.
- The applicability of both mathematical models of Shields [22] and van Rijn [20] can be described by the shear Reynolds number, but not by grain sizes. This also supports the fact that the shear Reynolds number-based classification of the sediment transport is expected to be more reliable than a grain-size-based method.
- The investigation of the measurements, for which the van Rijn bed load formula was validated and calibrated [20,57,58], suggests that the sand bed load can occur at lower Re^* range, $Re^* < 400$. Complementing this by the statement of Parker [8], it is assumed that in rougher hydraulic regime ($Re^* > \sim 1000$), the sand transport appears only in suspended form, while the bed load consists mainly of gravel.
- The Re^*-based assessment of the laboratory experiments of Wilcock et al. [26] shows that the Re^* range ~3–400 seems to be a critical zone, where the yield of the sand and gravel are roughly the same. A characteristic grain size cannot refer to such phenomena.

Accordingly, the following conjecture was stated: The Re^* parameter is a more suitable parameter for the prediction and distinction of the sand- or gravel-dominated bed load transport, than either simply a characteristic grain size (d_{90} or d_{50}), or the Re_p. Moreover, Re^* between ~300 and 400 seems to be a critical zone: above-gravel, below-sand domination can be expected in the bed load transport.

2.4. Description of the Validation Measurements

In order to investigate accuracy of the conjecture, an assessment considering characteristic local bed grain sizes, local bed shear stress values as well as calculated Reynolds numbers was performed using datasets from recent laboratory and field experiments of the authors. First, the results from the laboratory experiments of Török et al. [59] were used. In this experiment, morphological processes (scouring, bed armoring, and aggradation) were investigated around a single groin, using mixed-size bed material with an initial d_{50} of 5.16 mm. The local grain size distributions were determined by an automated image-based grain detection software tool BASEGRAIN [60]. Local bed shear stress values were estimated according to the TKE-Method [61] using the near-bed point-velocity measurements, carried out with a 3D Acoustic Doppler Velocimeter (ADV) [62]. The measurements were carried out at quasi-equilibrium bed geometry. Then, negligible bed changes were measured in the flume ($\Delta z < \pm 1$ mm/h); however, weak bed load transport ($Q_b < 0.3$ kg/h) of mainly the sand particles ($d_{50} = 1.3$ mm) could still be observed at the outlet of the flume, where the d_{50} of the bed material ranged between 4 and 6 mm [59].

Second, field data from a section of a large river was also assessed. The field experiments were carried out in the Hungarian section of the Danube River at mean flow regime, where the main characteristic morphological parameters were already introduced in the Introduction point. Forty-seven bed material samples were taken by a drag-bucket sampler, for which the grain-size distributions of the samples were determined by sieving analysis [63]. Parallel to the local bed material samplings, time-averaged flow velocity profiles were measured by an acoustic Doppler current profiler (ADCP) (WorkHorse Rio Grande 1200 kHz) at the sampling locations. Based on these long-term, fixed-boat measurements, the local bed shear stress values could be estimated using the turbulent wall law [64]. No bed load sampling could be performed during the two measurement campaigns. However, the bed load rating curve at the closest monitoring station (6 km upstream) [17] represents that low, but not negligible, bed load transport took place during both measurements ($Q_b < 0.1$ kg/s).

The data regarding the field and laboratory measurements can be found in Appendix A (Table A1).

2.5. Measurement Data Processing for Validation Purpose of the Stated Conjecture

Field and laboratory data from the above-introduced reach of the Danube River in Hungary and laboratory experiments were conducted to confirm the stated conjecture. In order to investigate the shear Reynolds number-based classification potential, the samples were sorted into three classes according to their origin place (Figures 5 and 6): the black squares refer to the sandy bed material samples, which were collected from the near-bank zones. Here, the bed material consists predominantly of silt and sand, gravel transport can be expected only during flood waves. The red stars originate from the side bars (e.g., bed material illustrated in Figure 2). As it was introduced in Chapter 2 and based on Figure 2, the dominant grain size in the bed load is variable (both sand and gravel can dominate) and it is strongly flow-dependent. Finally, the samples from the main channel are depicted by the blue diamonds. Based on the bed material analysis, only gravel transport can occur at this region of the channel, because sand can be vanishingly found. However, as the measurements indicate stable bed surface, pure gravel transport can be expected at mean flow conditions. The investigation examines whether the shear Reynolds number is a proper variable to separate the different classes of the points.

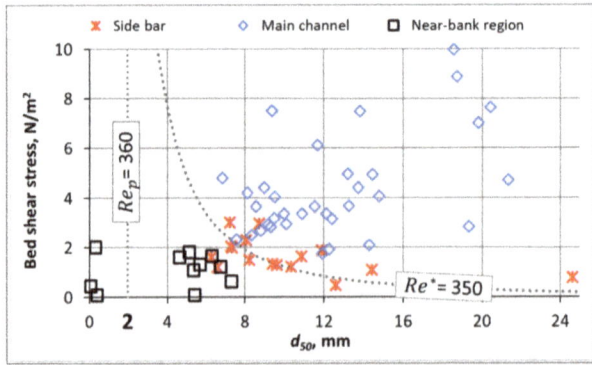

Figure 5. The bed shear stress as the function of the d_{50} grain size, regarding field and laboratory measurements [17,64].

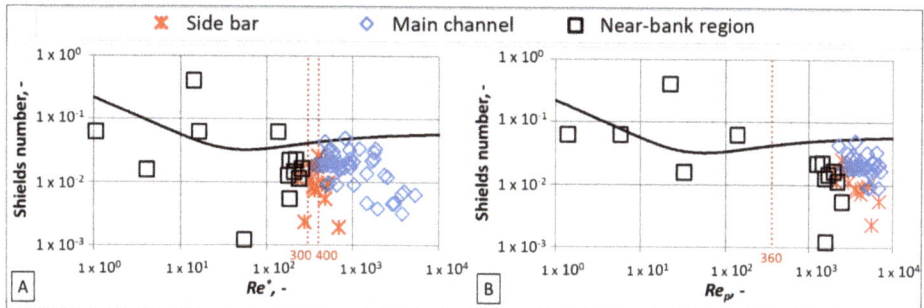

Figure 6. (**A**) Shields number as the function of the shear Reynolds number, based on laboratory and field data [17,59,64]. Shields curve is indicated with black line. (**B**) Shields number as the function of the explicit particle Reynolds number, based on laboratory and field data [17,59,64]. Shields curve is indicated with black line.

Furthermore, the comparison of the shear and particle Reynolds number-based investigations was also examined. First, the scattering of the three point classes as the function of Re^* and Re_p were compared. Based on this comparison, a more accurate separation method becomes outlined visually.

Finally, in order to perform a quantitative assessment on the performance of the Re^*- and Re_p-based approaches, a statistical method was applied, which provides information on the representativity of the Reynolds number ranges for the characteristic bed load transport processes. Here, the log-normal distributions of the Re^* and Re_p values were calculated, separately for the three above-distinguished groups. The reason for using the log-normal distribution was the fact that asymmetric distributions were expected (based on the scattering of the points), where Re numbers cannot take negative values [65]. The probability density and the cumulative distribution functions are described in the Appendix A. Based on the cumulative distribution functions, the Re^*- and Re_p-based separatenesses of the different point groups become qualitatively characterizable.

3. Results

First, the field measurement data were analyzed in Figure 5, separately. The scattering of the points reflects a clear tendency: the bed shear stress increases with higher grain sizes. Nevertheless, when indicating a separation line, which represents a Re^* of 350, the points from the three well-distinguishable regions indeed show different features. Points representing the shallower, sandy regions fall below the line. As most of these sampling locations can be characterized with mean grain sizes in the gravel range ($d_{50} > 2$ mm), the Shields–Parker diagram would consequently suggest clear gravel transport. However, considering the relatively low bed shear stress values (<2 N/m^2), the Shield curve predicts only sand motion (see the findings based on Figure 2, where $\tau^* = 0.05 \to \tau \approx 2$ N/m^2). Indeed, the d_{50} of the bed material samples at the outlet of the flume (4–6 mm) would refer to gravel domination in the bed load, but the corresponding d_{50} of the bed load samples show clear sand domination (1.3 mm) [59].

Second, the points distributed along the line belong to the bars. At this part of the points, the domination of sand or gravel cannot be well distinguished, but most probably a similar amount of them is transported. Accordingly, two phenomena can occur: (i) the bed shear stress values are low and the bed materials are quite uniform and coarse, resulting in resting particles (red stars, $\tau < \sim 1$ N/m^2); (ii) higher flow transport capacity can cause almost the same order of magnitude in the sand and gravel discharge, if the bed material contains both fractions (red stars, where $\tau > \sim 2$ N/m^2 and see also Figure 4).

Finally, the blue diamonds located above the line represent the stable [17,59,64] main stream with coarse grains. Because of the armored bed surface, weak sediment transport occurs during mean flow discharge. Only higher flood waves can result in significant bed load transport [64], containing of local bed material, i.e., only gravel grains.

The constant particle Reynolds number of 360 values are indicated with a vertical dotted line. Unlike the separation line of $Re^* = 350$, the particle Reynolds number does not distinguish the different point classes.

Assessing the relationship between local Shields number and the shear Reynolds number for both the laboratory and field experiments (Figure 6A), the statements in the previous paragraphs are corroborated regarding the Re^*-based classification of the landforms. Again, the $Re^* \approx 300$–400 range outlines as a suitable indicator for the identification whether the gravel or the sand particles dominate.

Contrarily, no clear distinction of the dominating morphodynamic processes can be made if the particle Reynolds number is considered (Figure 6B). Most of the points, representing all three different transport modes, range between $1000 < Re_p < 10,000$, clearly below the critical condition provided by Shields. Accordingly, a stable bed surface is expected both in the laboratory and in the field study. However, this is only the case for the armored sections of the study sites (blue diamonds) [17,59,64], whereas the bed load measurements of the laboratory experiments and the bed load rating curve at the study site in the Danube River demonstrate that weak bed load transport took place in both cases. Moreover, the bed load material investigation of the laboratory experiment suggested that the bed contained mainly sand fractions. Overall, despite the fact that the Shields curve would suggest stable bed, weak and sand-dominated bed load transport was monitored during the field and laboratory measurements.

Finally, the results of the log-normal-distribution-based probability examination are presented. As a summary of the calculations, the probability data was summarized for the six log-normal distributions, considering $Re^* = 300$, $Re^* = 400$, and $Re_p = 360$ as critical values.

Table 2 shows that 5% of the near-bank points are in the $Re^* > 300$ range, 60% of the gravel bar points take Re^* between 300 and 400, while only 3% of the points from the main channel are expected below 400. It can also be seen that, below the critical particle Reynolds number value ($d = 2$ mm; $Re_p = 360$), only the fine sediment transport nature can be expected. However, according to the log-normal distribution, 85% of the near-bank points are over 360. Thus, the $Re_p = 360$ indicates well that below this value mainly finer sediment transport is expected. In turn, the sediment transport nature cannot be determined obviously above this border.

Table 2. Probability values of the shear Reynolds number and the particle Reynolds number, calculated based on the fitted log-normal distribution for grouped data.

Landform Classes	Re^*			Re_p	
	<300	< >	>400	<360	>360
Sand aggradation, near-bank points	95%	5%	0%	15%	85%
Side bar	5%	60%	35%	0%	100%
Main channel, bed armor	0%	3%	97%	0%	100%

The most important outcome of the probability analysis is that the Re^* between 300 and 400 is indeed a critical range: below 300 the sand transport dominates, above 400 the gravel motion dominates.

The probability values indicated for the three main morphodynamic processes emphasize the suitability of Re^* for the determination of the locally dominant grain size range in the bed load transport. On the other hand, the application of Re_p suggests no clear classification for the same categorization, when spatially varied bed material is present, such as at the investigated river reach.

4. Discussion

Previous studies report that the dominant sediment transport nature in rivers with uniform bed material can be reliably determined as the function of the so-called explicit particle Reynolds number (Re_p), but yields less accurate estimations in cases when the bed material composition shows strong variability even along shorter reaches.

Based on the available and accessible data, namely 70 bed material samples and related local bed shear stress values gathered both from recent own laboratory and field experiments, it could be confirmed that, instead of the utilization of the explicit particle Reynolds number (Re_p), or the application of characteristic grain sizes, e.g., d_{50} or d_{90}, the shear Reynolds number (Re^*) is a more adequate parameter to assess the locally dominant sediment transport nature. The probability calculations indicate that the sand-dominant bed load is estimable by 95% accuracy, while the gravel majority transport by 97%. In contrast, the Re_p-based former method shows significantly greater inaccuracy. Although the gravel-dominated locations are estimated precisely, the method calculates the prediction of fine sediment majority by 85% inaccuracy.

Besides the authors' own experimental data, it could also be shown that the Re^*-based approach works reliably for the laboratory and field data based on which the widely used van Rijn [20,57,58] and the Wilcock and Crowe [26] bed load formulas were validated. The method is able to indicate the fine particle domination for all of the validation and calibration data of the van Rijn model correctly. Moreover, the transitional range between the sand and gravel majority becomes detectable in the case of sand–gravel mixture bed material measurements, with which the Wilcock and Crowe model was validated.

As another example, the authors assessed the datasets, consisting of 45 coupled bed shear stress—d_{50} characteristic grain size value pairs from field measurements, published by Mueller et al. [66]. In that study, the median surface grain sizes of the gravel-bed streams and rivers were reported to vary between 0.027 and 0.21 m, which indicates quite coarse bed material and armored bed surface. Accordingly, the calculated Re^* are consequently high, all estimated Re^* values exceed 400.

Based on the data assessment introduced above, the following classification could be set up:

- $Re^* < \sim 300 \rightarrow$ sand transport dominates;
- $Re^* > \sim 400 \rightarrow$ gravel transport dominates;
- $\sim 300 < Re^* < \sim 400 \rightarrow$ gravel accumulating and side bar formation is expected.

Besides the promising validation results, the theory must be further proven by direct bed load field measurements. Also noted, that the local bed load content depends not only on the local bed material, but also on the arriving bed load material [67]. This obvious effect is not taken into account by the introduced method. Additionally, the benchmark measurements were carried out at weak bed load discharges. Thus, the validation needs to be extended to a wider range of the flow regime. However, the authors believe that the presented classification indicates meaningful progress in the investigation methods of the nonuniform bed load.

5. Conclusions

The herein-presented classification method of the locally dominant fraction in bed load transport can contribute to better implementation of different sediment transport investigation methods applied in large rivers. For instance, a well-known issue related to field sediment transport monitoring is the high uncertainty of direct bed load sampling methods. Having information on the local bed material and on the flow field, the offered Re^*-based approach can suggest where and what sort of sampling techniques would be the most suitable to collect reliable sediment information.

The results can also contribute to the development of improved computational modeling tools. Indeed, instead of applying one specific empirical sediment transport formula in a simulation to calculate local bed load transport, the Re^*-based approach can be utilized to distinguish between several formulas, each of them having a certain application range. Thus, the combined use of several formulas becomes possible.

Author Contributions: G.T.T. conceived the study, collected the data, conducted the data analysis, preformed the statistical analysis and drafted the manuscript. G.T.T. and S.B. discussed the results. S.B. worked on subsequent drafts of the manuscript. S.B. and J.J. participated in the design and coordination of the study.

Funding: This research was supported by the ÚNKP-17-3 and 17-4 New National Excellence Program of the Ministry of Human Capacities, which is highly appreciated, and was partly supported by MTA TKI of the Hungarian Academy of Sciences. Support of grant BME FIKP-VÍZ by EMMI is also kindly acknowledged. The authors acknowledge the funding of the OTKA FK 128429 grant.

Conflicts of Interest: The authors declare no conflict of interest.

Appendix A

The probability density function for the log-normal distribution is expressed as:

$$P(x) = \frac{1}{x\sigma\sqrt{2\pi}} e^{-\frac{1}{2}\left(\frac{\ln x - u}{\sigma}\right)^2}, \tag{A1}$$

where the parameters are $-\infty < \mu < \infty$ (scale parameter) and $\sigma > 0$ (shape parameter), can be obtained by the fitting of the function to the known points. The integration of the log-normal distribution curves results in the cumulative distribution function, which is:

$$F(x) = \Phi\left(\frac{\ln x - u}{\sigma}\right), \tag{A2}$$

where Φ is the Laplace integral [68,69].

Table A1. Measured morphodynamic data of field [17,64] (at the introduced reach of the Danube River in Hungary) and laboratory [59] measurements and of the referenced laboratory experiments [17,59,64].

	Main Channel, Bed Armor					Sand Aggradation, Near-Bank Points					Side Bar				
	u_*	τ^*	Re^*	Re_p	d_{50}	u_*	τ^*	Re^*	Re_p	d_{50}	u_*	τ^*	Re^*	Re_p	d_{50}
Danube, rkm 1798–1795	0.054	0.020	498	3546	9.2	0.025	0.005	185	2512	7.3	0.022	0.002	276	5705	12.6
	0.053	0.009	1031	10,835	19.4	0.010	0.001	55	1604	5.4	0.035	0.007	364	4242	10.4
	0.046	0.009	657	6886	14.3	0.022	0.062	1	1	0.0	0.028	0.002	689	15,567	24.6
	0.042	0.009	500	5252	11.9	0.129	0.062	16	6	0.1	0.033	0.006	478	6995	14.5
	0.058	0.019	633	4583	10.9	0.128	0.062	136	140	1.1	0.036	0.008	347	3802	9.6
	0.061	0.020	701	5005	11.6	0.045	0.399	14	23	0.3	0.041	0.009	443	4577	10.9
	0.066	0.030	598	3429	9.0	0.010	0.016	4	33	0.4	0.037	0.009	347	3676	9.4
	0.044	0.010	540	5481	12.3						0.044	0.010	519	5231	11.9
	0.069	0.014	1465	12,559	21.4										
	0.058	0.003	3745	65,995	64.6										
	0.100	0.033	1855	10,182	18.6										
	0.064	0.026	605	3737	9.5										
	0.064	0.017	946	7251	14.8										
	0.069	0.043	475	2283	6.9										
	0.094	0.029	1766	10,314	18.7										
	0.084	0.022	1659	11,230	19.8										
	0.087	0.049	813	3658	9.4										
	0.056	0.021	535	3715	9.5										
	0.048	0.019	365	2657	7.6										
	0.087	0.049	813	3658	9.4										
	0.053	0.019	499	3631	9.3										
	0.058	0.021	581	4026	10.0										
	0.077	0.007	3955	46,694	51.3										
	0.049	0.004	1956	31,713	39.6										
	0.079	0.012	2543	23,129	32.1										
	0.061	0.005	3042	44,346	49.5										
	0.082	0.006	5374	67,130	65.3										
	0.060	0.005	2896	42,850	48.4										
	0.073	0.006	3714	46,211	50.9										
	0.048	0.005	1449	21,195	30.3										
	0.066	0.020	913	6486	13.7										
	0.079	0.013	2345	20,525	29.6										
Laboratory measurements	0.065	0.032	528	2951	8.1	0.040	0.022	186	1262	4.6	0.035	0.011	228	2155	6.6
	0.078	0.032	915	5087	11.7	0.033	0.013	177	1570	5.3	0.055	0.026	398	2481	7.2
	0.070	0.023	930	6118	13.2	0.036	0.014	207	1727	5.7	0.055	0.021	478	3287	8.7
	0.060	0.026	518	3194	8.6	0.043	0.022	219	1474	5.1	0.045	0.017	329	2537	7.4
	0.086	0.033	1197	6548	13.8	0.041	0.016	257	2015	6.3	0.039	0.011	320	3016	8.3
	0.058	0.017	704	5374	12.1	0.035	0.011	237	2220	6.7	0.045	0.017	330	2506	7.3
	0.061	0.017	805	6151	13.3						0.048	0.018	387	2914	8.1
	0.070	0.021	1018	7012	14.5						0.040	0.016	253	2005	6.3
	0.056	0.016	697	5558	12.4										
	0.054	0.018	549	4084	10.1										
	0.087	0.023	1785	11,739	20.4										
	0.050	0.019	417	3066	8.3										
	0.052	0.019	459	3339	8.8										

References

1. Dey, S. Fluvial Processes: Meandering and Braiding. In *Fluvial Hydrodynamics: Hydrodynamic and Sediment Transport Phenomena*; Springer: Berlin/Heidelberg, Germany, 2014; pp. 529–562.
2. Leopold, B.L.; Wolman, M.G. *River Channel Patterns: Braided, Meandering and Straight*; Geological Survey Professional Paper; U.S. Government Printing Office: Washington, DC, USA, 1957; Volume 282.
3. Lane, E.W. *A Study of the Shape of Channels Formed by Natural Streams Flowing in Erodible Material*; U.S. Army Corps of Engineers Report; The Division: Omaha, NE, USA, 1957; Volume 141.
4. Leopold, L.B.; Wolman, M.G. River meanders. *Geol. Soc. Am. Bull.* **1960**, *71*, 769–794. [CrossRef]
5. Schumm, S.A. Evolution and response of the fluvial system, sedimentologic implications. *SEPM Spec. Publ.* **1981**, *31*, 19–29.

6. Jaeggi, M.N.R. Effect of Engineering Solutions on Sediment Transport. In *Dynamics of Gravel-Bed Rivers*; Billi, P., Hey, R.D., Thorne, C.R., Tacconi, P., Eds.; John Wiley & Sons Ltd.: Hoboken, NJ, USA, 1992.
7. García, M.H. Sediment Transport and Morphodynamics. In *Sedimentation Engineering: Processes, Measurements, Modeling, and Practice, Manuals and Reports on Engineering Practice*; ASCE: Reston, VA, USA, 2008; pp. 21–163.
8. Parker, G. Transport of Gravel and Sediment Mixtures. In *Sedimentation Engineering*; Garcia, M., Ed.; American Society of Civil Engineers: Reston, VA, USA, 2008; ISBN 978-0-7844-0814-8.
9. Schumm, S.A.; Dumont, J.F.; Holbrook, J.M. *Active Tectonics and Alluvial Rivers*; Cambridge University Press: Cambridge, UK; New York, NY, USA; Melbourne, Australia, 2000; ISBN 0-521-66110-2.
10. Edward, J. Hickin. Chapter 4. In *Fluid Mechanics*; Online course note; Simon Fraser University and R.S. Graphics and Printing: Burnaby, BC, Canada, 2009.
11. Piton, G.; Recking, A. The concept of travelling bedload and its consequences for bedload computation in mountain streams. *Earth Surf. Process. Landf.* **2017**, *42*, 1505–1519. [CrossRef]
12. Farkas-Iványi, K.; Guti, G. The Effect of Hydromorphological Changes on Habitat Composition of the Szigetköz Floodplain. *ACTA Zool. Bulg.* **2014**, *7*, 117–121.
13. Hankó, Z.; Starosolszky, Ö.; Bakonyi, P. Megvalósíthatósági tanulmány a Duna környezetének és hajózhatóságának fejlesztésére (Danube Environmental and navigation Project, Feasibility Study). *Vízügyi Közlemények* **1996**, *78*, 291–315.
14. Rákóczi, L. A Duna-meder sorsa Szap és Szob között (Destiny of the Danube channel between Szap and Szob). *Vízügyi Közlemények* **2000**, *82*, 262–280.
15. Goda, L. A Duna gázlói Pozsony-Mohács között (Shallows of the River Danube between Pozsony, Bratislava and Mohács. *Vízügyi Közlemények* **1995**, *77*, 71–102.
16. Baranya, S.; Józsa, J.; Török, G.T.; Ficsor, J.; Mohácsiné Simon, G.; Habersack, H.; Haimann, M.; Riegler, A.; Liedermann, M.; Hengl, M. A Duna hordalékvizsgálatai a SEDDON osztrák-magyar együttműködési projekt keretében (Introduction of the joint Austro-Hungarian sediment research under the SEDDON ERFE-project). *Hidrológiai Közlöny* **2015**, *95*, 41–46.
17. Török, G.T.; Baranya, S. Morphological investigation of a critical reach of the upper Hungarian Danube. *Period. Polytech. Civ. Eng.* **2017**, *61*. [CrossRef]
18. Észak-Dunántúli Vízügyi Igazgatóság. *Nagyvízi mwe haederkezelési terv (High Water River Management Plan) 01.nmt.02. (egyeztetési terv)—Duna 1809,76—1786,00 fkm*; 2014; Professional report; Észak-Dunántúli Vízügyi Igazgatóság: Győr, Hungary, 2014.
19. Liedermann, M.; Gmeiner, P.; Pessenlehner, S.; Haimann, M.; Hohenblum, P.; Habersack, H. A Methodology for Measuring Microplastic Transport in Large or Medium Rivers. *Water* **2018**, *10*, 414. [CrossRef]
20. Van Rijn, L.C. Sediment Transport, Part I: Bed Load Transport. *J. Hydraul. Eng.* **1984**, *110*, 1431–1456. [CrossRef]
21. Wilcock, P.R.; Crowe, J.C. Surface-based transport model for mixed-size sediment. *J. Hydraul. Eng.* **2003**, *129*, 120–128. [CrossRef]
22. Shields, A. Application of Similarity Principles and Turbulence Research to Bed-Load Movement. *Mitt. Preuss. Versuchsanst. Wasserbau Schiffbau* **1936**, *26*, 47.
23. Bergillos, R.J.; Rodríguez-Delgado, C.; Millares, A.; Ortega-Sánchez, M.; Losada, M.A. Impact of river regulation on a Mediterranean delta: Assessment of managed versus unmanaged scenatios. *Water Resour. Res.* **2016**, *52*, 5132–5148. [CrossRef]
24. Bergillos, R.J.; Ortega-Sánchez, M.; Masselink, G.; Losada, M.A. Morpho-sedimentary dynamics of a micro-tidal mixed sand and gravel beach, Playa Granada, southern Spain. *Mar. Geol.* **2016**, *379*, 28–38. [CrossRef]
25. Fischer-Antze, T.; Rüther, N.; Olsen, N.; Gutknecht, D. Three-dimensional (3D) modeling of non-uniform sediment transport in a channel bend with unsteady flow. *J. Hydraul. Res.* **2009**, *47*, 670–675. [CrossRef]
26. Wilcock, P.R.; Kenworthy, S.T.; Crowe, J.C. Experimental Study of the Transport of Mixed Sand and Gravel. *Water Resour. Res.* **2001**, *37*, 3349–3358. [CrossRef]
27. Wilcock, P.R. A two-fraction model for the transport of sand/gravel mixtures. *Water Resour. Res.* **2002**, *38*, 1–12. [CrossRef]
28. Sziło, J.; Bialik, R.J. Grain Size Distribution of Bedload Transport in a Glaciated Catchment (Baranowski Glacier, King George Island, Western Antarctica). *Water* **2018**, *10*, 360. [CrossRef]

29. Bialik, R.J. Numerical study of saltation of non-uniform grains. *J. Hydraul. Res.* **2011**, *49*, 697–701. [CrossRef]
30. Einstein, H.A. *The Bed-Load Function for Sediment Transportation in Open Channel Flows*; Department of Agriculture, Soil Conservation Service: Washington, DC, USA, 1950.
31. Ashida, K.; Michiue, M. Study on hydraulic resistance and bedload transport rate in alluvial streams. *Trans. Jpn. Soc. Civ. Eng.* **1972**, *206*, 59–69. [CrossRef]
32. Meyer-Peter, E.; Müller, R. Formulas for Bed-Load Transport. In Proceeding of the IAHSR 2nd Meeting, Stockholm, Sweden, 7–9 June 1948.
33. Wiberg, P.L.; Smith, J.D. Model for calculating bed load transport of sediment. *J. Hydraul. Eng.* **1989**, *115*, 101–123. [CrossRef]
34. Török, G.T.; Baranya, S.; Rüther, N. 3D CFD Modeling of Local Scouring, Bed Armoring and Sediment Deposition. *Water* **2017**, *9*, 56. [CrossRef]
35. Gaeuman, D.; Andrews, E.D.; Krause, A.; Smith, W. Predicting fractional bed load transport rates: Application of the Wilcock-Crowe equations to a regulated gravel bed river. *Water Resour. Res.* **2009**, *45*, 1–15. [CrossRef]
36. Janssen, S.R. Testing Sediment Transport Models under Partial Transport Conditions. Master's Thesis, University of Twente, Enschede, The Netherlands, 2010.
37. Török, G.T.; Baranya, S.; Rüther, N. Three-dimensional numerical modeling of non-uniform sediment transport and bed armoring process. In Proceedings of the 18th Congress of the Asia & Pacific Division of the International Association for Hydro-Environment Engineering and Research, Jeju, Korea, 19–23 August 2012.
38. Rüther, N.; Olsen, N.R.B. Modelling free-forming meander evolution in a laboratory channel using three-dimensional computational fluid dynamics. *Geomorphology* **2007**, *89*, 308–319. [CrossRef]
39. Rüther, N.; Olsen, N.R.B. 3D modeling of transient bed deformation in a sine-generated laboratory channel with two different width to depth ratios. In Proceedings of the International Conference on Fluvial Hydraulics, Lisbon, Portugal, 6–8 September 2006; Taylor & Francis: New York, NY, USA, 2006.
40. Rüther, N.; Olsen, N.R.B. Three-dimensional modeling of sediment transport in a narrow 90° channel bend. *J. Hydraul. Eng.* **2005**, *131*, 917–920. [CrossRef]
41. Olsen, N.R.B.; Melaaen, M.C. Three-dimensional calculation of scour around cylinders. *J. Hydraul. Eng.* **1993**, *119*, 1048–1054. [CrossRef]
42. Bihs, H.; Olsen, N.R.B. Numerical Modeling of Abutment Scour with the Focus on the Incipient Motion on Sloping Beds. *J. Hydraul. Eng.* **2011**, *137*, 1287–1292. [CrossRef]
43. Deltares. *3D/2D modelling suite for integral water solutions. Delft3D-FLOW, User Manual*; Deltares: Delft, The Netherlands, 2014; Volume 710.
44. Skinner, C.J.; Coulthard, T.J.; Schwanghart, W.; Van De Wiel, M.J.; Hancock, G. Global sensitivity analysis of parameter uncertainty in landscape evolution models. *Geosci. Model Dev.* **2018**, *11*, 4873–4888. [CrossRef]
45. Unal, N.E. Shear Stress-Based Analysis of Sediment Incipient Deposition in Rigid Boundary Open Channels. *Water* **2018**, *10*, 1399. [CrossRef]
46. Habersack, H.; Kreisler, A. Sediment Transport Processes. In *Dating Torrential Processes on Fans and Cones*; Beniston, M., Ed.; Springer: Dordrecht, The Netherlands; Heidelberg, Germany; New York, NY, USA; London, UK, 2013; ISBN 9789400743359.
47. Zanke, U.C.E. *Hydromechanik der Gerinne und Küstengewässer*; Blackwell; Vieweg+Teubner Verlag: Berlin, Germany, 2002; ISBN 978-3-322-80212-5.
48. Parker, G.; Klingeman, P.C. On why gravel rivers are paved. *Water Resour. Res.* **1982**, *18*, 1409–1423. [CrossRef]
49. Cushman-Roisin, B. Rivers and Streams, Chapter 15. In *Environenmental Fluid Mechanics*; John Wiley & Sons, Inc.: Hanover, NH, USA, 2019.
50. Parker, G. Surface-based bedload transport relation for gravel rivers. *J. Hydraul. Res.* **1990**, *28*, 417–436. [CrossRef]
51. Dey, B.S.; Papanicolaou, A. Sediment Threshold under Stream Flow: A State-of-the-Art Review. *KSCE J. Civ. Eng.* **2008**, *12*, 45–60. [CrossRef]
52. Dwivedi, A.; Melville, B.W.; Shamseldin, A.Y.; Guha, T.K. Analysis of hydrodynamic lift on a bed sediment particle. *J. Geophys. Res.* **2011**, *116*. [CrossRef]
53. Morsi, S.A.; Alexander, A.J. An investigation of particle trajectories in two phase flow systems. *J. Fluid Mech.* **1972**, *55*, 193–208. [CrossRef]

54. Rubinow, S.I.; Keller, J.B. The transverse force on a spinning sphere moving in a viscous fluid. *J. Fluid Mech.* **1961**, *11*, 447–459. [CrossRef]
55. Saffman, P.G. The lift on a small sphere in a slow shear flow. *J. Fluid Mech.* **1965**, *22*, 385–400. [CrossRef]
56. Dey, S. Sediment threshold. *Appl. Math. Model.* **1999**, *23*, 399–417. [CrossRef]
57. Fernandez Luque, R. Erosion and Transport of Bed-load Sediment. BSc Thesis, Delft Technical University, Delft, The Netherlands, 1974.
58. Fernandez Luque, R.; van Beek, R. Erosion and Transport of Bed-load Sediment. *J. Hydraul. Res.* **1976**, *14*, 127–144. [CrossRef]
59. Török, G.T.; Baranya, S.; Rüther, N.; Spiller, S. Laboratory analysis of armor layer development in a local scour around a groin. In Proceedings of the International Conference on Fluvial Hydraulics, RIVER FLOW 2014, Lausanne, Switzerland, 3–5 September 2014; Taylor and Francis Group: Lausanne, Switzerland, 2014; pp. 1455–1462.
60. Detert, M.; Weitbrecht, V. User guide to gravelometric image analysis by BASEGRAIN. In *Advances in River Sediment Research*; Fukuoka, S., Nakagawa, H., Sumi, T., Zhang, H., Eds.; CRC Press: Boca Raton, FL, USA, 2013; ISBN 9781138000629.
61. Kim, S.C.; Friedrichs, C.T.; Maa, J.P.Y.; Wright, L.D. Estimating bottom stress in tidal boundary layer from Acoustic Doppler velocimeter data. *J. Hydraul. Eng.* **2000**, *126*, 399–406. [CrossRef]
62. Cea, L.; Puertas, J.; Pena, L. Velocity measurements on highly turbulent free surface flow using ADV. *Exp. Fluids* **2007**, *42*, 333–348. [CrossRef]
63. Przyborowski, Ł.; Loboda, A.M.; Bialik, R.J. Experimental investigations of interactions between sand wave movements, flow structure, and individual aquatic plants in natural rivers: A case study of Potamogeton Pectinatus, L. *Water* **2018**, *10*, 1166. [CrossRef]
64. Török, G.T. Methodological Improvement of Morphodynamic Investigation Tools for Rivers with Non-Uniform Bed Material By. Ph.D. Thesis, Budapest University of Technology and Economics, Budapest, Hungary, 2018.
65. Koris, K. *Hidrológia II. (Hydrology II)*; 1. kiadás.; Class note (Egyetemi jegyzet): Budapest, Hungary, 2014; ISBN 978-963-12-0752-1.
66. Mueller, E.R.; Pitlick, J.; Nelson, J.M. Variation in the reference Shields stress for bed load transport in gravel-bed streams and rivers. *Water Resour. Res.* **2005**, *41*. [CrossRef]
67. Vowinckel, B.; Jain, R.; Kempe, T.; Fröhlich, J. Entrainment of single particles in a turbulent open-channel flow: A numerical study. *J. Hydraul. Res.* **2016**, *54*, 158–171. [CrossRef]
68. *EasyFit Professional, Version 5*; MathWave Technologies: Washington, DC, USA, 2015.
69. Aristizabal, R.J. Estimating the Parameters of the Three-Parameter Lognormal Distribution. FIU Electronic Theses and Dissertation, Florida International University, St. Miami, FL, USA, 2012.

© 2019 by the authors. Licensee MDPI, Basel, Switzerland. This article is an open access article distributed under the terms and conditions of the Creative Commons Attribution (CC BY) license (http://creativecommons.org/licenses/by/4.0/).

Article

Modelling of Soil Erosion and Accumulation in an Agricultural Landscape—A Comparison of Selected Approaches Applied at the Small Stream Basin Level in the Czech Republic

Jiří Jakubínský [1,*], Vilém Pechanec [2], Jan Procházka [2] and Pavel Cudlín [1]

1 Global Change Research Institute CAS, Bělidla 986/4a, 603 00 Brno, Czech Republic; cudlin.p@czechglobe.cz
2 Department of Geoinformatics, Palacký University Olomouc, 17. listopadu 50, 771 46 Olomouc, Czech Republic; vilem.pechanec@upol.cz (V.P.); jan.prochazka07@upol.cz (J.P.)
* Correspondence: jakubinsky.j@czechglobe.cz

Received: 28 December 2018; Accepted: 20 February 2019; Published: 26 February 2019

Abstract: This article deals with the modelling of erosion and accumulation processes in the contemporary cultural landscape of Central Europe. The area of interest is the headwater part of the small stream catchment—the Kopaninský Stream in central Czech Republic. It is an agricultural and forest–agricultural landscape with a relatively rugged topography and riverbed slope, which makes the terrain very vulnerable to water erosion. The main aim of this article is to compare the results of four selected soil erosion and sediment delivery models, which are currently widely used to quantitate the soil erosion and sediment accumulation rates, respectively. The models WaTEM/SEDEM, USPED, InVEST and TerrSet work on several different algorithms. The model outputs are compared in terms of the total volume of eroded and accumulated sediment within the catchment per time unit, and further according to the spatial distribution of sites susceptible to soil loss or sediment accumulation. Although each model is based partly on a specific calculation algorithm and has different data pre-processing requirements, we have achieved relatively comparable results in calculating the average annual soil loss and accumulation. However, each model is distinct in identifying the spatial distribution of specific locations prone to soil loss or accumulation processes.

Keywords: soil loss; sediment delivery; erosion modelling; environmental change; agriculture; Czech Republic

---

1. Introduction

Soil erosion is a physical process occurring almost everywhere in the world, even in landscapes with low slopes. Its main cause is, in addition to natural terrain properties, the character of the soil substrate and local climatic conditions, especially the prevailing land-use type and intensity of the anthropogenic pressure to which the given landscape is subjected. According to Pimentel [1], about 10 million hectares of fertile land are eroded each year, and about 115 million hectares of soil are affected by erosion in Europe [2]. Generally, this topic has been dealt with in the long term by both the professional community and political decision makers – as evidenced by the fact that soil erosion, as the main soil degradation process, has been identified as a key priority within the Soil Thematic Strategy of the European Commission [3]. In Europe, a number of research projects have been carried out in the past focusing on soil erosion and its modelling, and the importance of the theme is also illustrated by the fact that the European Soil Data Centre (ESDAC) was established within the Joint Research Centre in Ispra, Italy [4].

It is obvious that humans—as the initiators of soil loss, increasing the susceptibility of the soil to erosion—act directly through specific anthropogenic influences most often connected with land-use changes, as well as indirectly, usually in connection with the impacts of climate change, which is also conditioned by human activities to a certain extent. Indirect impacts of human activities affecting the soil erosion rates are most often manifested in the form of extreme hydrometeorological phenomena (flash floods or drought episodes). The relationship between the impacts of climate change and the soil loss potential is being given increasing attention, e.g., [5–8]. This is also closely related to the evapotranspiration rates and plant biomass production, among other factors [9]—it is a mechanism to reduce biomass production that could lead to long-term increases in soil erosion due to sparser vegetation cover [10].

When the upper layer of the soil cover is dredged from the agricultural land, there is a rapid decrease in the fertility, which leads to an obvious decrease in the yields of the crops grown there. An easily erodible soil substrate on sloping areas also presents a certain risk during significant precipitation episodes, where specific flood events in the form of debris flow (a fast-moving landslide made up of liquefied, unconsolidated, and saturated sediment mass) may occur. De Vente et al. [11] consider such debris flow an off-site effect of soil erosion. This phenomenon is manifested, for example, in the clogging of riverbeds, increased flood risk, and reduced lifetime of water reservoirs [12]. This makes soil erosion a critical problem [13] that requires holistic solutions involving physical and socioeconomic approaches. For these reasons, accurate quantification of soil loss and sediment delivery rates is very important, and information on the spatial distribution of these phenomena in a given landscape is also crucial. In addition to processes characterizing the current state, the possibility of predicting soil erosion–accumulation rates in the near future also becomes significant. Modelling of these phenomena takes into account, in particular, the expected impacts of climate change on the current cultural landscape as well as locally specific conditions that usually reflect the decision-making of regional stakeholders—usually landowners or local governments. The outlined factors influencing the prediction of soil erosion are, on the practical level, reflected through the land-use changes.

The Universal Soil Loss Equation (USLE) [14] and its more recent version, the Revised Universal Soil Loss Equation (RUSLE), revised by Renard et al. [15], are applied for the purposes of determining the volume of soil erosion almost everywhere in the world. Algorithms of a large number of soil loss and sediment delivery models and procedures are based on these equations—the models applied in this study are no exception. In the long term, soil erosion issues are highly relevant, especially in the Mediterranean region [16], which is, according to de Vente et al. [17], particularly sensitive to erosion due to its climate characterized by dry summers followed by intense autumn rainfall and often a steep topography with fragile soils. These findings illustrate the soil erosion risk maps created, for example, in Italy [18], Greece [19] or Spain [20]. Similar conditions, when the erosion–accumulation phenomena in the landscape were additionally amplified by the impacts of climate change starting in the late 1980s, were analyzed and subsequently described also in North Africa, as exemplified by Hallouz et al. [21] in a case study from Algeria. In Central European conditions, the increased susceptibility of soil to erosion is still mainly due to the use of inappropriate farming methods on sloping land, but the influence of climatic conditions and extreme hydrometeorological situations is increasing, which greatly increases the area of cultivated land with potential for higher soil erosion risk (e.g., [22–24]). The choice of specific crops grown on those lands plays a key role in the erosion rate risk —in Central Europe, this issue was dealt with, e.g., by Jones et al. [25] or Lieskovský and Kenderessy [26].

In the Czech Republic, soil erosion is one of the most widespread types of soil degradation [27], with more than 50% of agricultural land at present being threatened by water erosion and more than 10% by wind erosion [28]. The modelling of erosion processes in the Czech landscape was dealt with, e.g., by Dostál et al. [29], Krása et al. [30], Van Rompaey et al. [31] or Konečná et al [32]. Although the currently used approaches allow relatively accurate identification of sites subject to erosion risk only based on modelling techniques (i.e., without the need for field measurements), there is still considerable uncertainty associated with the determination of the volume of sediment eroded. This is also evidenced by the relatively different results achieved by individual models applied to the same territory within this study. The aim of this article is to determine the main causes of the differences between four selected, commonly used soil loss and sediment delivery models. Specifically, it is a comparison of the models in terms of their ability to identify sites with the potential for soil erosion or accumulation, as well as the exact quantification of the volume of transported material within the analyzed area. Based on this information, it is possible to determine the range of the most suitable application tasks in which the models achieve the highest-quality results.

2. Materials and Methods

2.1. Case Study Area

Selected soil loss and sediment delivery models were applied to the study area located in central Czech Republic in the Bohemian-Moravian Highlands, approximately 90 km southeast of Prague. This is the headwater area of the Kopaninský Stream (IV. order stream) with an average altitude of 565 m, with an area of 7.1 km², diverting water to the Želivka River (the left-side tributary of the Sázava River). The exact position of the selected watershed is shown in the map in Figure 1. The area is characterized by rugged terrain and a stream from its headwater area to the final profile near Velký Rybník Village at a distance of 4.3 km, surpassing an elevation of 108 m. Long-term average discharge in the final profile (between 1991 and 2000) was 0.027 m^3 s^{-1} [33]. The basin of the Kopaninský Stream is classified as a slightly warm and damp climatic area, characterized by an average annual temperature of 7.1 °C. In terms of precipitation totals, the study area is located in a rather drier region (average annual rainfall is 665 mm per year) under the influence of the continental climate.

Figure 1. Kopaninský Stream catchment and its location within the Czech Republic.

The soil type is cambic hyperskeletic Leptosol, with extremely permeable shallow loamy sands. Along the streambed, there are mainly Gleysols and Pseudogleysols in terms of soil cover. From the granulometric point of view, clay soils predominate, supplemented with sandy loam along the watercourse. The catchment area is approximately half covered with forests, and the rest is agricultural land, which is prone to erosion due to the relatively rugged relief (see Figure 2). The erosive potential is increased by the fact that these are compact and large-scale blocks of soil cover, without fragmentation made by line vegetation cover, etc. When evaluating the vegetation cover, the most common are the "distance to nature habitats" and "alien habitats," which together cover approximately 660 hectares. Selected hydrometeorological and geomorphological characteristics of the area, important also because of their influence on the formation of rainfall-runoff processes in the landscape, are given in Table 1. For the catchment, an agricultural-forest landscape is typical, with approximately 50% of forest stands—these are mainly coniferous forest with an area of nearly 260 hectares. The rest of the territory comprises non-irrigated arable land and also land principally occupied by agriculture, with significant areas of natural vegetation. The composition of agricultural crops corresponds to natural conditions. Grains and oilseed rape predominate; there is also a significant proportion of maize and peas, and some root crops.

Figure 2. Typical agricultural landscape in the area of interest (photo: O. Cudlín, 2016).

Table 1. Selected hydrometeorological and geomorphological characteristics of the study area.

	Average Annual Values			Average Terrain Slope (°)	Average Stream Slope (°)
Precipitation (mm)	Discharge in Closing Profile ($m^3\,s^{-1}$)	Specific Outflow ($l.s^{-1}\,km^{-2}$)	Air Temperature (°C)		
665.0	0.027	3.802	7.1	18.3	2.6

2.2. Data Sources

The quality of soil loss and sediment accumulation model outputs is directly dependent on the spatial resolution of the topography used in the form of the digital elevation model (DEM). In particular, the identification of sites with significant terrain slope and their extent, as well as the selection of various relief microforms that can influence the drainage processes at a given river basin by their morphometry, are very important. Often, these are anthropogenic forms of relief, such as road and rail embankments, flood levees, various water passes and other forms of relief, which, moreover, often occur in an agricultural landscape. However, the location and spatial extent of these forms are now relatively easy and reliably identifiable due to the high accuracy of commonly available terrain models based on the application of modern Earth remote sensing methods (e.g., LIDAR). This study used the digital elevation model of the Czech Republic of the 5th generation (DEM 5G) with a spatial resolution of 5 × 5 m and an average height error of 0.18 m in the exposed terrain and 0.30 m in forest terrain. The model was created from data acquired by airborne laser scanning in 2009–2013 and the resulting data are provided by the Czech Office for Surveying, Mapping and Cadastre (COSMC). DEM is the basic input layer of all soil loss and sediment delivery models based on the USLE equation. Based on these altitudinal data, the slope length and steepness factor (LS factor) can be estimated, which is one of the most important input parameters affecting the results of erosion process calculations. All models used work with a DEM in the form of a raster file.

Another key input for erosion–accumulation models is the data on the nature of prevailing land-use types. In this study, data from the ZABAGED®dataset (COSMC), supplemented by data from the Land Parcel Identification System (LPIS), and subsequently verified on the basis of the current aerial images of the catchment and its field survey, were used for the purposes of characterizing the landscape coverage. The land cover was categorized based on the requirements of individual models applied in this study—because of different working requirements, each model worked with a different number of land-use categories. The values of the individual factors of the USLE equation are also crucial inputs of the models—they are commonly used in its basic (USLE) or revised form (RUSLE) for determining the vulnerability of agricultural soils to water erosion. This approach is given by the following equation [14]:

$$A = R \times K \times LS \times C \times P, \tag{1}$$

where A is the long-term average annual soil loss [t ha^{-1} year^{-1}]; R is a rainfall erosivity factor expressed in terms of kinetic energy, total precipitations and intensity of high precipitations; K represents the soil erodibility factor expressed depending on the texture and structure of soil, organic matter content in the regolith layer and permeability of the soil profile; LS is the topographic factor expressing the effect of uninterrupted slopes and steepness on soil loss amount; C is cover and the management factor expressed depending on the state of vegetation cover and agricultural technologies used; and P represents the support practice factor.

These parameters are given in the form of raster files and in some cases as constant values for the entire area of interest. All models used in the study work on the basis of the RUSLE equation, but some parameters are not applicable for the conditions of the Czech Republic according to the RUSLE methodology because of a lack of data. The equations used are therefore a combination of RUSLE and USLE approaches. The specific values used for the individual equation factors and their sources are presented in Table 2. The spatial distribution of some factors (i.e., K, LS and C factor) within the basin is shown in the maps in Figure 3.

Table 2. Factors of the (R)USLE equation, its values and data sources used for the models applied.

Factor	Factor description	Used Values	Unit	Data Source
R	Rainfall erosivity factor	40.00	MJ ha^{-1} cm h^{-1}	Janeček et al. [34]
K	Soil erodibility factor	0.32–0.48[1]	t ha^{-1}	
LS	Topographic factor (slope length and steepness)	0.03–50.47[1]	-	Own computation based on Desmet and Govers [35]
C	Cover and management factor	0.00–0.65[1]	-	Janeček et al. [34]
P	Support practice factor	1.00	-	

[1] The values of the given factor are spatially variable and the actual distribution within the area of interest is shown in Figure 3.

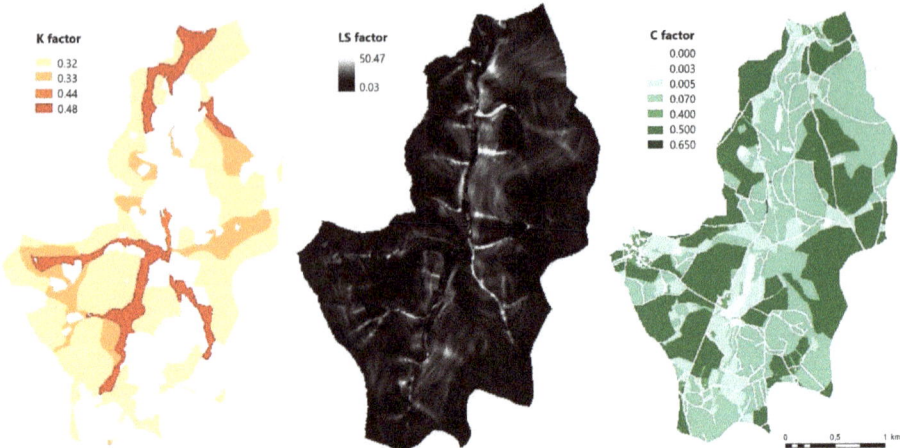

Figure 3. Spatial distribution of the K, LS and C factor values in the study area.

2.3. Approaches Used to Soil Loss and Sediment Accumulation Modelling

The following paragraphs provide basic information on the individual models applied within this article. However, this is not a detailed description of the algorithms and equations on which the models are based; this information is contained in the original studies published by authors of the given models or in the cited articles, the authors of which have also applied the models in practice. For the purposes of this study we selected a total of four models, which can be categorized as soil loss and sediment delivery models according to their official descriptions. According to available documentation, each model uses the (R)USLE equation parameters to calculate the soil loss amounts and other different algorithms to calculate the soil accumulation. In all the models applied, this is the basic USLE equation (see Equation (1)), which determines the potential soil loss from the area of 1 ha per year (it means the long-term average value valid at each site of the river basin). The USLE equation is modified for use in the GIS environment by replacing the "L" and "S" factors by the combined topographic "LS" factor derived directly from the digital elevation model. Modification of the "R" factor is not applicable here because all models consistently expect a direct input of the rainfall erosivity value, not a sub-calculation of the rain erosion efficiency based on meteorological measurements. For ease of comparison, the support practice factor "P" is not considered, the value 1 is explicitly used. An overview of the models used, their selected parameters and input data requirements is provided in the Table 3.

Table 3. Selected parameters of the models applied.

Model		InVEST	USPED	TerrSet	WaTEM/SEDEM
Basic input parameters	DEM	Raster file	Raster file	Raster file	Raster file
	R factor	Raster file	Numerical value	Raster file	Numerical value
	K factor	Raster file	Raster file	Raster file	Raster file or value
	LS factor	Numerical value	Raster file	Numerical value	Numerical value
	C factor	Numerical values (for each land-use category)	Raster file	Raster file	Raster file or value
	P factor	Numerical value	Numerical value	Raster file	Not included

Table 3. Cont.

Model	InVEST	USPED	TerrSet	WaTEM/SEDEM
Other inputs (compulsory)	River basin layer (shapefile), Biophysical parameters, Land-use and land cover data, Flow accumulation threshold, Calibration parameters	Not included	Terrain properties to identify areas with similar erosion rates, SDR values	Watercourse layer (shapefile), Land-use and land cover data, Ptef, Parcel Connectivity Data, Calibration parameters
Optional inputs	Information on drainage systems	Calibration parameters	No data value, Units used, Land fragmentation	Algorithm of LS factor computation, Retention ponds, Ploughing direction, Soil roughness, Units used
Geodata format	Common raster file, Shapefile (.SHP), .CSV file (biophysical parameters only)	Common raster file	Idrisi file (.RST)	Idrisi file (.RST)

2.3.1. InVEST

InVEST is a large set of open-source tools (models) designed to map and evaluate ecosystems and natural processes. There is a wide range of tools such as models for carbon modelling, crop pollination modelling, sediment transport modelling, but also even for fisheries or recreation modelling. The individual tools are installed in the form of one package, but each is started separately. InVEST is developed within the Natural Capital Project, which is based on a partnership between several of the world's leading academic institutions—Stanford University, the Chinese Academy of Sciences, the University of Minnesota, The Nature Conservancy and the World Wildlife Fund. InVEST is one of the few in the range of used models to be further developed [36]. Earlier versions of the entire InVEST package are also implemented in TerrSet model as an Ecosystem Service Modeler tool.

Outputs from the sediment transport model include the sediment load delivered to the stream at an annual time scale (i.e., "sediment accumulation"), as well as the amount of sediment eroded in the catchment and retained by vegetation and topographic features. The sediment delivery module (used for computation of the sediment accumulation amount) is a spatially explicit model working at the spatial resolution of the input digital elevation model raster. For each pixel, the model first computes the amount of annual soil loss from that pixel, and then computes the sediment delivery ratio (SDR), which is the proportion of soil loss actually reaching the stream. Once sediment reaches the stream, we assume that it ends up at the catchment outlet, thus no in-stream processes are modelled. This approach was proposed by Borselli et al. [37] and has received increasing interest in recent years [38–40]. The amount of annual soil loss on pixel is given by the revised universal soil loss equation (RUSLE).

Based on the work by Borselli et al. [37], the model first computes the connectivity index (IC) for each pixel. The connectivity index describes the hydrological linkage between sources of sediment (from the landscape) and sinks (like streams). Higher values of IC indicate that source erosion is more likely to make it to a sink (i.e., is more connected), which happens, for example, when there is sparse vegetation or higher slope. Lower values of IC (i.e., lower connectivity) are associated with more vegetated areas and lower slopes. The sediment delivery ratio for a pixel is then derived based on the connectivity index IC, maximum theoretical SDR and calibration parameters, defining the shape of the SDR-IC relationship (which is an increasing function). The value of sediment transport from a pixel is the amount of sediment eroded from that pixel that actually reaches the stream [38].

2.3.2. USPED

USPED (Unit Stream Power-Based Erosion/Deposition Model) is a two-dimensional model of erosion and accumulation based on the RUSLE equation. This model is oriented to the spatial distribution of areas with topographical potential for soil erosion and accumulation. The model was developed in the USA in collaboration with experts from the U.S. Army Construction Engineering Laboratories, Illinois Natural History Survey, and the Faculty of Science of Comenius University in Bratislava, Slovakia. The first results were obtained by using this model in Central Illinois and Yakima Ridge, Washington, USA [41]. USPED has been applied to the basins of various scales in the USA [42] and in Italy [43,44]. Unlike the one-dimensional revised universal soil loss equation (RUSLE) models, which assume erosion mainly depends on rainfall detachment capacity, USPED model assumes that soil erosion and accumulation mainly depend on the sediment transport capacity of the surface runoff. If soil particles are already detached by rain, but there is not enough runoff to transport the soil particles because of the terrain shape or vegetation effect, the actual amount of erosion will be significantly reduced. The total surface runoff through a given pixel determines the sediment transport capacity of this area. USPED assumes that both surface runoff and sediment transport are continuous variables. The net erosion/accumulation of sediment can be positive, indicating soil deposition, or negative, indicating soil loss. The sediment transport capacity calculation is approximated using a static sediment flow rate.

In the Central European conditions, the model was used, for example, in a Slovak study by Dotterweich et al. [45] or in the Czech Republic by Bek [46] or more recently by Vysloužilová and Kliment [47]. The major advantage of the USPED model is that it does not require the installation of any special software and can be easily implemented in common GIS environments. Generally, there are a number of interrelated raster data operations that were implemented in ESRI ArcGIS 10.2 (using Raster Calculator and Spatial Analyst Tools). The result is a raster file (the size of the grid corresponding to the input data parameters) giving information about the nature of the phenomena prevailing in each cell of the given raster—positive values indicate predominant sediment delivery and negative values indicate soil loss. A detailed procedure for implementing the USPED model is described in [47] or [43].

2.3.3. TerrSet

TerrSet is commercial raster-based software that includes a large number of individual modelling tools. It also includes tools for modelling the impacts of climate change, and integrates a whole set of InVEST models (older versions of tools) or tools for processing of the remote sensing data. The software is developed and distributed by Clark Labs at Clark University in Worcester, MA, USA [48]. The calculation of soil loss and sediment accumulation was done using the RUSLE model combination, followed by the SEDIMENTATION tool from the IDRISI GIS Analysis tool kit. The main differences from other software are based on the fact that the actual calculation is performed on raster data, but all the analyses take place on the most homogeneous surfaces, the parameters of which are chosen when starting the RUSLE model and are viewed as mutually isolated patches.

According to Clark Labs [48] the SEDIMENTATION tool utilizes patch-level output from the RUSLE modelling tool to evaluate the net soil movement (erosion or deposition) within patches. The Sediment Delivery Ratio (SDR) is determined from the average annual sediment yield. Sediment yield (as in the case of InVEST model) is the amount of soil loss (erosion) that reaches a stream and is transported within the waterway. Determining net erosion or deposition begins with the total soil loss by patch produced from the RUSLE module. To determine net soil loss or deposition in each patch, SEDIMENTATION tool first determines the average elevation for each patch. Then the highest elevation in the river basin or the highest elevation in each field is located. The direction of movement of the soil is then established by the relative elevation differences between contiguous patches. Movement is always in the downslope direction. The amount of soil loss that moves into the surrounding lower patches is proportional to the length of the common boundary between the higher patch and the lower patches. Next, the net soil loss or soil accumulation in all lower patches is calculated as follows. Using the output from RUSLE, the proportional soil loss for the higher patch is compared to the soil loss for the lower patch. The difference between the amounts of soil loss from the higher patch to the lower patch represents the net soil loss or accumulation in the lower patch.

2.3.4. WaTEM/SEDEM

WaTEM/SEDEM is a specific tool developed at KU Leuven in Belgium, designed for modelling soil erosion and transport processes. The software is freely distributed and is based on the IDRISI GIS analysis tool and also uses its RST file format, which is related to the need to convert input data. The model can be used to determine soil loss and accumulation values due to water erosion, to identify areas of agricultural land susceptible to water erosion processes or to simulate the effects of different soil conservation scenarios [49,50]. The water erosion component of WaTEM/SEDEM uses an adapted version of the Revised Universal Soil loss equation (RUSLE) to calculate mean annual soil loss values. Runoff patterns are calculated with a flow algorithm that takes into account field borders, tillage direction and road infrastructure. Sediment is routed along these flow paths to the nearest river using a transport capacity term that is proportional to the potential rill erosion rate. The boundaries of the land and the topology of the river network are taken into account, which makes it suitable for modelling larger river basins.

The sediment production values do not correspond with erosion values as calculated with RUSLE, the latter being higher. RUSLE only predicts soil loss per pixel, not sediment deposition. On pixels where sediment deposition is lower than RUSLE erosion, netto erosion will be the case. It is the latter value which is used to calculate sediment production (soil accumulation). If sediment accumulation is higher than RUSLE soil loss, there will be netto sediment accumulation and this value will be used to calculate the catchment's total sediment accumulation value. This model uses the R, K, C and P factors of the RUSLE equation. The topographical L and S factors are automatically calculated from the DEM. In order to adapt the RUSLE to a two-dimensional landscape, the upslope length is replaced by the unit contributing area, i.e., the upslope drainage per unit of contour length. The use of contributing areas implies that the effects of flow convergence and divergence on topographically complex landscapes are explicitly accounted for [51].

The erosion rate is considered equal to the sum of the potential erosion unless the local transport capacity is exceeded. The transport capacity coefficient describes the proportionality between the potential for rill erosion and the transport capacity of the overland flow. The transport capacity is the maximum sediment mass that can be transported by the overland flow. If the sediment production is higher than this transport capacity, sediment will be deposited. Thus, the higher the transport capacity coefficient, the more sediment can be transported downslope. For each land use type, transport capacity can be different. The transport capacity on a given slope segment was considered to be directly proportional to the erosion potential. If the sediment inflow exceeds the transport capacity, deposition occurs, so that the amount of material equals the transport capacity. WaTEM/SEDEM version 2004, launched through the 2006 version, is used in this study, which brings together all available versions of the given software and corrects some of their errors. The algorithms used are listed in the program manual [52]. This software generally belongs to relatively frequently used tools—in Central Europe, its outputs were presented, for example, by Bezak et al. [53] or in the Czech Republic by Van Rompaey et al. [31] and by Krása [54], which describes in more detail the individual algorithms of the model.

3. Results

3.1. Analysis of Soil Loss Caused by Water Erosion

Selected models were applied within the area of interest to determine the soil loss amount caused by water erosion. Each model works on a slightly different principle of soil loss estimation.

The WaTEM/SEDEM (W/S) model calculates the "netto erosion" (the results range from negative values indicating soil erosion to positive values that represent the sediment accumulation). This principle is also applied within the USPED model. The InVEST model provides results in the form of "average erosion," which can be understood as a direct output from the USLE equation (the results range from zero to positive values that represent soil loss). In contrast, the TerrSet model calculates both of these options ("netto erosion" and "average erosion"). Comparison of the results of the four used models in terms of spatial distribution of erodible sites within the catchment is provided by the maps in Figures 4 and 5. From these outputs, a different principle of calculation is again evident—whereas the TerrSet model refers to the soil erosion amounts on individual patches corresponding to the sites with a homogeneous land-use type (the basic spatial unit is therefore a unitary land-use category), the W/S predicts the resulting soil erosion in a resolution corresponding to the extent of the input data, i.e., individual pixels with an area of 100 m^2. On the same principle as W/S, other applied models also work. For this reason, TerrSet's results may be somewhat less accurate in terms of identifying specific sites with the potential to soil loss. It can be concluded that in terms of spatial distribution, comparable results are achieved in particular by InVEST and TerrSet model, which identify almost identical sites, but differ in the estimated soil erosion volumes slightly (with total soil loss in the whole basin ranging from 43.8 to 55.4 tonnes per year). In the case of the USPED model, it is obvious that, as well as sediment accumulation modelling, soil loss volumes are significantly higher than in the case of other models, although significant soil erosion is expected in only a few localities within the catchment. When comparing the results of all four applied models, we can conclude that the models differ especially in terms of the total soil erosion volumes expected throughout the catchment—the values vary from 42.0 to 76.0 tonnes per year. This variance is largely influenced by the different area of the sites with soil erosion potential (see Table 4), which is also the result of a different approach to the areas where soil erosion and sediment accumulation were not identified. While W/S assigns a "0" value and includes these areas in following calculations, all other models assigns a "No data" attribute to that territory and do not work with it in any further steps. The difference in the number of pixels included in the calculations for each model is a key factor affecting the overall results.

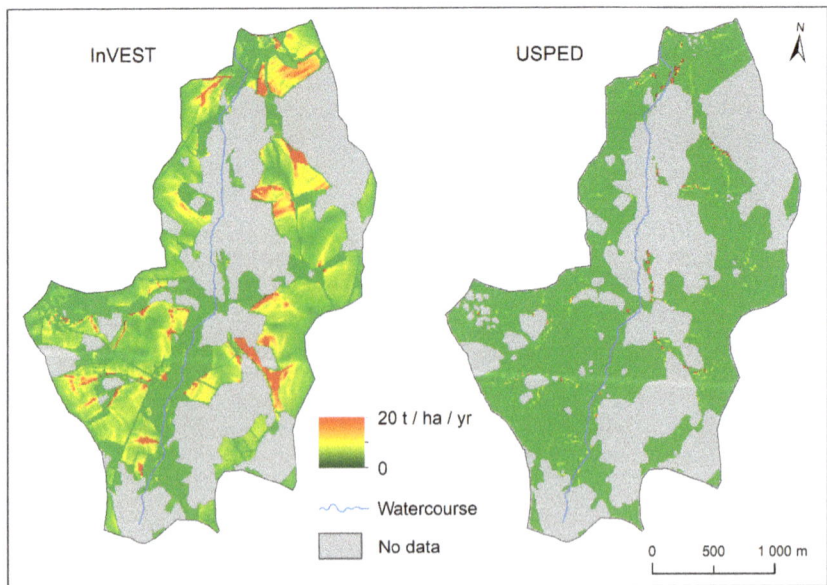

Figure 4. Soil loss in the study area based on the results of the InVEST and USPED models.

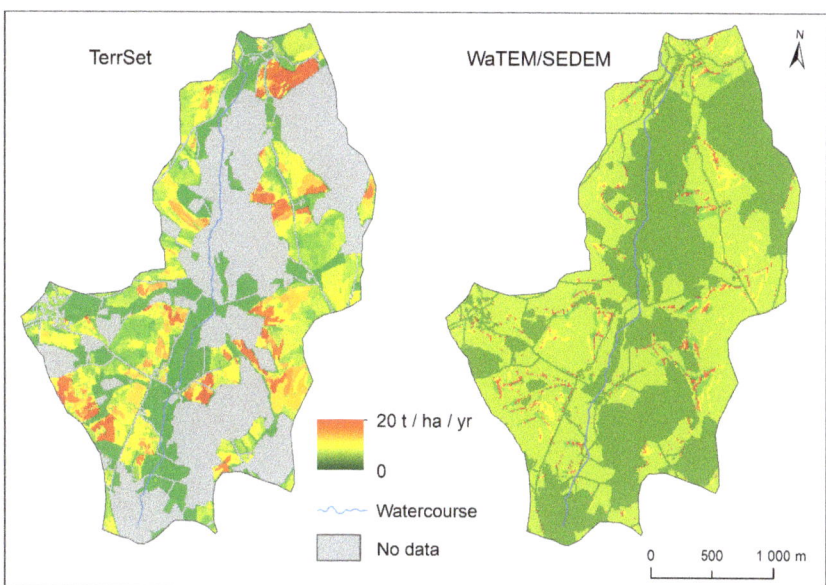

Figure 5. Soil loss in the study area based on the results of the TerrSet and WaTEM/SEDEM models.

Table 4. Comparison of the basic soil loss characteristics of the four models applied.

Model Used	InVEST	USPED	TerrSet	WaTEM/SEDEM
Total soil loss in the basin (t year^{-1})	43.78	76.03	55.38	41.96
Total area with soil loss (ha)	349.38	186.2	354.77	305.10
Average soil loss (t ha^{-1} year^{-1})	0.062	0.107	0.078	0.059

The calculated soil loss values and the respective areas of the sites where the erosion takes place are shown in the graph in Figure 6. In principle, this figure captures the dependence between the number of pixels and the corresponding soil erosion values of each output raster file. The resulting chart points to the fact that in terms of the nature of the relationship between the two variables studied, there are significant differences between the models applied. The W/S model achieves the lowest soil losses, with the values found in areas ranging from 0.01 ha to almost 10 hectares (according to the number of pixels with given values).

Relatively similar results are achieved by the InVEST model, which has slightly increased soil loss values and pixels with very low erosion values are not present in such amount as compared to W/S. The curve pattern of the InVEST model outputs indicates that there is almost the same number of pixels with heterogeneous values of soil loss. A similar situation with slightly higher values (with an area covering of about 3 ha) is also the case for the USPED model. Significantly different values that occur in the TerrSet model output are mainly due to the fact that the model works with larger patches characterized by a homogeneous and higher average value.

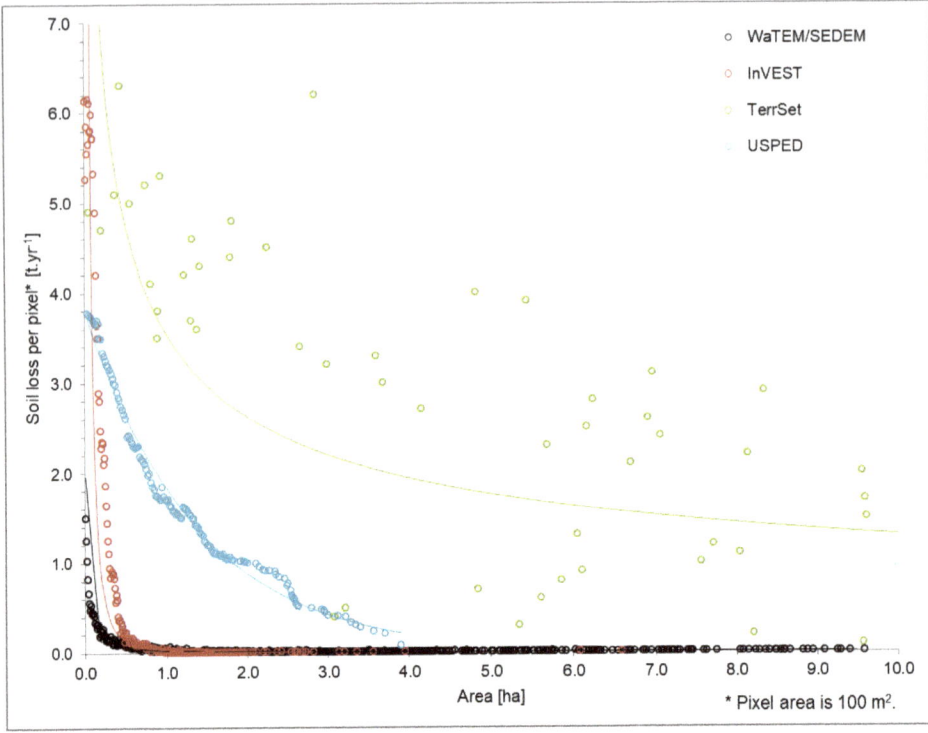

Figure 6. Relationship between the soil loss values and the corresponding number of pixels identified within the catchment where soil erosion takes place based on the results of the four models applied.

3.2. Analysis of Sediment Accumulation as a Consequence of Soil Loss by Water Erosion

In addition to the soil erosion, all of the applied models also allow the computation of sediment accumulation, and since the study area is a coherent catchment (its upper part)—a hydrologically correctly defined spatial unit—the volume of sediments should correlate with the amount of eroded material to some extent (taking into account that part of the sediment volume is transported outside the area of interest through the closing profile).

Each model has achieved relatively different results not only in terms of total volumes of the sediment accumulation (Table 5), but also when comparing the spatial distribution of the observed phenomenon (Figures 7 and 8). The outputs of all the models have been unified for ease of interpretation in terms of the sediment accumulation amounts displayed, ranging from 0 to 20 tonnes per hectare per year. Although it is not obvious from map visualization, the highest values of sediment accumulation in the study area are predicted by the USPED model, which assumes soil accumulation of more than 0.11 tonnes per hectare per year. The main reason is the accumulation of extremely high amounts in the small area, which roughly corresponds to the results of the other models applied. In terms of total soil accumulation in the basin, the results of the InVEST and TerrSet models are relatively similar. However, since InVEST assumes sedimentation on a much larger area, it is clear from the comparison of average sedimentation (tonnes per hectare per year) that the outputs for InVEST and USPED are more comparable.

Based on the spatial pattern of soil accumulation values, a slightly different calculation algorithm is also apparent when comparing the three models—InVEST, USPED (in Figure 7) and WaTEM/SEDEM (on the right in Figure 8) with the TerrSet model outputs (on the left in Figure 8). While these three models counts soil accumulation (as well as soil loss) in each pixel alone and do not communicate with their surroundings ("per pixel" method), the remaining TerrSet model counts using "per patch" method. This means that at the beginning the area is divided into patches, defined on the basis of the homogeneous terrain properties and landscape cover. This approach creates a group of adjacent pixels that together make up one area in which the calculation takes place. The maximum length of drainage path is given by the user's value or by landscape elements interrupting the drainage path. Individual patches communicate with each other, which mean that they are receiving information about computed values in adjacent areas. This aspect has a significant effect on the L factor (slope length factor), since in the case of patches the uninterrupted length of drainage path is much larger compared to the size of the per pixel models (i.e., the length of the drainage line in each pixel is always the same).

Table 5. Comparison of the basic soil accumulation characteristics of the four models applied.

Model Used	InVEST	USPED	TerrSet	WaTEM/SEDEM
Total soil accumulation in the basin (t year^{-1})	21.24	80.84	22.42	39.86
Total area with accumulation (ha)	320.58	123.87	93.10	46.65
Average soil accumulation (t ha^{-1} year^{-1})	0.030	0.114	0.032	0.056

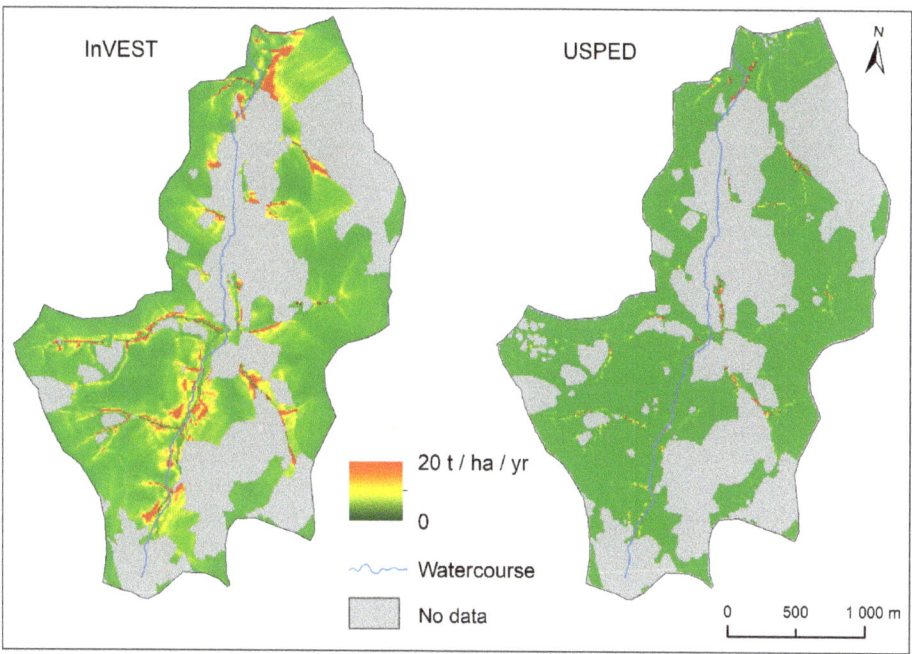

Figure 7. Sediment accumulation as a result of soil loss caused by water erosion in the study area—comparison of InVEST and USPED outputs.

When assessing the spatial distribution of the sediment accumulation capacity of the catchment, it is obvious that large sediment accumulation-prone areas are predicted, especially by the InVEST and TerrSet model, which expects the sediment accumulation to occur at the foot of practically all slopes (this is also reflected in the sediment accumulation area, achieving the highest extent from all the models in case of the InVEST). However, based on the data in Table 5, it is evident that InVEST identifies large areas susceptible to sediment accumulation, but at the same time the model assigns relatively low values to those sites, compared to other models (up to 269 tonnes/hectare/year). By comparing the graphical outputs of all models, the influence of a different calculation algorithm is obvious. While the second pair of models (Figure 8) shows that the key factor for the prediction of the spatial distribution of sediment accumulation is the nature of land-use (the individual areas identified correspond to the extent and boundary position of the land-use units), for the first pair of models (Figure 7) the terrain characteristics play a crucial role—topography is projected onto the results, especially in the case of the InVEST model. Specific to the InVEST and USPED models, the calculation is also carried out on road surfaces, which the TerrSet and WaTEM/SEDEM models exclude from the calculation.

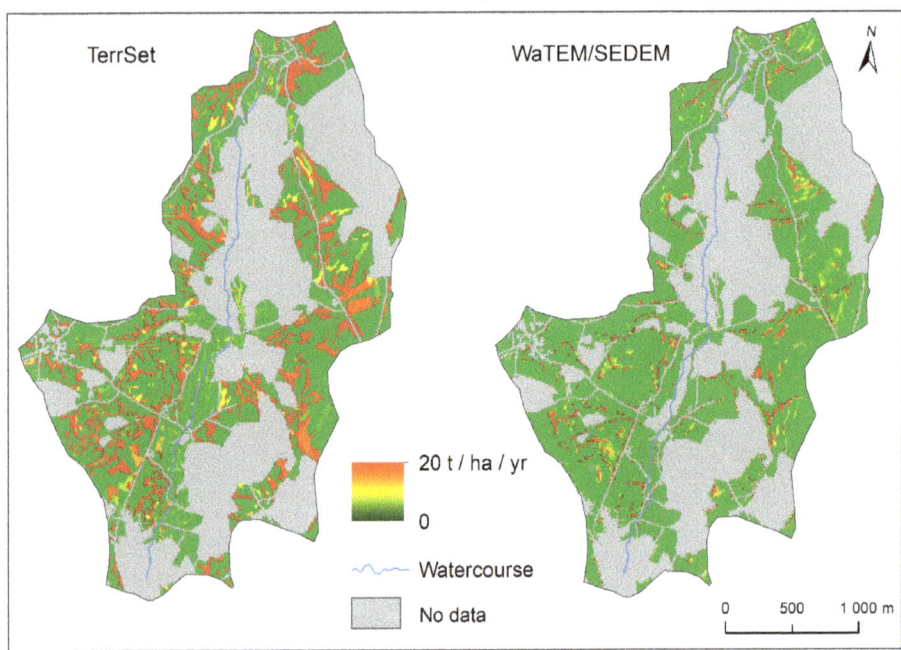

Figure 8. Sediment accumulation as a result of soil loss caused by water erosion in the study area—comparison of TerrSet and WaTEM/SEDEM outputs.

The above-described differences in the results of the analyzed models are evident, especially in the detailed view of the graphical outputs, showing the spatial distribution of the observed phenomena (Figure 9). The W/S model attribute high sediment accumulation values to only a few relatively small areas and minimal accumulation values that are homogeneous on a relatively large area. In contrast, the InVEST model allows for much more extensive areas, characterized by increased, but not extreme, sediment accumulation. From the anthropogenically conditioned forms of relief, the road network plays a very important role in soil erosion and sediment accumulation modelling. This element influences the results of the models by modifying the morphological parameters of the terrain, which is then reflected in input data for modelling (DEM), but also in the manner in which a given model processes this type of landscape cover. The TerrSet and WaTEM/SEDEM models used exclude the road network from calculation (they gave the "No data" attribute to the pixels), which is reflected especially when comparing the average sediment accumulation or soil loss for the whole catchment area—each model therefore calculates with a little different area affected by the number of pixels with the "No data" attribute.

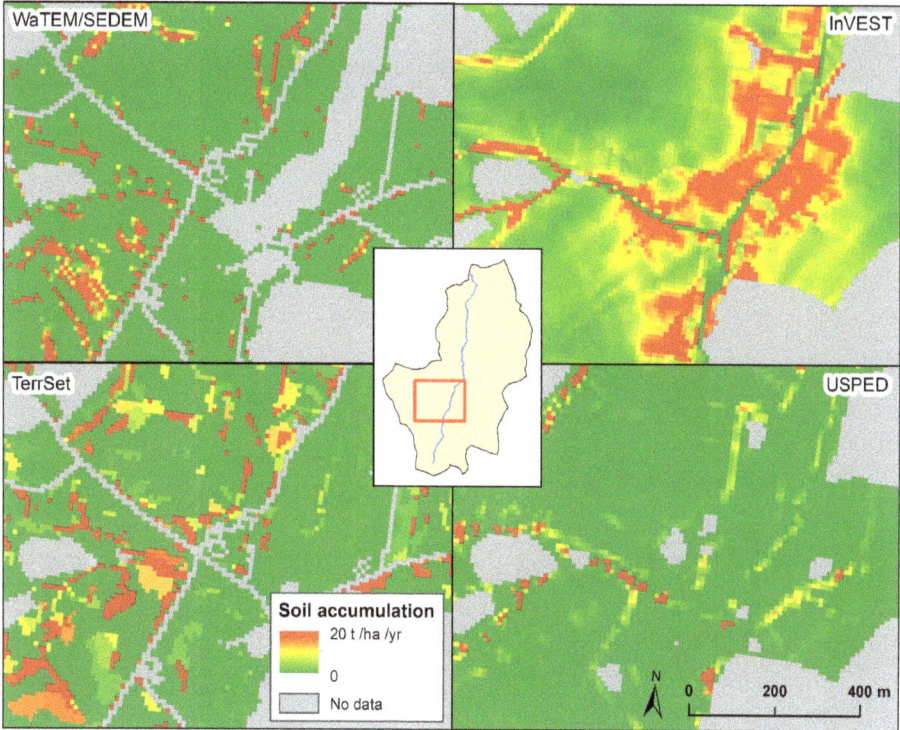

Figure 9. Detailed view on the selected site within the area of interest—comparison of the soil accumulation modelling results.

For the efficient interpretation of differences in the results of the analyzed models, graph (Figure 10) was created. The graph show the course of the dependencies between the amount of sediment accumulation (i.e., soil accumulation per an area of one hectare) and an area that is characterized by given specific soil accumulation capabilities. Based on Figure 10 it is obvious that the analyzed models have different results in terms of their quality. From the distribution of values within the correlation plot, it can be stated that the greatest differences is observed between the W/S and the USPED model, where W/S assigns a large number of pixels with very low soil accumulation values, the USPED assumes the occurrence of pixels with significantly higher values. The TerrSet and InVEST models have a very similar distribution of values within the correlation plot.

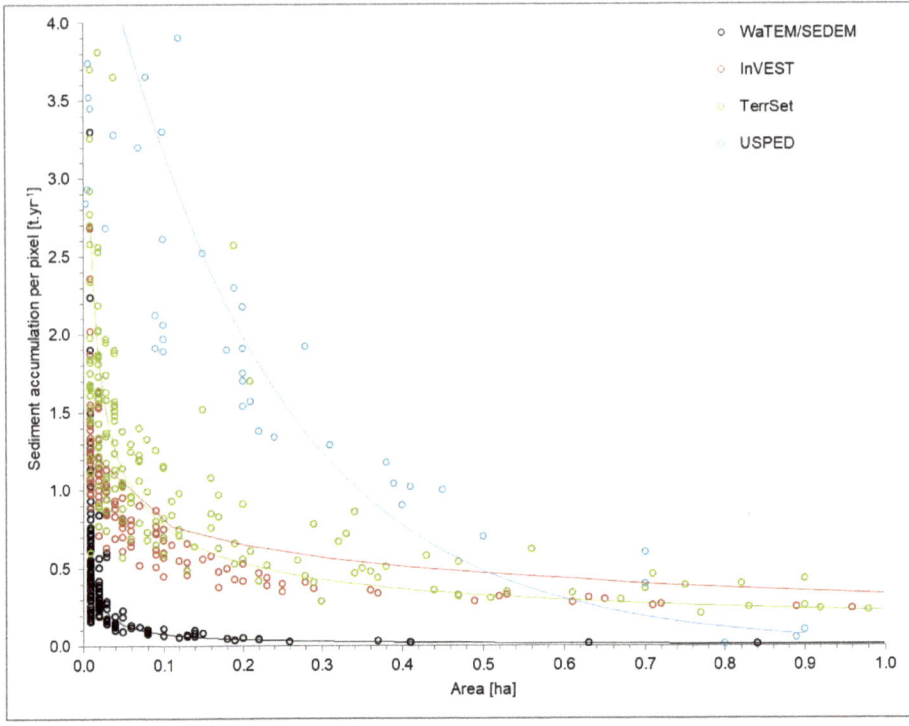

Figure 10. Relationship between the sediment accumulation values and the corresponding number of pixels (pixel area is 100 m^2) identified within the area where the accumulation takes place based on the results of the four models applied.

4. Discussion and Conclusions

Based on the application of the four soil loss and sediment accumulation models on a selected small stream catchment in a typical Central European landscape, it can be stated that there are obvious differences between their outputs, both in terms of the total expected soil erosion and sedimentation volumes and spatial distribution of these phenomena within the catchment. Different model results, however, cannot be considered as a symptom of a certain error in the processing, but rather as a result of different algorithms and approaches used to process the input data (which were otherwise identical for each model). One of the main causes of differences in the results is the fact that each model actually works with a different number of pixels within the used raster files. This fact is dependent on different approaches to the classification of the land cover, for example, the W/S model also includes areas where there is no soil erosion (assigns a "0" value), whereas the other applied models refer to these areas as "No data" and exclude them. Furthermore, the TerrSet and W/S models also earmark the road network from the final calculation. This different fragmentation of the study area and the continuity of individual patches are key factors influencing the resulting soil erosion and sediment deposition values, respectively. Another factor influencing soil loss values is the runoff algorithm (flow accumulation algorithm) used—while W/S and TerrSet use the "Multiple Flow Direction" algorithm, USPED and InVEST have only a "Single Flow Direction" mechanism integrated. The details about their differences and influences on the runoff values are described in Wolock and McCabe [55].

Apart from the above-described differences between the results of each modelling process and its probable causes, it is also necessary to take into account the data input quality, which can greatly influence the overall accuracy of the results. It is mainly the spatial and temporal uncertainty of the input data that affects the quality of the result. An example is the "cover and management factor" (C factor), where the same value for each crop for the whole year was applied to all the models. These values are based on widely used tabular reports valid for the territory of the Czech Republic [34]. A similar principle is often used in other European countries. However, it is evident that the use of a single C factor value is largely inaccurate, especially in regions with alternating seasons, as it fluctuates during the course of the year, depending on the actual growth phase of vegetation. Much more accurate results that could subsequently be used for other decision-making processes in a given area can be achieved by determining the C factor value by means of a series of several soil erosion partial calculations, for different conditions, depending on a particular growing season (for more information on this issue, see Pechanec et al. [56]). However, to assess the quality of model outputs, the use of a single value is sufficient, especially with regard to the necessity of affecting the differences in the actual soil loss or sediment delivery calculation algorithm, not in the quality of input data.

The models used can be further compared according to their user requirements and input data requirements—in this regard, the InVEST model is probably the most user-friendly. It has a relatively purely processed user interface and has the lowest input and pre-processing requirements. The WaTEM/SEDEM model offers a fairly good user interface, but the need for high-quality input data and its pre-processing is considerable. Additionally, the need to work with data in IDRISI format can now be considered as a certain disadvantage. In contrast, the USPED model does not have any custom interface—it is essentially a series of calculations (mathematical operations with raster files) that can be done in any GIS software. Overall, this approach can be considered relatively easy; however, due to the large number of intermediate stages, it is a rather time-consuming process. The advantages and disadvantages of all analyzed models with respect to their practical applicability are summarized in Table 6.

Table 6. Rating of the models in terms of their demands on the input data and previous user experiences. "++" means the most positive rating – user-friendly model, low input data requirements, etc.; "–" indicates the most negative rating; and 0 means that the model does not have a user interface.

Model Used	InVEST	USPED	TerrSet	WaTEM/SEDEM
Data pre-processing requirements	++	-	+	-
User-friendly interface	++	0	+	+
Claims for prior user experiences	+	++	-	-
Failure sensitivity (including the difficulty of error detecting)	+	+	-	-

Another possibility to evaluate the quality of the outputs of individual models is offered based on their comparison with real soil erosion values, observed and measured in the field directly within the area of interest. Since the field measurement process of soil loss is a relatively demanding issue, it is not always possible to perform such a comparison and thus verify the results of model simulations. The advantage of the area studied in this article is the fact that it is a long-term observation area, which has been investigated in the past by a number of other Czech authors (i.e., Doležel et al. [57] or Fučík et al. [58]—among other measurements, soil erosion has been measured over a certain time period. Pavlík et al. [59] carried out field measurements in the Kopaninský Stream catchment in 2005 and 2006 and found out that the average annual soil erosion amounts vary in the range between 3.4 and 4.4 t ha^{-1} year^{-1}. According to the Czech Hydrometeorological Institute (CHMI) in the area of interest the average annual precipitation amounts recorded in these years (2005 and 2006) were 798 resp. 826 mm, which is a significantly above-average state (long-term average annual value is 665 mm year^{-1}). The outputs can also be compared with the study realized by Panagos et al [8], which predicts a soil erosion of approximately 1.0–8.9 t ha^{-1} year^{-1} in the soil erosion map of Europe and soil database (ESDAC) for the given area of interest. Based on these results, it is evident that most models applied achieve highly underestimated values when compared with the outputs mentioned.

However, it is very important to emphasize that the comparison of the model outputs with the real values obtained by field measurements is only for the idea of how much the model outputs can differ from the real situation. But it is not possible to consider the models as inoperative because of the differences, as the main reason for these differences is the fact that the models do not capture the current conditions in the locality, but only the typical average values valid for a wider area. One of the key elements affecting significantly the outputs is the choice of the C factor (cover and management factor). In this study we were used year round constant value recommended for the Czech Republic by Janeček et al. [34]. Another reason for discrepancies between model outputs and the real situation may also be the unavailability of information on the existence of possible erosion control measures in the area of interest. For this reason, a constant value of "1" for the P factor (Support practice factor) is entered into the models, but for example, in the above-mentioned comparison, we do not have any detailed information on the particular state of the erosion control measures at the site where the measurements took place in the past. An important finding from the comparison of the four selected models is that all models achieve very similar results in assessing the overall average soil loss (the values found range from 0.125 to 0.156 tonnes per hectare per year). Thus, by applying these models, relatively similar results can be obtained, but in terms of the spatial distribution of the particular soil loss or accumulation areas, the outputs can vary considerably and they cannot be used for precise identification of specific locations. For the purposes of precise identification of such soil erosion-prone areas as is required for example in landscape planning or flood protection solutions, it is appropriate to use differently designed models, described, for example, in Pechanec et al. [60].

Overall, it can be summarized that for the purposes of application tasks the InVEST model is probably the most acceptable from the four models compared within this article. This model is less demanding in terms of claims on the input data and is characterized by a relatively high-quality user interface, which makes it usable even for less experienced users. The USPED model is also easy to apply, but the ease lies mainly in the fact that it is a sequence of operations feasible in any GIS software, therefore the steps are clear and the user can easily verify the quality of the partial results. The opposite is the W/S model, which works in the form of a black box (the user does not see the algorithms of the individual calculations) and the input data need to be entered in a specified format. However, all the models applied are characterized by the demands on a certain user experience in the field of environmental modelling and, in particular, the availability of the most accurate data, which completely determines the quality of the outputs.

Author Contributions: J.J. wrote the main paper and performed the USPED model computations. V.P. and J.P. performed the remaining model computations. V.P. also discussed the results and commented on the manuscript at all stages. P.C. discussed the results and commented on the manuscript in its initial phase.

Funding: This study was supported by the Ministry of Education, Youth and Sports of CR within the National Sustainability Program I (NPU I), grant number LO1415 and also by the project "LAND4FLOOD: Natural Flood Retention on Private Land", grant number LTC18025 and project "Small watercourses and riparian ecosystem management to mitigate the impacts of environmental change (SMART2Envi)," grant number LTC18069.

Acknowledgments: The authors would like to thank Ing. Petr Fučík, Ph.D. for his initial idea to carry out this type of research in the given area of interest and also to Ing. Ondřej Cudlín, Ph.D. for providing photos from the area of interest shown in the article.

Conflicts of Interest: The authors declare no conflict of interest.

References

1. Pimentel, D. Soil erosion: A food and environmental threat. *Environ. Dev. Sustain.* **2006**, *8*, 119–137. [CrossRef]
2. Kibblewhite, M.G.; Miko, L.; Montanarella, L. Legal frameworks for soil protection: Current development and technical information requirements. *Curr. Opin. Environ. Sustain.* **2012**, *4*, 573–577. [CrossRef]
3. Communication from the Commission to the Council, the European Parliament, the European Economic and Social Committee and the Committee of the Regions—Thematic Strategy for Soil Protection. Brussels 2006.

Available online: https://eur-lex.europa.eu/legal-content/EN/TXT/?uri=CELEX:52006DC0231 (accessed on 17 November 2018).
4. Panagos, P.; Van Liedekerke, M.; Jones, A.; Montanarella, L. European Soil Data Centre (ESDAC): Response to European policy support and public data requirements. *Land Use Policy* **2012**, *29*, 329–338. [CrossRef]
5. Morgan, R.P.C.; Quinton, J.N. Erosion modeling. In *Landscape Erosion and Evolution Modeling*; Springer: Boston, MA, USA, 2001; pp. 117–143.
6. Nunes, J.P.; Seixas, J.; Pacheco, N.R. Vulnerability of water resources, vegetation productivity and soil erosion to climate change in Mediterranean watersheds. *Hydrol. Process. Int. J.* **2008**, *22*, 3115–3134. [CrossRef]
7. Lorencová, E.; Frélichová, J.; Nelson, E.; Vačkář, D. Past and future impacts of land use and climate change on agricultural ecosystem services in the Czech Republic. *Land Use Policy* **2013**, *33*, 183–194. [CrossRef]
8. Panagos, P.; Borrelli, P.; Poesen, J.; Ballabio, C.; Lugato, E.; Meusburger, K.; Montanarella, L.; Alewell, C. The new assessment of soil loss by water erosion in Europe. *Environ. Sci. Policy* **2015**, *54*, 438–447. [CrossRef]
9. Nearing, M.A.; Pruski, F.F.; O'Neal, M.R. Expected climate change impacts on soil erosion rates: A review. *J. Soil Water Conserv.* **2004**, *59*, 43–50.
10. Pruski, F.F.; Nearing, M.A. Runoff and soil-loss responses to changes in precipitation: A computer simulation study. *J. Soil Water Conserv.* **2002**, *57*, 7–16.
11. De Vente, J.; Poesen, J.; Verstraeten, G.; Van Rompaey, A.; Govers, G. Spatially distributed modelling of soil erosion and sediment yield at regional scales in Spain. *Glob. Planet. Chang.* **2008**, *60*, 393–415. [CrossRef]
12. Verstraeten, G.; Poesen, J. The nature of small-scale flooding, muddy floods and retention pond sedimentation in central Belgium. *Geomorphology* **1999**, *29*, 275–292. [CrossRef]
13. Montgomery, D.R. Soil erosion and agricultural sustainability. *Proc. Natl. Acad. Sci. USA* **2007**, *104*, 13268–13272. [CrossRef] [PubMed]
14. Wischmeier, W.H.; Smith, D.D. *Predicting Rainfall Erosion Losses—A Guide to Conservation Planning*; Science and Education Administration, USDA: Hyattsville, MD, USA, 1978; 62p.
15. Renard, K.G.; Foster, G.R.; Weesies, G.A.; McCool, D.K.; Yoder, D.C. *Predicting Soil Erosion by Water: A Guide to Conservation Planning with the Revised Universal Soil Loss Equation (RUSLE)*; United States Department of Agriculture: Washington, DC, USA, 1997.
16. Guerra, C.A.; Maes, J.; Geijzendorffer, I.; Metzger, M.J. An assessment of soil erosion prevention by vegetation in Mediterranean Europe: Current trends of ecosystem service provision. *Ecol. Indic.* **2016**, *60*, 213–222. [CrossRef]
17. Grimm, M.; Jones, R.J.; Rusco, E.; Montanarella, L. Soil erosion risk in Italy: A revised USLE approach. *Eur. Soil Bureau Res. Rep.* **2003**, *11*, 23.
18. Van Rompaey, A.; Bazzoffi, P.; Jones, R.J.; Montanarella, L. Modeling sediment yields in Italian catchments. *Geomorphology* **2005**, *65*, 157–169. [CrossRef]
19. Kouli, M.; Soupios, P.; Vallianatos, F. Soil erosion prediction using the revised universal soil loss equation (RUSLE) in a GIS framework, Chania, Northwestern Crete, Greece. *Environ. Geol.* **2009**, *57*, 483–497. [CrossRef]
20. García-Ruiz, J.M. The effects of land uses on soil erosion in Spain: A review. *Catena* **2010**, *81*, 1–11. [CrossRef]
21. Hallouz, F.; Meddi, M.; Mahé, G.; Toumi, S.; Rahmani, S.E.A. Erosion, Suspended Sediment Transport and Sedimentation on the Wadi Mina at the Sidi M'Hamed Ben Aouda Dam, Algeria. *Water* **2018**, *10*, 895. [CrossRef]
22. Boardman, J.; Poesen, J. Soil erosion in Europe: Major processes, causes and consequences. *Soil Erosion Eur.* **2006**, *4*, 477–487.
23. Scholz, G.; Quinton, J.N.; Strauss, P. Soil erosion from sugar beet in Central Europe in response to climate change induced seasonal precipitation variations. *Catena* **2008**, *72*, 91–105. [CrossRef]
24. Mullan, D. Soil erosion under the impacts of future climate change: Assessing the statistical significance of future changes and the potential on-site and off-site problems. *Catena* **2013**, *109*, 234–246. [CrossRef]
25. Jones, R.J.; Spoor, G.; Thomasson, A.J. Vulnerability of subsoils in Europe to compaction: A preliminary analysis. *Soil Tillage Res.* **2003**, *73*, 131–143. [CrossRef]
26. Lieskovský, J.; Kenderessy, P. Modelling the effect of vegetation cover and different tillage practices on soil erosion in vineyards: A case study in Vráble (Slovakia) using WATEM/SEDEM. *Land Degrad. Dev.* **2014**, *25*, 288–296. [CrossRef]

27. Novotný, I. *Příručka Ochrany Proti Vodní Erozi*; Ministerstvo Zemědělství: Praha, Czech Republic, 2014; 73p. (In Czech)
28. Novotný, I. *Příručka Ochrany Proti Erozi Zemědělské Půdy*; Ministerstvo Zemědělství a Výzkumný ÚSTAV MELIOrací a Ochrany Půdy, v.v.i.: Praha, Czech Republic, 2017; 86p. (In Czech)
29. Dostál, T.; Krása, J.; Váška, J.; Vrána, K. The map of soil erosion hazard and sediment transport in scale of the Czech Republic. *Vodní Hospodářství* **2002**, *52*, 46–48. (In Czech)
30. Krása, J.; Dostál, T.; Vrána, K.; Plocek, J. Predicting spatial patterns of sediment delivery and impacts of land-use scenarios on sediment transport in Czech catchments. *Land Degrad. Dev.* **2010**, *21*, 367–375. [CrossRef]
31. Van Rompaey, A.; Krása, J.; Dostál, T. Modelling the impact of land cover changes in the Czech Republic on sediment delivery. *Land Use Policy* **2007**, *24*, 576–583. [CrossRef]
32. Konečná, J.; Karásek, P.; Fučík, P.; Podhrázská, J.; Pochop, M.; Ryšavý, S.; Hanák, R. Integration of Soil and Water Conservation Measures in an Intensively Cultivated Watershed – A Case Study of Jihlava River Basin (Czech Republic). *Europ. Countrys.* **2017**, *1*, 17–28.
33. Doležal, F.; Kvítek, T.; Soukup, M.; Kulhavý, Z.; Tippl, M. Czech highlands and peneplains and their hydrological role, with special regards to the Bohemo-Moravian Highland. *IHP HWRP-Berichte* **2004**, *2*, 41–56.
34. Janeček, M. *Ochrana Zemědělské Půdy Před Erozí. Metodika*; Česká Zemědělská Univerzita, Fakulta Životního Prostředí: Praha, Czech Republic, 2012; 113p. (In Czech)
35. Desmet, P.J.J.; Govers, G. A GIS procedure for automatically calculating the USLE LS factor on topographically complex landscape units. *J. Soil Water Conserv.* **1996**, *51*, 427–433.
36. Natural Capital Project. Available online: https://www.naturalcapitalproject.org/ (accessed on 5 May 2018).
37. Borselli, L.; Cassi, P.; Torri, D. Prolegomena to sediment and flow connectivity in the landscape: A GIS and field numerical assessment. *Catena* **2008**, *75*, 268–277. [CrossRef]
38. Cavalli, M.; Trevisani, S.; Comiti, F.; Marchi, L. Geomorphometric assessment of spatial sediment connectivity in small Alpine catchments. *Geomorphology* **2013**, *188*, 31–41. [CrossRef]
39. Lopez-Vicente, M.; Poesen, J.; Navas, A.; Gaspar, L. Predicting runoff and sediment connectivity and soil erosion by water for different land use scenarios in the Spanish Pre-Pyrenees. *Catena* **2013**, *102*, 62–73. [CrossRef]
40. Sougnez, N.; Wesemael, B.; Van Vanacker, V. Low erosion rates measured for steep, sparsely vegetated catchments in southeast Spain. *Catena* **2011**, *84*, 1–11. [CrossRef]
41. Mitášová, H.; Hofierka, J.; Zlocha, M.; Iverson, L.R. Modelling topographic potential for erosion and deposition using GIS. *Int. J. Geogr. Inf. Syst.* **1996**, *10*, 629–641.
42. Warren, S.D.; Mitášová, H.; Hohmann, M.G.; Landsberger, S.; Iskander, F.Y.; Ruzycki, T.S.; Senseman, G.M. Validation of a 3-D enhancement of the Universal Soil Loss Equation for prediction of soil erosion and sediment deposition. *Catena* **2005**, *64*, 281–296. [CrossRef]
43. Pistocchi, A.; Cassani, G.; Zani, O. Use of the USPED model for mapping soil erosion and managing best land conservation practices. In Proceedings of the 1st International Congress on Environmental Modelling and Software, Lugano, Switzerland, 24–27 June 2002.
44. Pelacani, S.; Märker, M.; Rodolfi, G. Simulation of soil erosion and deposition in a changing land use: A modelling approach to implement the support practice factor. *Geomorphology* **2008**, *99*, 329–340. [CrossRef]
45. Dotterweich, M.; Stankoviansky, M.; Minár, J.; Koco, Š.; Papčo, P. Human induced soil erosion and gully system development in the Late Holocene and future perspectives on landscape evolution: The Myjava Hill Land, Slovakia. *Geomorphology* **2013**, *201*, 227–245. [CrossRef]
46. Bek, S.; Chuman, T.; Šefrna, L. The Usability of Contours in Erosion Modelling: A Case Study on ZABAGED, Czech Republic. *Acta Universitatis Carolinae* **2008**, 1–2, 77–86.
47. Vysloužilová, B.; Kliment, Z. Modelování erozních a sedimentačních procesů v malém povodí. *Geografie* **2012**, *2*, 170–191. (In Czech)
48. CLARK LABS. Available online: https://clarklabs.org/about/ (accessed on 12 May 2018).
49. Verstraeten, G.; Van Oost, K.; Van Rompaey, A.; Poesen, J.; Govers, G. Evaluating an integrated approach to catchment management to reduce soil loss and sediment pollution through modelling. *Soil Use Manag.* **2002**, *18*, 386–394. [CrossRef]

50. WaTEM/SEDEM Homepage. Available online: http://geo.kuleuven.be/geography/modelling/erosion/watemsedem/index.htm (accessed on 5 May 2018).
51. Hamel, P.; Falinski, K.; Sharp, R.; Auerbach, D.A.; Sánchez-Canales, M.; Dennedy-Frank, P.J. Sediment delivery modeling in practice: Comparing the effects of watershed characteristics and data resolution across hydroclimatic regions. *Sci. Total Environ.* **2017**, *580*, 1381–1388. [CrossRef] [PubMed]
52. Notebaert, B.; Vaes, B.; Verstraeten, G.; Govers, G. *WaTEM/SEDEM Version 2006 Manual*; KU Leuven, Physical and Regional Geography Research Group: Leuven, Belgium, 2006.
53. Bezak, N.; Rusjan, S.; Petan, S.; Sodnik, J.; Mikoš, M. Estimation of soil loss by the WaTEM/SEDEM model using an automatic parameter estimation procedure. *Environ. Earth Sci.* **2015**, *74*, 5245–5261. [CrossRef]
54. Krása, J. *Hodnocení Erozních Procesů Ve Velkých Povodních za Podpory GIS. Dizertační Práce*; České Vysoké Učení Technické v Praze, Fakulta Stavební, Katedra Hydromeliorací a Krajinného Inženýrství: Praha, Czech Republic, 2004; 176p. (In Czech)
55. Wolock, D.M.; McCabe, G.J. Comparison of Single and Multiple Flow Direction Algorithms for Computing Topographic Parameters in TOPMODEL. *Water Resour. Res.* **1995**, *5*, 1315–1324. [CrossRef]
56. Pechanec, V.; Mráz, A.; Benc, A.; Cudlín, P. Analysis of spatiotemporal variability of C-factor derived from remote sensing data. *J. Appl. Remote Sens.* **2018**, *12*, 016022. [CrossRef]
57. Doležal, F.; Kulhavý, Z.; Kvítek, T.; Soukup, M.; Čmelík, P.; Fučík, P.; Novák, J.; Peterková, E.; Pilná, P.; Pražák, M.; et al. Hydrologický výzkum v malých zemědělských povodích. *J. Hydrol. Hydromech.* **2006**, *54*, 217–229. (In Czech)
58. Fučík, P.; Kaplická, M.; Kvítek, T.; Peterková, J. Dynamics of Stream Water Quality during Snowmelt and Rainfall—Runoff Events in a Small Agricultural Catchment. *Clean Soil Air Water* **2012**, *40*, 154–163. [CrossRef]
59. Pavlík, F.; Dumbrovský, M.; Podhrázská, J.; Konečná, J. The influence of water erosion processes on sediment and nutrient transport from a small agricultural catchment area. *Acta Univ. Agric. Silvic. Mendel. Brun.* **2012**, *60*, 155–165. [CrossRef]
60. Pechanec, V.; Burian, J.; Kiliánová, H.; Němcová, Z. Geospatial analysis of the spatial conflicts of flood hazard in urban planning. *Morav. Geogr. Rep.* **2011**, *19*, 41–49.

© 2019 by the authors. Licensee MDPI, Basel, Switzerland. This article is an open access article distributed under the terms and conditions of the Creative Commons Attribution (CC BY) license (http://creativecommons.org/licenses/by/4.0/).

MDPI
St. Alban-Anlage 66
4052 Basel
Switzerland
Tel. +41 61 683 77 34
Fax +41 61 302 89 18
www.mdpi.com

Water Editorial Office
E-mail: water@mdpi.com
www.mdpi.com/journal/water

www.ingramcontent.com/pod-product-compliance
Lightning Source LLC
LaVergne TN
LVHW071945080526
838202LV00064B/6680